INTRODUCTION
to
FOOD
BIOTECHNOLOGY

CRC Series in
CONTEMPORARY FOOD SCIENCE

Fergus M. Clydesdale, Series Editor
University of Massachusetts, Amherst

Published Titles:

CRC Series in
CONTEMPORARY FOOD SCIENCE

INTRODUCTION
to
FOOD
BIOTECHNOLOGY

Perry Johnson-Green

Department of Biology
Acadia University
Wolfville, Nova Scotia
Canada

CRC PRESS

Boca Raton   London   New York   Washington, D.C.

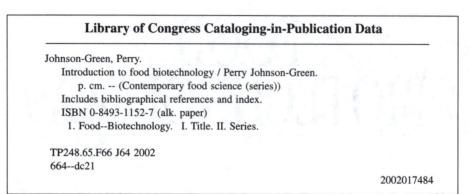

Library of Congress Cataloging-in-Publication Data

Johnson-Green, Perry.
    Introduction to food biotechnology / Perry Johnson-Green.
        p. cm. -- (Contemporary food science (series))
    Includes bibliographical references and index.
    ISBN 0-8493-1152-7 (alk. paper)
        1. Food--Biotechnology. I. Title. II. Series.

TP248.65.F66 J64 2002
664--dc21

                                                                    2002017484

This book contains information obtained from authentic and highly regarded sources. Reprinted material is quoted with permission, and sources are indicated. A wide variety of references are listed. Reasonable efforts have been made to publish reliable data and information, but the author and the publisher cannot assume responsibility for the validity of all materials or for the consequences of their use.

Neither this book nor any part may be reproduced or transmitted in any form or by any means, electronic or mechanical, including photocopying, microfilming, and recording, or by any information storage or retrieval system, without prior permission in writing from the publisher.

The consent of CRC Press LLC does not extend to copying for general distribution, for promotion, for creating new works, or for resale. Specific permission must be obtained in writing from CRC Press LLC for such copying.

Direct all inquiries to CRC Press LLC, 2000 N.W. Corporate Blvd., Boca Raton, Florida 33431.

Trademark Notice: Product or corporate names may be trademarks or registered trademarks, and are used only for identification and explanation, without intent to infringe.

Visit the CRC Press Web site at www.crcpress.com

Preface

Biotechnology is becoming increasingly important to food. In some industries (e.g., brewing), it is part of a process that has deep roots in human society, whereas many other applications of biotechnology are new to food production and processing systems. Food biotechnology is also new to consumers; its introduction sometimes leads to opposition from consumer groups and antibiotechnology activist groups. In some cases, opposition has been strong enough to influence government policy toward regulation of biotechnology.

However, many aspects of food biotechnology are virtually invisible to the consumer. Microbial products are increasingly common ingredients in processed foods, and the diagnostic tools used by the food industry to maintain food safety often have a biotechnological component. Consumers are becoming more aware of nutraceuticals and functional foods, and have enthusiastically embraced this aspect of biotechnology.

Food scientists, nutritionists, dietitians, and agricultural professionals must have a rich understanding of food biotechnology, because biotechnology has the potential to be used as a tool within each of these disciplines. For example, transgenic plant biotechnology can be used to modify food or to improve its performance as a component of a processed food. Plants can also be modified to have higher nutrient or vitamin contents, increased levels of health-promoting compounds, or decreased levels of toxins or allergens. Transgenic plant technology has already been used extensively to improve the efficiency of food production, and there will be more applications of this sort in the future. One of the main objectives of this book is to lay a solid foundation in all areas of food biotechnology that can also be used as a springboard to careers in biotechnology. Readers will acquire an understanding of the language used in biotechnology, as well as the biological and chemical concepts that are important in each field. One of the major themes is biological diversity — the fount of most biotechnological innovation. Biotechnologists need to appreciate how the natural world has provided important tools to enhance technology. Another theme is the frequent use of examples. Some examples are currently used in the food industry, whereas others are taken from the research literature.

Food professionals also need to be aware of the controversial aspects of food biotechnology. The final chapter reviews ethical and regulatory issues, but an effort has been made to discuss them throughout the book. For example, Chapter 4 includes a discussion of the potential of transgenic plants to harm nontarget insects such as the monarch butterfly. Chapters 3 and 7 also have sections devoted to specific controversies in food biotechnology (bovine growth hormone and eosino-philia–myalgia syndrome, respectively).

Each chapter closes with a list of recommended reading. These are a mixture of general sources which provide a wide range of supporting material for topics covered in the chapter and, more specific, which support examples used in the book. The order of the lists corresponds to the sequence of topics in the chapter.

This book has benefited greatly from interactions and feedback with students in Food 3413 over the years. I am also indebted to Sheila Potter for training in Corel Draw and Krista Patterson for administrative support. Finally, I thank Julia Green-Johnson for her continual encouragement and advice.

Author

Perry Johnson-Green has taught a senior course in food biotechnology since 1995, as well as courses in food microbiology, sensory science, and human biology at Acadia University in Wolfville, Nova Scotia. He has been involved in a wide range of research, covering neuroscience, plant–microbe interactions, and the potential use of plant-derived antimicrobial compounds as food preservatives. Current research topics include interactions between probiotic yeast and mammalian cell function. A member of the Canadian Institute of Food Science and Technology, the Canadian Botanical Association, and the Institute of Food Technologists, he frequently participates in public discussions on consumer issues in food biotechnology.

Table of Contents

1 The Scope of Food Biotechnology

I. OVERVIEW

Food is central to human society. Food production and processing are vital components of global and local economies, and all aspects of the food industry rely heavily on technology. The development of food technology has been one of the great success stories of the 20th century. In the western world, we enjoy the luxury of year-round access to a dizzying diversity of fresh and processed foods, due to technological improvements in our ability to grow, store, and process plant and animal foods. In developing countries, food production has generally increased as populations increase, largely because of the development of high-yielding seeds.

However, from a global perspective, it is easy to find fault with current food technology. In industrialized countries, food production relies heavily on fossil fuels and products derived from them, such as synthetic pesticides and chemical fertilizers (production of fertilizers requires large inputs of energy, usually derived from fossil fuels). Hunger and malnutrition are still common in the developing world, despite attempts to increase the efficiency of food production in those regions. Throughout the world, arable land is becoming scarce; erosion and salinization are placing greater stresses on crops. The human population also continues to grow. Although the pace of growth has slowed considerably in the last 20 years, most forecasts call for continued population growth, particularly in the developing world. Increasing food production to accommodate for this growth is a challenge.

Many problems are felt on a smaller scale, in relation to the conversion of food commodities into valuable products. The incidence of foodborne illness is increasing in many parts of the world, and this increase has partly been blamed on more complex food processing systems. Increased consumer demand for relatively unprocessed foods has also contributed to the risk of food-borne illness. Recent cases of *Escherichia coli* O157:H7 infection from unpasteurized apple cider are an example of this phenomenon. Consumers are more aware of the dangers of food-borne pathogens and are more concerned about the potential health risks associated with pesticide residues on food.

Biotechnology certainly has the potential to alleviate many of these problems. Such efforts may lead to more efficient and sustainable agricultural systems and reduced reliance on chemical pesticides and fertilizers. One objective of this book is to develop an understanding of the approaches that can be taken to improve food production systems. For example, the application of **recombinant DNA technology** to directly modify an organism's genetic structure has provided producers of corn, cotton, and potato an alternative to synthetic pesticides to control insect pests. The ability to transfer DNA from a bacterium (*Bacillus thuringiensis*) to a plant, a process that was unthinkable 40 years ago, is crucial to the development of these improved plants. *B. thuringiensis* produces toxic proteins that kill a narrow range of insects. When the gene for this type of toxic protein is transferred to a plant, thus making a **transgenic plant**, the plant becomes toxic to susceptible insects. This technology can also be used to improve the nutritional qualities of plant foods and improve processing characteristics (e.g., modifying the characteristics of high-molecular glutenins in wheat to increase the rising power of bread doughs).

Biotechnology also has the potential to improve food safety, through the development of enhanced **diagnostic** systems to detect pathogens and toxins in food. Microbes also play a positive role in the food industry; -**microbial biotechnology** is an increasingly important part of food processes. A variety of valuable enzymes, amino acids, and polysaccharides can be obtained from bacteria and fungi, and their use is steadily increasing in the food industry.

Why, then, is food biotechnology so controversial? Consumers, particularly in Europe, Britain, and Japan, are particularly hostile to transgenic crops. Many consumer and environmental groups believe that these crops are potentially dangerous to humans and the environment and that these risks have been insufficiently assessed. There is also a common perception that food biotechnology is beneficial only to the companies involved in its development, and not to the general public. It is also a favorite target of antiglobalization activists, who view biotechnology as a means to exploit producers in the developing world.

Public concern about biotechnology increased in the fall of 1999, when large-scale contamination of "Starlink" corn was found in taco shells and various other foodstuffs destined for human consumption. This variety of corn contains a gene that produces a protein toxic to the European corn borer, a destructive pest. However, it was never approved for human consumption; prior to the Starlink scandal, however, it could be used in animal feed. The reason for this regulatory strategy was a concern that the novel protein had the potential to provoke an allergic response in humans, because it was somewhat resistant to breakdown by digestive enzymes. As a result of this contamination, several companies suffered substantial fiscal damage, and public distrust of food biotechnology grew.

Soon after the Starlink scandal, Swiss biotechnologists announced the development of "golden rice," a variety of rice that has much higher levels of β-carotene than normal rice. The human body can use β-carotene to synthesize vitamin A. Vitamin A deficiency is widespread in the developing world and is the leading cause of noninfectious blindness. The most common reason for this deficiency is overreliance on rice, which is often the only food available to people living in dire poverty and to subsistence farmers in southeast Asia.

It is uncertain how effective golden rice will be in combating vitamin A deficiency, and it is not a permanent solution to global malnutrition, but it will probably lead to improvements in vitamin A nutrition of some of the world's poor. The biotechnology industry has been quick to capitalize on golden rice, using it as an example of the potential of biotechnology to beneficially affect human society. This use of golden rice as a public relations tool has been widely criticized, especially by the developers of golden rice. The pathway of development of golden rice has been very different from that of commercial transgenics. Funding for development came from the Swiss government and an American foundation (the Rockefeller Foundation). The developers of golden rice have always been adamant that golden rice seed would be freely available to farmers in the developing world and that they would not seek intellectual property rights for this transgenic plant. In contrast, companies such as Monsanto have successfully obtained patents for their transgenic crops, and patent rights are vigorously defended. A number of high-profile court cases in the U.S. and Canada have demonstrated that the biotechnology industry considers intellectual property of transgenic crops to be essential.

Consumers throughout the world, then, are faced with contradictory information about biotechnology. Numerous surveys have also found that consumers tend to misunderstand the nature of biotechnology. For example, a survey of Australian consumers in 2000 found that most do not have a clear understanding of the potential risks and benefits of food biotechnology, yet they feel that the risks are greater than the benefits. Thus, consumers are suspicious of the entrance of biotechnology into the food industry, but this suspicion is not driven by specific knowledge. Consumers in the rest of the world are likely in a similar state. This is a crucial time for food biotechnologists; the public is generally hostile to biotechnology, but there seems to be room for improvement in attitudes. To encourage this change, the biotechnology industry needs to show consumers that biotechnology has the potential to improve their lives, without undue risk to the environment or to human health.

Most consumers are unaware of the less controversial aspects of food biotechnology, such as microbe-derived food additives (e.g., xanthan gum) and DNA-based diagnostic techniques. These types of biotechnology offer few potential risks to consumers and the environment and have generally been ignored by antibiotechnology activist groups.

The main objective of this book is to achieve an understanding of all aspects of food biotechnology. This will enable the reader to focus on specific technologies with more advanced readings, as needed. Risks and benefits will also be explained, with reference to numerous examples, allowing a thorough understanding of the controversies that surround food biotechnology.

II. WHAT IS BIOTECHNOLOGY?

Because of the wide range of processes that have been described as biotechnology, it is difficult to concisely define the term. However, as applied to food, biotechnology usually describes one of the following processes:

- The direct modification of DNA of plants, animals, or microbes that are used as food (popularly known as "genetically modified organisms," or GMOs)
- The use of microbes or microbial products as food or food additives
- DNA- or protein-based methods for the detection or identification of microbes or microbial products in food

This list is not exclusive; many aspects of traditional plant breeding include processes that are commonly considered to be biotechnology. For example, in their search for new useful traits, plant breeders often attempt to cross crop plants with their wild relatives. This is usually facilitated by various techniques of cell culture, which is generally considered to be a biotechnological process.

Directed use of microbes in agriculture is also becoming more common. *Rhizobium,* a bacterium that is able to transfer atmospheric nitrogen to certain plants, thus decreasing reliance on fertilizers, is commonly inoculated onto seeds. Microbial biocontrol of insect and weed pests is also increasingly popular.

Functional foods and nutraceuticals are sometimes considered to be part of biotechnology. Functional foods are usually defined as foods that are part of a normal human diet that provide benefits to health beyond the supply of basic nutrients. Nutraceuticals are sometimes defined as the specific compounds within functional foods that are responsible for their health-promoting effects; more commonly, though, the term "nutraceutical" is restricted to the marketing of food components that improve health and are sold in a purified form (e.g., as a tablet). Because functional foods are usually derived from common, unmodified plants produced through conventional agriculture, they do not fit well with other forms of food biotechnology. For this reason, they will not be discussed extensively in this book. However, the use of recombinant DNA technology to increase levels of vitamins is clearly biotechnology, but also has functional food implications, because many vitamins have health-promoting effects (e.g., as antioxidants) that are distinct from their traditional role as vitamins. Many researchers in the biotechnology community are very interested in the potential of recombinant DNA technology to increase levels of other food components that improve health or prevent disease.

III. RECOMBINANT DNA TECHNOLOGY

A. GENE CLONING

The major breakthrough in the development of recombinant DNA technology was the ability to clone genes. This refers to the process of isolating a specific **gene** from an organism's **genome** (the entire set of genetic information in an organism). In Chapter 3, we will discuss in detail how this is done, but in general terms, genes are usually cloned by inserting fragments of a genome into a **vector** (Figure 1.1). A vector is an agent that can be used to move DNA segments from one organism to another. **Plasmids**, small circular double-stranded DNA molecules that are capable of replication within their host cell, are commonly used as vectors. Once a plasmid vector has been inserted into a cell, the cell that contains the desired gene can be located and separated from cells that contain other fragments of DNA.

Gene cloning allows careful study of a gene's sequence and properties, and it also allows the gene to be transferred to a wide variety of organisms. Thus, a gene isolated from a bacterium can be transferred to another bacterium, a plant, or an animal. In some cases, gene transfer is relatively easy; in others (e.g., inserting a gene into a multicellular animal), it is much more challenging and complex. A defining feature of molecular biotechnology is the ability to transfer specific genes

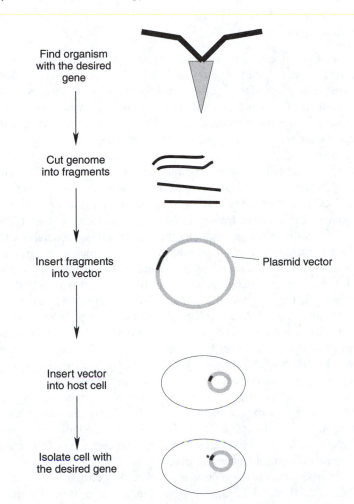

Find organism with the desired gene

Cut genome into fragments

Insert fragments into vector — Plasmid vector

Insert vector into host cell

Isolate cell with the desired gene

FIGURE 1.1 Overall process of gene cloning. In this example, a desirable gene is present in a carrot cultivar. DNA from the carrot is cut into small fragments and inserted into a plasmid vector, which is then inserted into a host cell. The cell with the desired gene is then located and isolated.

from organism to organism without the restrictions of incompatibility that otherwise apply (e.g., animals will breed successfully only with animals of the same species).

The basic techniques of gene cloning were developed in the mid-1970s. A product of direct gene transfer is considered to be **recombinant**, because its genome now consists of DNA from different organisms. The transfer process is known as **genetic engineering**, and in the popular media, the products are known as **genetically modified organisms (GMOs)**. In this book "GMOs" will be used in general reference to organisms (plants, animals, and microbes) that have been modified using recombinant DNA technology, and the terms "**transgenic plants**," "**transgenic animals**," and "**recombinant microbes**" will be used when referring to specific types of GMOs.

B. TRANSGENIC PLANTS

Sometime after the last glaciation (~11,000 years ago), humans began to cultivate plants and herd animals for food. They probably also began to breed these crops and animals. Breeding is fundamentally a simple process: parents with superior traits are allowed to mate (for plants, this means that pollen from one parent is used to pollinate flowers of the other parent), and offspring that have desirable traits are then selected and allowed to breed. This process continues until the desired improvement is obtained.

Breeding is still an effective mechanism for improvement and is frequently used to increase such traits as yield and disease resistance, as well as characteristics important to food processors (e.g., sucrose levels in potatoes) and nutritionists (e.g., levels of β-carotene in carrots). However, the great disadvantage that breeders face is the lack of control over the gene mixing that occurs during normal sexual reproduction. When a nucleus from a pollen grain fertilizes a nucleus in an ovule all of the chromosomes of the pollen nucleus are mixed with all of the chromosomes of the egg cell in the ovule. In many cases undesirable traits are passed to the egg cell along with desirable traits. In the mid-1970s, plant scientists were quick to see the potential of recombinant DNA technology to revolutionize plant breeding. Instead of being dependent on characteristics found in other cultivars, or closely related species, plant breeders could transfer genes from virtually any other organism. Furthermore, the wholesale mixing of genomes that occurs during conventional (traditional) breeding, does not occur with recombinant DNA technology. Only the desired gene and one or two other genes, depending on the methodology used, are transferred to the recipient plant. This means it is less likely that valuable traits will be lost during the gene transfer. In the early 1980s, the first studies of transgenic plants were published, and in the early 1990s, several companies took initial steps toward commercialization of transgenic plants.

Transgenic crop plants have received intense public scrutiny and remain highly controversial. Since 1994, 51 transgenic crop varieties have been released into the U.S. (Figure 1.2). Most of these were released between 1995 and 1998, and most confer benefits to producers; they provide herbicide resistance, contain genes that produce proteins that are toxic to insect pests, or give resistance to viruses that are plant pathogens (Table 1.1). Relatively few of the released crops are directly relevant to consumers, except for varieties with delayed ripening. Delayed-ripening tomatoes offer two potential advantages: a longer period of ripening in the field, without the softening that normally accompanies vine ripening, and an increased content of solids, due to the decreased rate of pectin breakdown. The first characteristic allows better transport of vine-ripened tomatoes, and the second allows the production of improved tomato paste.

In the future, there will likely be an increase in development and release of transgenic plants that have direct benefits to the food industry and to consumers (Table 1.2). Currently, few such transgenics exist. Why? One reason is that modifications aimed at processing or consumer problems often are technically difficult to achieve. For example, the suppression of lipoxygenase (LOX) activity in peas and other legumes improves flavor and aroma. However, it may also lead to decreased

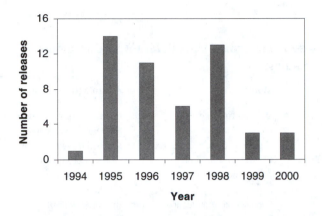

FIGURE 1.2 Commercial releases of trangenic crops in the U.S. between 1994 and 2000.

TABLE 1.1
Introduced Traits in Recombinant
Crops Released in the U.S.
between 1994 and 2000

Trait	Number of Releases
Herbicide resistance	23
Insect resistance	14
Sterility/fertility	8
Delayed ripening	6
Virus resistance	3
Modified lipid profile	2
Improved nutritional content	1

Note: Some crop varieties contain more than one introduced trait.

---

plant resistance against various stresses, such as insect attack. Therefore, such modification must be done with a thorough understanding of its implications.

Few of the world's crops have been modified through recombinant DNA technology (Table 1.3). Rice and wheat have been particularly difficult to transform, partly because of difficulties in growing rice and wheat cells *in vitro*. However, recent advances have made the transformation of wheat and rice relatively straightforward; the number of recombinant wheat and rice releases is likely to increase in the next five years. Commercial development of transgenics of minor crops (e.g., most vegetable crops) is unlikely until there is greater public acceptance of transgenic plants.

Recombinant crops have proven to be very popular with producers. Every year, the Economic Resource Service (ERS) of the U.S. Department of Agriculture

TABLE 1.2
Potential Uses of Transgenic Plants Relevant to Consumers and the Food Industry

Plant	Modification	Advantage
Tomato and other plants	Delayed ripening	Easier transport of fruits; improved quality
Tomato	Increased chitinase	Less post-harvest spoilage
Corn	Control over starch structure	Fewer requirements for starch conversion[a]
Corn, canola, etc.	Control over lipid profile	Oils that promote human health
Various plants	Addition of phytase	Decreased antinutritional compounds (phytate)
Legumes	Suppression of protease inhibitors	Increased digestibility
Soybean	Suppression of lipoxygenase	Improved flavor
Wheat	Increased HMW[b] glutenin	Improved bread quality
Barley	Increased β-glucanase levels	Fewer haze problems in beer
Various plants	Modification of enzyme activity	Increased antioxidants
Peanut and other plants	Elimination of allergens	Less allergenicity
Rice, tomato, etc.	Increased provitamin A	Increased vitamin A supply[c]

[a] Starch is normally converted to a range of products (e.g., maltodextrins) through the use of microbial enzymes.
[b] High molecular weight.
[c] Transgenics with increased vitamin E have also been developed.

TABLE 1.3
Recombinant Crops Released in the U.S. between 1994 and 2000

Crop	Number of Releases
Corn	13
Canola	10
Tomato	6
Cotton	5
Potato	4
Soybeans	3
Squash	2
Cantaloupe, flax, papaya, rice	1 each

(USDA) surveys a random selection of farmers. These surveys indicate progressive increases in acreage of recombinant crops between 1996 and 2000, with some recombinant varieties achieving close to 50% of total acreage (e.g., herbicide-resistant soybeans). Insect-resistant and herbicide-resistant crops have been particularly successful, in some cases because of decreased costs of chemical control of insect or weed pests.

Such large plantings means that much of the corn, soybean, and canola used in food processing in North America are transgenic. However, because only a few transgenic crops (one cultivar of insect-resistant corn and one cultivar of herbicide-resistant soybeans) can currently be imported and sold in food in Europe, some of the North American–grown transgenic cultivars must be separated. This process is known as crop **segregation**, and for some commodities (e.g., soybeans in the U.S.), it is difficult because such segregation was not necessary before the introduction of transgenic crops. Thus, systems for separate handling of different cultivars have to be introduced into current systems, which may be an expensive process. Many farmers, particularly after the Starlink scandal, fear that such requirements for segregation may impede international trade in transgenic crops. Consequently, they must balance any benefits of growing recombinant crops against potential difficulties in selling their crop. Problems related to contamination of GMO-free crops by transgenic seed have also occurred in the last 5 years. This, along with the Starlink scandal, demonstrates that it is technically difficult to completely segregate commodities based on their recombinant status. The contamination problem is likely to hinder commodity exports in the future, unless global consumer distrust of recombinant crops decreases in intensity. Recently released (July 2001) proposed regulations governing the use of transgenic crops in food in the European Union (EU) state that any food containing more than 1% transgenic products must be labeled to indicate that it contains GMOs. Thus the presence of low levels of transgenic seed in bulk commodity shipments of non-GMO seed could cause problems.

In Europe, no new transgenic plants will be approved until 2002. However, the process for approval of new transgenic cultivars is still under development, and the EU is not expected to implement the new process until 2003. Therefore, it is unlikely that large increases in the sowing of transgenic crops in Europe will occur in the near future. Anti-biotechnology activist groups are much more vocal and successful in Europe than in North America, and the European public remains unconvinced of the human and environmental safety of recombinant foods. Attitudes toward transgenic crops appear to be improving in Britain, although the widespread antagonism toward experimental plots shows that the atmosphere is still not encouraging for producers who wish to grow transgenic crops.

C. RECOMBINANT MICROBES

To date, the use of recombinant microbes in food production and processing has been limited to recombinant microbial enzymes and a recombinant hormone (**bovine growth hormone** [BGH]) to boost milk production. Recombinant **chymosin** (rennet) is an example of a recombinant enzyme produced by microbes. The bovine chymosin gene was transferred to several fungal species via recombinant DNA technology in the 1980s. Recombinant chymosin is now widely used throughout the world in cheese making. Chymosin increases the rate of curd formation during initial fermentation of milk by lactic acid bacteria (Figure 1.3). Traditionally, chymosin was obtained from the stomach of slaughtered calves, but the supply from this source is somewhat unstable. In contrast, recombinant chymosin does not have this instability, because it can be produced through growth of recombinant yeasts in large bioreactors

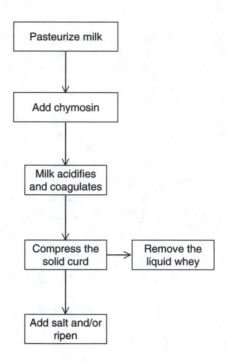

FIGURE 1.3 The process of cheese making, illustrating the importance of chymosin.

(vessels used for large-scale growth of cells). Interestingly, recombinant chymosin is currently triggering hostility between the U.S. and the EU. Proposed EU regulations governing labeling of food containing GMOs exempt cheese made using recombinant chymosin, because such cheese, when it is sold to consumers, contains negligible amounts of recombinant chymosin. However, the U.S. government maintains that this constitutes an unfair trade practice, because it implies that European products such as cheese are exempt from labeling, whereas foods containing small amounts (>1%) of transgenic plants (mostly imported from the U.S.) must be labeled.

The other major use of recombinant microbes in food is recombinant bovine growth hormone (rBGH) by dairy farmers. The gene for BGH was transferred to *E. coli*, and large-scale culture of this recombinant bacterium yields large amounts of rBGH. This is then injected into cows, which increases milk production.

The American public largely opposed the introduction of rBGH into the dairy industry (see Chapter 3) even though it was deemed to be safe by the U.S. Food and Drug Administration (FDA). It cannot be used legally in Canada, Europe, and many other countries. In some cases, this is due to animal welfare concerns, and in others, concern over potential effects on human health. It is worth noting that most of the animal and health concerns are linked to the hormone itself and its effect on bovine metabolism rather than to the fact that rBGH is recombinant. In other words, injection of natural BGH into cows would lead to similar safety concerns.

D. TRANSGENIC ANIMALS

Given the commercial success of transgenic plants, the lack of commercially available transgenic animals is surprising. There are several reasons for this. Many crops had readily identifiable problems (e.g., corn and the European corn borer) that could be attacked with single genes (e.g., a gene that produces a protein that is toxic to the borer). Such simple problems are less common in animal production systems. For example, feed efficiency (the ability of an animal to convert feed into tissue) is tremendously important; if an animal can gain weight with less feed, the producer can reap large savings. Unfortunately, a number of factors control feed efficiency, and single-gene solutions are unlikely. Increased animal (e.g., pig) growth hormone improves feed efficiency but can also seriously affect animal health (e.g., abnormal skeletal development).

Fish are an exception. Transgenic salmon with boosted levels of growth hormone grow substantially faster than their nonrecombinant counterparts, apparently without major side effects. However, environmental concerns are substantial; transgenic fish would be raised in aquaculture, and in such systems it is difficult to prevent occasional escapes into the environment. Escaped fish could then interbreed with wild salmon populations, possibly resulting in decreased levels of overall fitness (see Chapter 5). Transgenic fish are usually sterile; thus escaping fish would be unable to breed with wild fish. Despite this protective feature, the release of transgenic fish remains a contentious issue.

The highly publicized success of Scottish researchers in producing an adult clone of a sheep ("Dolly") has focused attention on the possibility of using transgenic animals to produce human proteins in sheep. Such proteins (e.g., α_1-antitrypsin) have therapeutic uses. This "molecular pharming" will probably be the first commercial use of transgenic farm animals; it is debatable whether such use of animals is either a "food" or "agriculture." Nonetheless, the success or failure of molecular pharming in animals will have important reverberations on the development of transgenic animals with more conventional agricultural uses. Such uses include the modification of milk proteins or butterfat levels, the enhancement of resistance to viral or bacterial pathogens, and increased leanness of meat. Commercialization of transgenic animals is unlikely to happen before 2003, when the first transgenic salmon are expected to be ready for commercial use and sale, if regulatory approval is obtained.

Many animal reproductive technologies also fall under the rubric of biotechnology. Embryo cloning, preservation, and transfer are all part of modern animal husbandry and animal breeding programs. If an individual animal has particularly good characteristics, its offspring can be multiplied extensively using *in vitro* (cell culture) techniques, allowing the widespread dissemination of animals with good traits.

IV. MICROBIAL BIOTECHNOLOGY

A. PERSPECTIVES

Micro-organisms are extremely important in biotechnology. Bacteria and fungi can be grown in large scale using bioreactors, which are large vessels that typically allow

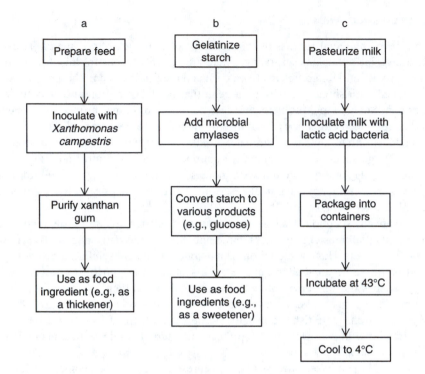

FIGURE 1.4 Examples of food processes that rely on microbes. (a) Production of xanthan gum by the bacterium *Xanthomonas campestris*. (b) Use of microbial enzymes (amylases) to convert starch into useful food ingredients. (c) Use of lactic acid bacteria to convert milk to yogurt.

close control of nutrient and oxygen levels, pH, and other environmental factors. In relation to the food industry, microbes are grown in bioreactors for one of four reasons: (1) they produce a compound that is useful, either as a food additive or as a food; (2) they produce enzymes that can be used to modify the properties of foods or food ingredients; (3) they transform a food into a different type of food; or (4) they transform waste products of the food industry (e.g., cellulose) into less environmentally harmful products (e.g., CO_2). **Xanthan gum** is an example of a microbe-derived food ingredient; this polysaccharide is a useful thickening agent used in salad dressings and other processed foods (Figure 1.4a). It is produced by *Xanthomonas campestris*, a relatively common bacterium. Modified starches are another group of common food ingredients. They are produced through the use of **amylases** and other microbial enzymes to alter the structure of corn or potato starch (Figure 1.4b). Common sweeteners such as glucose and fructose as well as various thickeners and fillers are obtained in this way. Yogurt is an example of a food that is made through microbial transformation of one food (milk) into another food (yogurt; Figure 1.4c). The brewing, wine-making, and distillery industries rely on the ability of the yeast *Saccharomyces cereviseae* to convert carbohydrates to ethanol.

We can make a distinction between traditional and modern microbial biotechnologies. Dairy and ethanol fermentations are the most common traditional biotechnologies; they have been practiced for millennia and are well accepted by consumers in most parts of the world. In contrast, xanthan gum and other microbial food additives and enzymes were not part of the food industry until the 20th century.

B. Traditional Microbial Biotechnology

All alcoholic beverages are produced using traditional biotechnological processes. **Yeasts** are added to carbohydrate substrates such as sucrose; the fungi use these substrates as a source of carbon and energy, and ferment them into ethanol and carbon dioxide. Yeasts are essential to the process. Without yeast, ethanol is not produced, and many of the flavors characteristic of each beverage are absent. Yeasts are also used in bread making. The production of carbon dioxide by yeast results in the formation of gas pockets, which drive the rising process.

Cheese, yogurt, and other fermented milk products are also made with the help of microbes (see Figures 1.3 and 1.4). In this case the **lactic acid bacteria** are vital. This group of bacteria produces energy through fermentation, and one of the main products of fermentation (often the sole product of fermentation) is lactic acid. This organic acid decreases the pH of milk, which causes thickening and coagulation of milk proteins, and creates an environment antagonistic to bacteria that are pathogens (e.g., *Salmonella* spp.) or that produce off flavors in milk (e.g., *Pseudomonas* spp.). The lactic acid bacteria (e.g., *Lactobacillus bulgaricus*) also produce volatile compounds that contribute to the flavor of fermented milk products.

C. Modern Microbial Biotechnology

Alcohol and lactic acid are not the only microbial products used in the food industry. Many of the enzymes, amino acids, and thickeners that are added to food are derived from a variety of microbes. In addition to xanthan gum, **glutamate** is also a microbial product that is a common food ingredient. It is added to many foods (e.g., dehydrated soups) to add "brothy" flavor and enhance other flavors. Glutamate is an amino acid and is therefore found in most proteins. However, it is easier to collect it from microbes than to separate glutamate in protein (e.g., soy protein) from other amino acids. *Corynebacterium glutamicum*, a Gram-positive bacterium, naturally has the ability to secrete large amounts of glutamate. This ability has been exploited by biotechnology companies, particularly in Japan.

V. DIAGNOSTIC BIOTECHNOLOGY

Food safety is an essential element of food security. In industrialized nations, food safety is currently of prime concern to the public and is vital to commercial success in the food industry. Pathogens such as *E. coli* O157:H7, *Campylobacter jejuni*, *Listeria monocytogenes*, and *Salmonella enteritidis* are increasingly frequent causes of outbreaks of food-borne illness, leading to great economic costs, and, in some cases, death. Overall rates of food-borne illness are also increasing, and combined with scandals such as the bovine spongiform encephalopathy (BSE) epidemic in the U.K., this has led to profound and widespread unease among consumers and governments and throughout the food industry.

The ability to detect and correctly identify contaminants in food is a vital part of the ongoing fight against food-borne illness. Traditional diagnostic methods for bacteria are effective but often time consuming. This limits their usefulness, because only limited numbers of samples can be taken, and often, contamination is undetected until illness occurs.

Many key diagnostic techniques have a biotechnological component. The most common of these is the use of mammalian **antibodies** to confirm the identification of bacterial species and to determine strain identity of bacteria. Such antibodies are produced through manipulation of the immune response in mammals such as mice and rabbits. They can also be obtained from **hybridoma** cells grown *in vitro*; (this term refers to the growth of cells outside of their normal location within a multicellular organism). Antibodies derived in this way are referred to as **monoclonal** antibodies. Antibody-based systems are also widely used clinically to identify viruses from samples of body fluids of people with food-borne illnesses.

DNA-based methods are also starting to make an impression in clinical and food-industry settings. The **polymerase chain reaction** (PCR) and the use of **labeled probes** can potentially increase the speed and sensitivity of methods in detecting and identifying pathogens. PCR is very sensitive; theoretically, the DNA of a single pathogen could be amplified and detected using PCR. One of the major problems with DNA-based methods is the complexity of foods; many compounds in most foods interfere with these methods. Nevertheless, we expect DNA-based diagnostics will be increasingly important in the food industry. At the present time, these methods are mainly used for the identification of particular strains of pathogens and the detection of transgenic plant DNA in commodities such as soybeans and corn.

VI. CONTROVERSIAL ASPECTS OF FOOD BIOTECHNOLOGY

Some aspects of food biotechnology are relatively free from controversy. Few consumers are aware of the use of microbial products such as recombinant chymosin in cheese or xanthan gum in salad dressings. The production of ethanol by yeasts is considered to be a safe technology, although ethanol abuse remains a troublesome issue for societies throughout the world. Diagnostic biotechnologies are seen as beneficial and powerful tools that can potentially decrease the incidence of food- and water-borne illness.

Transgenic plants and animals, however, have aroused the attention of activist and consumer groups, particularly in Britain and Europe. In these latter regions, biotechnologists have not been able to convince consumers that recombinant crops are safe, both for humans and for the environment. As a result, the EU and the U.K. have enacted legislation that requires labeling for foods containing recombinant crops. Nervousness among food processors and retailers over consumer distrust in biotechnology has resulted in the disappearance of foods containing recombinant crops from grocery shelves in the U.K. Although some consumer surveys have shown a slight lessening of public antipathy to food biotechnology, it is unlikely that recombinant foods will reappear in Europe in the near future.

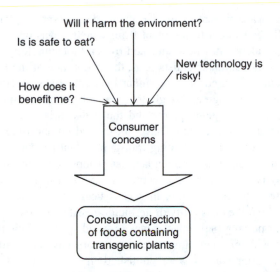

FIGURE 1.5 Factors leading to consumer hostility toward foods containing transgenic plants.

The situation is very different in the U.S., Canada, and Argentina, where large acreages of recombinant soybeans, corn, and canola are sown. These countries do not have labeling legislation, although they do permit, with some restrictions, voluntary labeling of foods containing recombinant crops. Antibiotechnology activist groups have had limited success in raising public opposition to food biotechnology, but numerous surveys have indicated that there is a substantial number of North Americans that are strongly opposed to the presence of transgenic plants in food, particularly if the food does not carry labels indicating the presence of GMOs. Most polls have also indicated a strong desire to introduce mandatory labeling of GMO-containing foods. So far, there has not been a ground swell of consumer activism to place pressure on the governments of the U.S. or Canada to introduce mandatory labeling, but that could change in the near future. Thus, the biotechnology industry in North America cannot afford to be complacent with respect to consumer activism and GMOs.

The main areas of concern for consumers and antibiotechnology activist groups are the potential risks for human and environmental safety (Figure 1.5). In terms of human safety, a common perception is that GMO-containing foods have been inadequately tested for the presence of unpredicted allergens or toxins. This has been exacerbated by the fact that many of the companies involved in commercialization of transgenic plants have not released to the public the data used to establish that their transgenic crops are safe for human consumption. Many groups also feel that recombinant DNA technology is intrinsically dangerous, because it constitutes "unnatural" mixing of genes that would never occur using normal reproductive processes. The biotechnology industry's attempts to reassure the public about human safety issues have largely been unsuccessful, partly because of continued insistence by the industry that human safety can be assessed using the process of **substantial equivalence**. This process involves a thorough comparison of levels of nutrients, toxins, and vitamins in a transgenic plant and the plant from which it was derived.

If these compounds are present in similar levels, the transgenic plant is considered to be substantially equivalent to the plant from which it was derived. The concept of substantial equivalence has been attacked by antibiotechnology activists and by a small number of influential scientists (e.g., the committee of the Royal Society of Canada that recently released an assessment of safety risks associated with food biotechnology). Scientists agree, though, that we lack alternative models for assessment of the safety of transgenic plants and that substantial equivalence has been successful to date. Transgenic plants have not caused human illness despite widespread consumption of foods containing transgenic plant products.

Another point of contention among activist groups and some scientists (particularly ecologists) is that the environmental safety of transgenic crops has been insufficiently assessed. For example, there has been much discussion in lay and scientific circles about the possibility of herbicide-resistant crops becoming troublesome weeds, or of passing their novel genes to weed species. This is often referred to as the **"superweed"** problem. Insect-resistant crops have come under much criticism, particularly after studies were published in 1999 and 2000 showing that **monarch** butterflies were harmed by the ingestion of large amounts of corn pollen from transgenic corn expressing an insecticidal protein. Most government regulatory agencies (e.g., the U.S. Environmental Protection Agency) contend that such risks are carefully assessed during the regulatory process. However, the question of environmental risks continues to be a potent source of consumer hostility to GMOs, particularly in Europe and the U.K.

The "What's in it for me?" question is also an important component of consumer hostility. Many people in the biotechnology industry hope that the introduction of transgenic crops targeted to increase consumer health (e.g., functional foods with enhanced levels of health-promoting compounds) will lead to greater public acceptance of GMOs. As an analogy, consider the microwave oven. It is unlikely that microwave ovens would have been embraced so quickly by consumers if they had not offered significant benefits.

VII. FOOD SECURITY

Food nourishes the body, and the production, processing, and distribution of food is crucial to global food security. It is also a crucial part of every nation's economy and political stability. Industrialized nations have an abundant supply of high-quality and diverse food throughout the year. As a consequence of this richness in food, questions of food security typically focus on three issues: (1) ensuring that food is safe and is not contaminated with pathogens or pesticides; (2) public education to encourage sound nutritional practices; and (3) alleviation of poverty.

However, in the developing world, food security is often low, because large numbers of people experience dire poverty. Food may be available, but that is irrelevant to those who lack the resources to buy it. This is a major problem in many countries that export food to richer nations. This illustrates the global inequity in food production and distribution that has been difficult to solve or alleviate, despite intense efforts in the latter half of the 20th century.

Biotechnology has the potential to increase food security. For example, many transgenic crops have decreased need of costly pesticides, because the crops themselves have been given the ability to fight off insects. Similar DNA-based technologies have also improved the range, sensitivity, and efficiency of diagnostic methods used to detect food-borne pathogens. An unfortunately small number of biotechnologists are developing low-cost diagnostic technologies that could be useful in developing countries.

Biotechnology can also help fight poverty and malnutrition. β-carotene-enriched golden rice is the best example of this. Vitamin A deficiencies are widespread in the developing world and cause a staggering array of public health problems; golden rice has the potential to alleviate this problem. Biotechnology could also decrease the reliance of producers on chemical fertilizers, while retaining the benefits of "western" agriculture — high yields with reduced labor inputs.

Although these beneficial results of biotechnology are real and substantial, the introduction of recombinant crops is a contentious issue throughout the developing world. Some countries (e.g., Sri Lanka) plan outright bans against the importation or planting of recombinant crops, whereas others (e.g., India) are attempting to develop their own biotechnology industry and are relatively receptive to recombinant crops.

The key question is whether the use of recombinant crops will accentuate the positive or negative aspects of the "green revolution," which led to enormous increases in agricultural productivity, but at the cost of increased economic disparity among farmers and increased reliance on technology and chemicals supplied by corporations from industrialized nations. Because most recombinant crops have been developed by corporate interests that are relatively uninterested in creating crops that are specifically tailored to agricultural problems in the tropics, where most developing nations exist, biotechnology may have a relatively small impact on this part of the world. However, increased western support of agricultural research in the tropics could lead to the development of transgenic crops targeted to specific agricultural problems in the developing world.

This idea is supported by the observation that traditional plant breeding has sometimes benefited poor farmers. For example, a recent initiative to introduce drought-resistant maize seeds to Zambian communities was successful in decreasing the effects of drought on village farmers. It is possible that creative minds, particularly those in the developing world, will find biotechnological solutions for food production problems that will have a similar positive impact.

RECOMMENDED READING

1. Dunwell, J. M., Transgenic crops: the next generation, or an example of 2020 vision, *Ann. Bot.*, 84, 269, 1999.
2. Daniell, H., Genetically modified food crops: current concerns and solutions for next generation crops, *Biotechnol. Genet. Eng. Rev.*, 17, 327, 2000.
3. Henry, R. J., Using biotechnology to add value to cereals, in *Cereal Biotechnology*, Morris, P. C. and Bryce, J. H., Eds., Woodhead Publishing, Cambridge, 2000, chap. 5.

4. Etherton, T. D. and Kris-Etherton, P. M., Recombinant bovine and porcine somatotropin: safety and benefits of these biotechnologies, *J. Am. Diet. Assoc.*, 93, 177, 1993.

5. Roller, S. and Dea, I. C. M., Biotechnology in the production and modification of biopolymers for foods, *Crit. Rev. Biotechnol.*, 12, 261, 1992.

6. Adams, M. R. and Moss, M. O., *Food Microbiology*, 2nd ed., Royal Society of Chemistry, Cambridge, 2000.

7. Glazer, A. N. and Nikaido, H., *Microbial Biotechnology: Fundamentals of Applied Microbiology*, W. H. Freeman, New York, 1995.

8. Hoover, D., Chassy, B. M., Hall, R. L., Klee, H. J., Luchansky, J. B., Miller, H. I., Munro, I., Weiss, R., Hefle, S. L., and Qualset, C. O., IFT expert report on biotechnology and foods: human food safety evaluation of rDNA biotechnology derived foods, *Food Technol.*, 54, 53, 2000.

9. Käferstein, F. K., Motarjemi, Y., Moy, G. G., and Quevado, F., Food safety: a worldwide public issue, in *International Food Safety Handbook: Science, International Regulation, and Control*, van der Heijden, K., Younes, M., Fishbein, L., and Miller, S., Marcel Dekker, New York, 1999, chap. 1.

10. Klijn, N., Weerkamp, A. H., and de Vos, W. M., Application of molecular detection and identification techniques in the study of the ecology of food associated microorganisms, in *Progress in Microbial Ecology*, Martins, M. E., Ed., Brazilian Society for Microbiology, Sao Paulo, Brazil, 1997, 214.

11. Wildman, R. E. C., Nutraceuticals: a brief review of historical and teleological aspects, in *Handbook of Nutraceuticals and Functional Foods*, Wildman, R. E. C., Ed., CRC Press, Boca Raton, FL, 2001, chap. 1.

12. Wenzel, G., The future role of bio-technology and genetic engineering, in *Food Security and Nutrition: The Global Challenge*, Kracht, U. and Schulz, M., Eds., St. Martin's Press, New York, 1999, chap. 22.

13. Swaminathan, M. S., Toward a food-secure world, in *Food Security: New Solutions for the Twenty-First Century*, El Obeid, A. E., Johnson, S. R., Jensen, H. H., and Smith, L. C., Eds., Iowa State University Press, Ames, 1999, chap. 6.

14. Domoney, C., Mullineaux, P., and Casey, R., Nutrition and genetically engineered foods, in *Nutritional Aspects of Food Processing and Ingredients*, Henry, C. J. K. and Heppel, N. J., Eds., Aspen Publishers, Gaithersburg, MD, 1998, chap. 6.

15. Ye, X., Al-Babili, S., Klöti, A., Zhang, J., Lucca, P., Beyer, P., and Potrykus, I., Engineering the provitamin A (beta-carotene) biosynthetic pathway into (carotenoid-free) rice endosperm, *Science*, 287, 303, 2000.

16. Bertini, C., Food security: international dimensions, in Food Safety, Sufficiency, and Security, Special publication no. 21, Council for Agricultural Science and Technology, Ames, Iowa, 1998, 38.

2 Tools of the Trade

I. THE HEART OF BIOTECHNOLOGY: CELL BIOLOGY

It is not easy being a biotechnologist. Biotechnologists need to be highly conversant in cell and molecular biology and comfortable with advanced topics specific to microbial, plant, or animal physiology. Many aspects of food biotechnology (e.g., functional food research) also require a sound understanding of food chemistry, as well as human nutritional epidemiology and physiology. As if that was not enough, food biotechnologists need to understand how market forces, regulatory bodies, and international trade can influence development of new technologies. It is beyond the scope of this book to supply all the necessary nutritional, food science, and biological background to become a proficient food biotechnologist. The emphasis in this chapter is on laying a solid foundation in microbiology and cell biology that will be helpful throughout the book. The reason for focusing on these topics is that microbes are tremendously useful to biotechnologists — they help us move bits of DNA from cell to cell, they produce many valuable biochemicals, and they are able to transform the ordinary into the extraordinary (e.g., grapes into wine). However, the main groups of microbes (viruses, bacteria, and fungi) have individual uses in biotechnology that are related to fundamental aspects of their structure and behavior. Hence, we will examine each group of microbes. Because DNA manipulation is so common in modern biotechnology, it is also important to understand the basics of how DNA works. We will also discuss some of the techniques that are used to manipulate DNA.

II. BACTERIA

A. BACTERIAL GROWTH

Bacteria are central to many aspects of food biotechnology (Figure 2.1). Bacterial enzymes, amino acids, vitamins, and polysaccharides are directly used in food processing, and they are also key participants in gene cloning and other molecular biological procedures. Bacteria are also the most frequent cause of food-borne illness, so development of efficient diagnostic tools for the detection of pathogenic bacteria is one of the major goals of food biotechnology. Thus, biotechnologists need to be aware of both the positive and negative aspects of bacteria.

Why do bacteria have this dual (good/bad) nature? One major reason (Table 2.1) is that bacteria tend to have simple growth requirements; many need only carbon (e.g., glucose), mineral nutrients (e.g., nitrogen), and water. Consequently, they grow very well in most foods. The presence of bacteria in food commodities, soil, air, and water sources makes it difficult to eliminate them from food. For this reason,

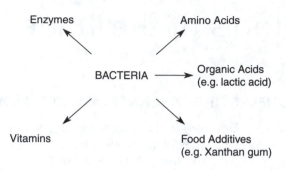

FIGURE 2.1 Useful products of bacterial metabolism.

TABLE 2.1

Characteristics of Bacteria that Are Relevant to Food Safety and Food Biotechnology

Characteristic	Relevance to Biotechnology	Relevance to Food Safety
Ease of growth	Cheap to grow in large scale	Can grow in most foods
Rapid growth	Cheap and easy to grow in large scale or small scale	Can quickly grow in many foods
Diverse physiology	Many opportunities for production of useful compounds	Difficult to assess safety
Toxin production	Opportunity for the development of diagnostic tools to detect toxins	Important cause of food-borne illness
Safe use in food	Some microbes can be used in food preparation and processing	Difficult to assess for novel microbial foods
Genetic structure	Useful for genetic manipulations	Dangerous traits can move among bacteria

food microbiologists and engineers have developed many strategies for inhibiting microbial growth in food, including the application of organic acids, salt, heat treatments, freezing, and refrigeration.

The ability to grow in simple media is also a blessing in certain contexts. Many bacteria can be easily grown at a laboratory scale, using nutrient broths and semisolid substrates such as agar or in large bioreactors (large-volume vessels specialized for the growth of microbes). Biotechnologists frequently use cheap "feeds" (materials used as a source of nutrients for the large-scale growth of microbes) such as molasses to supply all the nutrient requirements to bacteria grown in bioreactors. Thus, it is relatively easy to grow bacteria at a large scale. This partly explains the wide use of bacteria to produce products such as enzymes for the food industry.

Many bacteria are capable of rapid growth. This characteristic is mainly due to their small size and simple cellular structure. Bacteria lack organelles and generally make fewer compounds than eukaryotic cells; thus they can produce required structures and compounds more quickly than a typical eukaryotic cell. Bacteria such as

Escherichia coli can divide by binary fission in 30 min or less. This allows a single cell to produce a visible colony containing billions of cells in less than a day. Binary fission is simpler than mitosis, partly because of the presence of a single chromosome. Eukaryotic organisms, with their multiple chromosomes, require the complex process of mitosis to ensure that the cells resulting from division are genetically identical. In bacteria, this is a relatively simple process, allowing rapid growth. Again, this is both good and bad news for the food industry. It means that bacterial contamination of food can result in quick growth, resulting in large populations. This is a serious matter, particularly if the bacterium has pathogenic tendencies or produces toxins. On the other hand, the rapid growth rates of bacteria are useful to microbial biotechnologists interested in obtaining useful products from bacterial cultures.

Rapid growth makes it easier to study bacteria. Consider a tube with a suspension of bacteria (Figure 2.2). This suspension can be diluted and spread over an agar (semisolid) medium in a petri dish. If the dilution is correct (i.e., not too dilute and not too concentrated), then each cell will be well separated from other cells. If these cells are then allowed to grow, each will produce a visible colony. Within each colony, the cells are genetically identical "clones." This ability to quickly grow visible colonies from single cells is essential to many gene cloning procedures (see Chapter 3).

B. PHYSIOLOGICAL DIVERSITY

Another useful characteristic of bacteria is their great physiological diversity. Virtually all natural organic compounds can be used as a carbon source by one bacterium or another, although each species tends to specialize (e.g., *Erwinia carotovora* can degrade many plant cell components). This diversity is useful to biotechnologists because the enzymes that bacteria use to degrade organic compounds often can be used to transform foods. The most important example for the food industry is the transformation of plant-derived starch into an array of compounds with a wide variety of uses as food ingredients (Table 2.2). The enzymes that drive these transformations are obtained from bacteria or fungi.

Other products of bacterial metabolism, such as polysaccharides, vitamins, amino acids, organic acids, and lipids, are also useful. For example, the dairy industry depends heavily on the ability of lactic acid bacteria to efficiently ferment carbohydrates, releasing lactic acid as an end product. These bacteria are aerotolerant anaerobes (they can tolerate oxygen, but do not use it). They are completely dependent on fermentation to produce ATP (adenosine triphosphate) for cell growth and maintenance. As a consequence, they have become highly efficient at fermentation, and some strains, particularly those of *Lactococcus lactis*, are able to quickly produce enough lactic acid to drive the pH of milk from 6 to 4.5 within hours.

The metabolic diversity of bacteria is also exploited during cheese ripening and sauerkraut production. In both of these processes, there is a **succession** of bacterial growth; one bacterium (e.g., *Leuconostoc* in sauerkraut) will first decrease the pH of the substrate (shredded cabbage for sauerkraut production). This creates conditions favorable for the growth of other bacteria, which then flourish for a period of

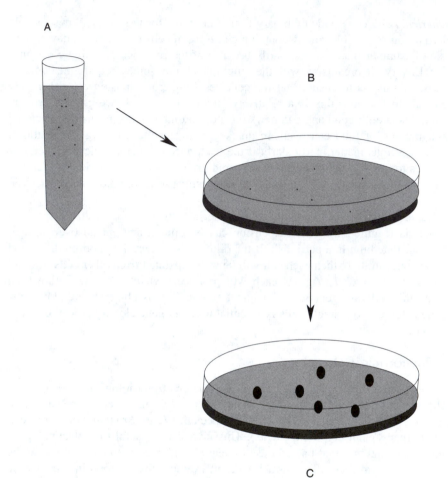

FIGURE 2.2 Growing bacterial clones (genetically uniform colonies) from a suspension of cells. A suspension of cells (A) is spread over an agar plate containing the appropriate nutrients (B). Each cell grows into a colony of genetically uniform bacteria (C).

TABLE 2.2

Commercial Uses of Starch and Glucose-Derived Products

Product	Use	Key Enzyme	Source of Enzyme	Microbe
Glucose	Sweetener	Glucoamylase	*Aspergillus niger*	Fungus
Fructose	Sweetener	Glucose isomerase	*Streptomyces* spp.	Bacterium
Maltodextrin	Thickener	α-Amylase	*Bacillus licheniformis*	Bacterium

time. With sauerkraut, *Lactobacillus plantarum* spp. grow, and this is a crucial part of sauerkraut production, because this bacterium produces compounds that contribute to the flavor of the final product. Many other traditional fermented foods have a similar dependence on numerous species of bacteria (and fungi).

For a bacterium to be exploited, it must be found and characterized. Traditionally, this has been achieved through screening processes, whereby a large number of bacterial isolates are tested for the presence of a desired attribute (e.g., production of a specific enzyme). Once a useful bacterium is isolated, it is identified and characterized further. For example, its ability to grow under a range of environmental conditions is assessed.

This screening process is tedious, time consuming, and expensive. However, a number of developments in the last 20 years have made the process of microbial discovery much more efficient. Genomics — the study of entire genomes of organisms — is progressing at a rapid rate, fueled by rapid improvements in DNA sequencing technology. The DNA of at least 30 bacterial species has been fully sequenced. As more and more bacterial genomes are sequenced and studied, it will become easier to identify bacteria that have biotechnological potential. **Bioinformatics** becomes important in this context. This word refers to the use of DNA or protein sequence data to understand structure–function questions. For example, if one needs to find an organism that produces a heat-stable enzyme that breaks down pectin, and if enough sequence information is available, one can simply use computer software and databases to find candidate organisms that are thermophilic and produce pectinases.

It is essential to have an orderly understanding of bacterial taxonomy to fully exploit their biotechnological potential. Current thinking divides all organisms into three domains: Archaea, Eubacteria, and Eukarya, which diverged early. All eukaryotic organisms are classified as Eukarya; Archaea and Eubacteria contain prokaryotic organisms that are fundamentally distinct from each other (e.g., cell wall structure is different in Archaea and Eubacteria). Most of the bacteria that are useful to biotechnologists are Eubacteria, but it is likely that Archaea will be used more in the future. Many Archaea are extremophiles that can tolerate extreme conditions of heat or salinity. Their enzymes often have similar extreme tolerances, which may prove to be useful in many biotechnological processes.

C. BACTERIAL GENETICS

The basic mechanisms of flow of genetic information (i.e., DNA replication, transcription, and translation) are similar in bacteria and eukaryotes but are less complex in bacteria. For example, eukaryotes have a set of chromosomes (one set for haploid cells and two sets for diploid cells); each chromosome is made up of a double-stranded DNA molecule as well as numerous proteins (e.g., histones). In contrast, bacteria have a single chromosome, with few associated proteins. Also, mechanisms of gene regulation are generally simpler and more fully understood than in eukaryotes. For these and other reasons, gene expression is more easily modified in bacteria. Another advantage is that **plasmids** (small, circular DNA molecules that are replicated by the host cell's DNA replication machinery) are common in bacteria but

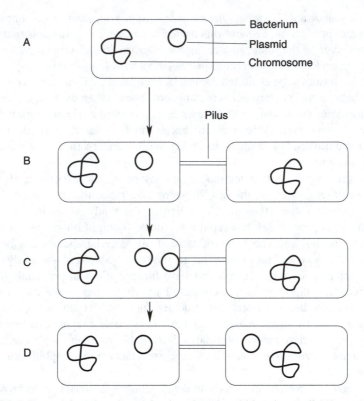

FIGURE 2.3 Bacterial conjugation. A conjugative plasmid in a host cell (A) contains genes that drive formation of a pilus (B) that joins the cell to another cell. A copy of the plasmid (C) is transferred to the other cell via the pilus (D).

rare in most eukaryotic cells. Plasmids are important to bacteria because they often contain useful genes (e.g., for antibiotic resistance), and certain plasmids, referred to as **conjugative plasmids**, contain genes that allow transfer of the plasmid from one cell to another through the process of **conjugation** (Figure 2.3). Initially, molecular biologists thought that conjugative plasmids would be useful tools for producing recombinant bacteria, but the problem with this approach is that recombinant genes may "jump" to other bacteria if the recombinant bacterium is released into the environment. For this reason, conjugating plasmids are not used for gene cloning. However, plasmids that are *incapable* of conjugation are useful DNA **vectors** in gene cloning. DNA from other sources can be easily inserted into plasmids, and the resulting **recombinant** plasmids can then be introduced into host bacteria such as *E. coli* (see Figure 1.1). The rapidity of bacterial cell division then allows the quick production of large amounts of plasmids, which can easily be purified from bacteria such as *E. coli*.

This is possible because plasmid vectors contain an *ori* sequence (usually about 300 bases long) that is recognized by proteins that initiate DNA replication. Plasmids that lack an *ori* will not be replicated, resulting in loss of the plasmid from the bacterial population, because the plasmid will not be passed on to dividing cells.

This is one of the reasons for the rarity of plasmids in eukaryotic cells — they are not usually recognized by the eukaryotic replication machinery. In bacteria, plasmids replicate *independently* of the chromosome; the frequency of plasmid replication varies widely. Thus, some plasmids are present in high numbers (**high-copy-number plasmids**) in each cell, whereas others are present in low numbers (**low-copy-number plasmids**). Generally, the former are preferred for cloning procedures. Because bacterial plasmids are recognized by bacterial polymerase, the cell must expend energy and materials toward plasmid replication. Thus, plasmids exert a **metabolic load** on host cells. Cells that have useless plasmids put their host cells at a disadvantage, compared to cells that lack plasmids. This leads to the loss of plasmids from bacterial populations, unless there is a gene on the plasmid that gives the host cell a competitive advantage over other bacteria. This is why plasmid vectors usually have **antibiotic-resistant genes**. If one wants to select cells that have a plasmid, exposing the cells to the antibiotic will result in death or growth suppression of cells that lack the plasmid. This strategy is central to most gene-cloning techniques (see Chapter 3, Section IV).

III. FUNGI ARE ALSO USEFUL AND VARIED

A. GENERAL CHARACTERISTICS OF FUNGI

"Fungus" describes a large range of physiologically and structurally diverse micro-organisms. Fungi are eukaryotic saprotrophs; they grow by breaking down organic compounds and utilizing the products of this decomposition as a source of carbon and energy. Fungi are directly important to the food industry (Table 2.3) as foods (e.g., mushrooms) and through the ability of fungi to transform food commodities. Economically, the most important industrial use of fungi is in the production of alcoholic beverages. *Saccharomyces cereviseae* is usually used for this purpose, although other yeasts, filamentous fungi, and bacteria share its ability to rapidly ferment carbohydrates to form ethanol. *S. cereviseae* is used to produce beer, wine, and virtually all distilled beverages. Its natural habitat is on the surface of sucrose-rich fruits (e.g., grapes), and it is well adapted for the fermentation of sucrose into ethanol and carbon dioxide.

Yeasts are unicellular fungi, and *S. cereviseae* is the best understood and most widely used yeast. Because of its extensive history as a model cell for research on eukaryotic cell biology and genetics, *S. cereviseae* is probably the best-understood eukaryotic organism. This, combined with its long history of safe food use, has led to numerous applications in the food industry. Its most useful characteristics are rapid growth under both anaerobic and aerobic conditions, efficient production of carbon dioxide and ethanol, and "clean" metabolism (i.e., it can be used in food and beverage production without adding strong off flavors or odors).

Various species of *Aspergillus* and *Penicillium* are extensively used in the food industry, because of their ability to transform basic commodities such as soybeans into value-added products such as soy sauce, and because they can secrete proteins in large amounts. These are filamentous fungi; they form extensive networks of filaments (hyphae) that have a width of one cell (Figure 2.4). Filamentous fungi are

TABLE 2.3
Importance of Fungi to the Food Industry

Fungus	Characteristic	Relevant Food
Positive Effects		
Agaricus bisporus and others	Fruit body formation	Food mushroom
Penicillium roquefortii and *Penicillium camembertii*	Ability to grow on cheese	Blue and camembert cheese
Kluyveromyces lactis and *Aspergillus nidulans*	Recombinant yeasts with chymosin gene	Production of chymosin (rennet) for cheese
Mucor miehei and others	Aspartic proteases	Cheese production
Kluyveromyces lactis	Lactase production	Alleviation of lactose intolerance
Saccharomyces cereviseae	Ethanol production	Fermented beverages
Saccharomyces cereviseae	CO_2 production	Bread making
Saccharomyces cereviseae	Rapid growth	Direct use as food
Aspergillus niger	Enzyme production	Starch processing, juice processing
Aspergillus niger	Citric acid production	Common food ingredient
Aspergillus oryzae	Flavor production	Soy sauce
Negative Effects		
Aspergillus flavus and others	Mycotoxin production	Contamination of various foods
Penicillium digitatum and others	Invasion of plant tissues	Cause of spoilage of fruits, grains, and vegetables
Botrytis cinerea and others	Invasion of plant tissues	Cause of plant disease in the field

efficient degraders of organic matter, because they can secrete hydrolytic enzymes and their hyphal colonies can quickly colonize organic matter. Amylases and proteases are two examples of useful enzymes that are produced by aspergilli.

Unfortunately, not all fungi are so benign or useful. Many fungi produce toxic metabolites (mycotoxins). Aflatoxins, produced by *Aspergillus flavus* and *Aspergillus parasiticus*, are the most worrisome of the mycotoxins, because of their potency as mutagens and carcinogens, even at low concentrations. *A. flavus* is a common contaminant on peanuts and cereal grains. Consequently, aflatoxin contamination of food and animal feeds is an important food safety issue. Mycotoxins are particularly troublesome in the developing world, because: (1) fungi grow particularly well under warm, moist conditions, which are common in the tropics; (2) storage technology (e.g., refrigeration) is less widespread than in industrialized nations; and (3) developing nations frequently lack the economic resources to use diagnostic technology to detect fungal and mycotoxin contamination.

Fungi are also a cause of great economic losses throughout the world, through the spoilage of stored food and the production of plant disease in the field. Historically, fungi have caused widespread epidemics of plant disease. For example, in the 1950s, North American wheat farmers faced large yield losses from wheat stem rust; this epidemic was finally stopped through the breeding of resistant wheat varieties, a process that continues to be important today. Fungal pathogens that attack field and harvested crops are currently an economic burden to both farmers and food

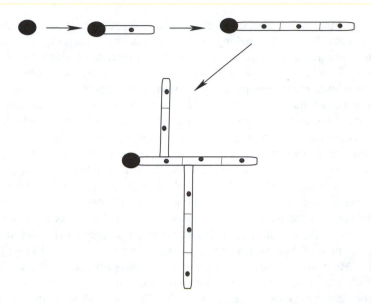

FIGURE 2.4 Formation of a fungal colony from a single spore. The spore germinates to form a one-cell-wide filament (hypha). The hypha elongates and forms branches, leading to an interconnected colony of hyphae. Clusters of intertwined hyphae are **mycelia**.

processors; fungicide application is frequently necessary to control diseases both pre- and post-harvest.

How can biotechnology help control these pathogens? Numerous efforts are underway to develop transgenic crops that have increased resistance to field diseases. This technology also has the potential to help protect plants from post-harvest diseases as well. Biocontrol, the use of organisms to prevent pathogens from attacking food plants, is enjoying increasing success and popularity partly because consumers throughout the world are pressing for decreased use of pesticides on food crops.

Biotechnology is also useful in the detection and identification of fungal plant pathogens and mycotoxins. Antibody-based technologies have been particularly successful in this context — a number of antibody-based tests for mycotoxins in foods are commercially available.

The development of diagnostic tests requires a thorough understanding of fungal taxonomy. The classification of fungi is complex, confusing, and currently in flux, but several basic divisions are well accepted. We now know that several of the groups traditionally included in the fungi are only distantly related to other fungi. This has led to a distinction between "pseudofungi," which are more closely related to certain protist groups, and "true fungi," which includes most of the economically important fungi. Both pseudofungi and true fungi can be troublesome plant pathogens. For example, the **oomycete** pseudofungus *Phytopthera infestans*, the biological cause of the Irish potato famine in 1850, is one of the most destructive plant pathogens. It is most closely related to heterokont protists, which includes diatoms and brown algae. In contrast, true fungi are phylogenetically more closely related to plants and animals than to pseudofungi.

The true fungi are divided into three divisions: **Zygomycota**, **Ascomycota**, and **Basidiomycota**. Most molds belong to Zygomycota or Ascomycota, whereas most mushrooms are members of Basidiomycota. Zygomycetes are quite different from other fungi. Their hyphae lack cross walls; this is one factor that allows them to grow rapidly over moist substrates. Zygomycetes are important agents of spoilage of fruits and other foods with high water activity. This group also contains useful fungi, particularly in the genera *Mucor* and *Rhizopus*. For example, *Rhizopus oligosporus* is used to transform soybeans into tempeh, a popular food in southeast Asia. Ascomycetes include most of the fungi important to food biotechnology (e.g., *S. cereviseae*), as well as a small number of prized food mushrooms (e.g., truffles and morels). Mycologists have also created an artificial group (Deuteromycota or Fungi Imperfecti) to contain isolates that lack a sexual stage in their life cycle. It has always been suspected that most of these species belong to either Ascomycota or Basidiomycota, and molecular techniques are increasingly used to determine the proper placement of species. Unfortunately, the current taxonomy of these fungi is confusing. Many species have two names: (1) the **anamorph**, the name classified under Deuteromycota, and (2) the **teleomorph**, the name that is given to the fungus if it can be classified under Ascomycota or Basidiomycota. This situation is relevant to food biotechnology because most of the fungi used in biotechnology belong to Deuteromycota. Thus, some species of *Penicillium* (anamorph) are also classified within the genus *Eupenicillium* (teleomorph). In general, though, anamorph names are usually used in reference to important fungi such as *Penicillium*, to minimize confusion.

B. THE USE OF FUNGI IN RECOMBINANT DNA TECHNOLOGY

Yeast and filamentous fungi are widely used in gene cloning, especially when the main aim of cloning the gene is to produce large amounts of recombinant proteins. Fungi are eukaryotes and share many of the organelles of plant and animal cells. These groups also share many intracellular processes. For example, messenger RNA (mRNA) transcripts in eukaryotes are often edited before translation (Figure 2.5), because eukaryotic genes typically contain noncoding regions. These introns are sequences of DNA within genes that are not translated into protein. Introns must be removed from mRNA transcripts before the mRNA leaves the nucleus and is translated in the cytoplasm. The remaining DNA sequences (exons) must also be spliced together to make an intact mRNA transcript before leaving the nucleus. This editing (removal of introns and splicing of exons) does not occur in eubacteria. This makes it difficult to clone eukaryotic genes in bacteria such as *E. coli*, although this difficulty can be alleviated through the use of cDNA ("copy" DNA) libraries (see Chapter 3, Section IV). Fungi can edit mRNA transcripts, but the editing process is slightly different than in mammals and plants. For this reason, cDNA libraries are frequently the method of choice, even when cloning a plant or animal gene into a yeast or a filamentous fungus.

Another fundamental difference between eukaryotic cells and eubacteria lies in the process of protein secretion. Protein secretion in eubacteria is relatively simple; a hydrophic **signal sequence** allows the protein to pass through the plasma membrane.

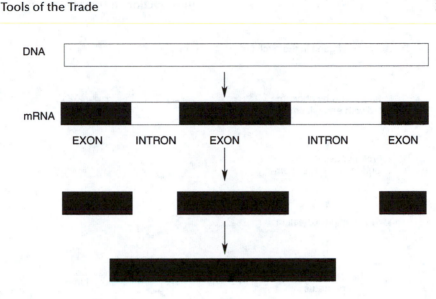

DNA

mRNA

EXON   INTRON   EXON   INTRON   EXON

Translation

FIGURE 2.5 Gene splicing in eukaryotes. A typical eukaryotic gene contains coding and noncoding regions. The mRNA transcript includes exons (coding regions) and introns (noncoding regions). Introns are cut out of the transcript, and the exons are then spliced together.

The signal sequence is then enzymatically removed. In contrast, proteins destined for secretion by eukaryotic cells are produced in ribosomes associated with the rough endoplasmic reticulum (RER). After translation, the proteins enter the RER and undergo processing, during which carbohydrate chains (glycosylation) are added to the polypeptide (Figure 2.6). This processing, which takes place in the RER and the Golgi apparatus, is usually essential for the function of the secreted protein. Secreted proteins travel from the RER to the Golgi apparatus within membrane-bound vesicles (transport vesicles) and are released from the Golgi in similar vesicles (secretory vesicles) that fuse with the cell membrane, releasing their contents outside of the cell.

Fungi are often able to process secreted proteins in a similar way as mammalian cells. Therefore, they are frequently a better choice for the **expression** (transcription + translation) of mammalian proteins than bacteria. Such production of recombinant mammalian proteins can be useful in the context of food production and processing; chymosin (rennet) and bovine growth hormone are two examples of this (see Chapter 3, Sections XI and XII). Some mammalian proteins are not processed properly in fungal cells, sometimes because of incorrect folding of the polypeptide after synthesis (in eukaryotes, specific proteins are involved in helping proteins achieve their proper, folded conformation). In such cases, the best alternative is to produce the protein in mammalian cell culture.

In comparison to bacteria, the main disadvantage of using fungi is that they are physiologically and genetically more complex. Getting DNA into fungi is also more difficult, and fewer plasmid vectors are available for the transformation of fungal cells. Despite these shortcomings, fungi are increasingly popular, both as tools of basic research and as key players in biotechnological development.

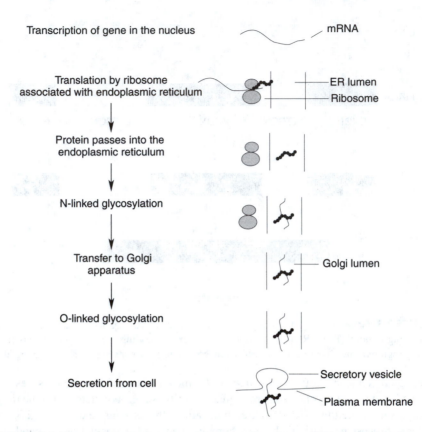

FIGURE 2.6 Processing of secreted proteins in eukaryotes. mRNA transcripts from the nucleus are translated by ribosomes associated with the rough endoplasmic reticulum (RER). The polypeptide then passes through the RER membrane into the lumen. Carbohydrate chains are added to NH_2 groups of asparagine residues (N-linked glycosylation). The polypeptide is then transported via a membrane-bound vesicle to the Golgi apparatus, where carbohydrate chains are added to hydroxyl groups of certain amino acids (e.g., serine). This is referred to as O-linked glycosylation. Finally, the polypeptide (a glycoprotein) is secreted from the cell. The protein is enclosed within a membrane-bound vesicle that is pinched off from the Golgi. This secretory vesicle then fuses with the plasma membrane, releasing the protein into the external environment.

IV. VIRUSES: USEFUL PARASITES

A. THE NATURE OF VIRUSES

Viruses are usually considered to be harmful parasites of cells, because of the enormous loss of life and economic productivity that they cause. Viruses are also important pathogens of crop plants and food animals, and the recent foot-and-mouth epidemic in Britain illustrates the ability of viruses to throw entire food production systems into chaos. As with bacterial and fungal pathogens, one of the benefits of biotechnology has been improved diagnostic tests for the detection and identification

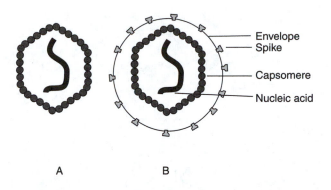

A   B

FIGURE 2.7 Structure of viruses. In nonenveloped viruses (A), the nucleic acid is surrounded by a capsid composed of protein subunits (capsomeres). In enveloped viruses (B), the capsid is enclosed by a membrane. The membrane usually has embedded protein spikes.

of viral pathogens. This is particularly relevant to food-borne viruses; despite widespread agreement that viruses are frequent causes of food-borne illness, most of the relevant viruses cannot easily be identified in food samples. Diagnostic biotechnology is expected to address this deficiency in the future.

Viruses are also useful. All organisms are hosts to viruses, and many viruses are adept at moving pieces of DNA from one host cell to another. Therefore, it is not surprising that viruses have been used in cloning projects, where the prime object is to move specific segments of DNA from one organism to another.

Viruses lack cellular structure and rely on host cells for reproduction. They can replicate only if they are able to infect the proper cell type and use the cell's machinery (enzymes and metabolites such as nucleotides and amino acids) to produce new virus particles. Viral particles (Figure 2.7) are made up of a protein **capsid** enclosing the viral **nucleic acid**, which can be RNA or DNA. The nucleic acid contains all the genetic information required for successful infection of a host cell. Viral infection usually follows the following pattern (Figure 2.8):

1. The virus attaches to a host cell. This is usually a highly specific interaction. The virus will attach to a host cell, but not usually to a nonhost cell. The virus usually binds to specific proteins (receptors) in the host cell membrane.
2. The virus enters the cell. With regard to bacterial viruses (bacteriophage), only the viral nucleic acid enters the cell. In viral infections of eukaryotic cells, the entire viral particle enters, often through endocytosis. After entry of an entire particle, the particle is "uncoated." This process results in breakdown of the capsid, releasing the viral nucleic acid into the cytoplasm.
3. The virus takes over the cell. Normal cellular metabolism is disrupted, and cellular enzymes, ribosomes, and metabolites are used to synthesize new viral proteins and nucleic acids. These components are then assembled to make new viral particles.

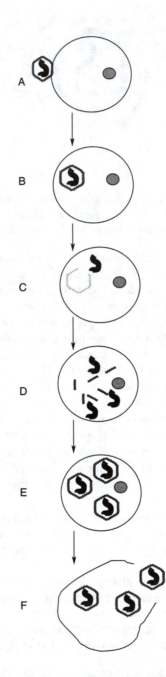

FIGURE 2.8 Infection cycle of a virus in a eukaryotic cell. The virus attaches to a host cell (A), enters the cell (B), is uncoated (C), directs the cell to make new viral components (D), assembles new particles (E), and causes lysis of the cell (F), thus releasing the new virus particles.

4. New virus particles are released from the cell. Bacteriophage are released through lysis of the cell, whereas viruses of eukaryotes are sometimes released through lysis and sometimes through budding, which may not be lethal to the host cell. Enveloped viruses acquire their envelope through budding — the viral particle is enclosed in part of the plasma membrane that is pinched off. This is broadly similar to the process involved in secretion of proteins through the formation of secretory vesicles (see Figure 2.6).

This infection cycle is often called the *lytic cycle*. The *lysogenic cycle* is similar, but includes a protracted stage when the virus is inactive and the viral nucleic acid is incorporated into the DNA of the host cell. One of the common points of confusion in biotechnology centers on the use of bacteriophage *lambda* (λ) in gene cloning. In introductory microbiology courses, this virus is usually taught as an example of a virus that has a lysogenic cycle. However, when λ is used in cloning procedures, genes required for lysogeny are usually removed. These strains quickly lyse susceptible cells after infection. See Chapter 3, Section VIII for an explanation of the use of viruses in gene cloning.

V. DNA: THE HEART OF BIOTECHNOLOGY

A. DNA STRUCTURE

Deoxyribonucleic acid (DNA) and its sister compound ribonucleic acid (RNA) are vital components of many biotechnological applications. The molecular biology revolution that has occurred in the last 20 years and created so many new biotechnological opportunities is fundamentally based on the ability to precisely manipulate DNA. Therefore, it is essential for biotechnologists to have a thorough understanding of DNA.

The prime role of DNA is to act as a reservoir of genetic information. This is possible because of the following structural features of DNA (Figure 2.9).

- DNA is a double helix made up of two antiparallel strands.
- Each strand is made up of a backbone of deoxyribose monosaccharides linked covalently through phosphate bridges.
- Each deoxyribose unit is linked covalently to a base consisting of either adenine (A), guanine (G), cytosine (C), or thymine (T).
- Two antiparallel strands, through hydrogen bonding between adjacent base pairs, can form a stable double helix.
- Hydrogen bonds form between complementary base pairs (C–G and A–T).
- Three linear bases on a strand code for a specific amino acid — this allows a linear sequence of bases on a strand of DNA to code for a linear sequence of amino acids on a polypeptide. Each group of three bases is a **codon**.

The antiparallel (upside down relative to each other) nature of the two strands is important in allowing a stable configuration of the double helix. The orientation of each strand is best described using 3′ and 5′ notation (see Figure 2.9). Close

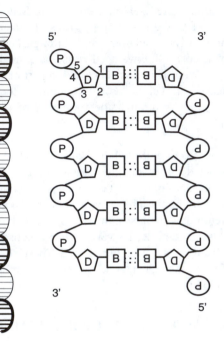

FIGURE 2.9 The structure of DNA. The backbone is made up of phosphate (P) and deoxyribose (D). The bases (B) interact through hydrogen bonds.

examination of the structure of the two strands shows that each phosphate molecule is attached to carbon 3(3′) of one deoxyribose and to carbon 5(5′) of the adjacent deoxyribose. At one end of a strand, there is a deoxyribose with a 3′ that is not attached through a phosphate molecule to another deoxyribose. At the opposite end of the strand, there is a deoxyribose with a 5′ end that is free. Thus, each strand has a 5′ end and a 3′ end. As we will see, the orientation of the strands is important to the function of DNA.

The DNA double helix can be reversibly separated into two single strands. This happens naturally during DNA replication and transcription and is driven by a specific enzyme. Heating also results in separation of the strands (**melting**), as does increasing the pH of the solution (e.g., through addition of NaOH). The latter process is referred to as **denaturation**. Both methods are used extensively in molecular biology. If either condition is relaxed (i.e., by cooling or by removing the NaOH), double-stranded DNA will re-form, through the association of complementary strands. This **re-annealing** of single-stranded DNA is called hybridization when the two strands are from different sources. Hybridization often involves of the use of a single-stranded DNA probe that anneals to a target strand of DNA. Hybridization can occur between two strands of DNA, two strands of RNA, or one each of DNA and RNA, as long as they are complementary. Hybridization of **primers** (short segments of single-stranded DNA) to longer strands of DNA is also an essential part of the polymerase chain reaction (PCR).

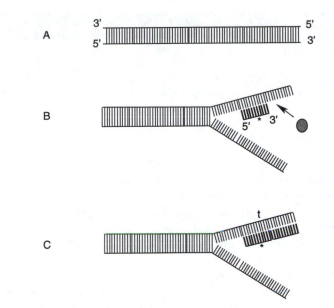

FIGURE 2.10 Replication of DNA. The double-stranded DNA (A) separates into single strands. A primer (*) anneals to one of the strands (B), and DNA polymerase (filled circle) extends the new strand (C), starting with the primer, and adding bases complementary to the template strand (t). The other original strand would also act as a template.

B. DNA REPLICATION

When a cell divides, it must ensure that both daughter cells have identical genetic structure; in other words, each cell must have the same sequence of DNA. This is achieved through DNA replication. In both bacteria and eukaryotes, this is accomplished through a suite of proteins and enzymes that separate the DNA into single strands, allowing DNA polymerase to use each strand as a template to build a new complementary strand (Figure 2.10). The following characteristics of DNA replication are particularly important; they are crucial to various techniques of molecular biology, including PCR:

- DNA polymerase cannot start to build a new strand of DNA unless it has an RNA primer that is complementary to a sequence on the template strand. DNA polymerase is able to extend a complementary DNA strand from this primer and later replaces the RNA primer with DNA. DNA primers are often used to control the starting point of DNA replication.
- The new DNA strand is always synthesized from the 5′ to the 3′ end.

As noted in the discussion of plasmids, DNA replication in cells also requires a specific DNA sequence that acts as an "origin of replication" (*ori*) on the chromosome (eukaryotic cells often have multiple sites of origin on each chromosome).

FIGURE 2.11 Structure of a eubacterial gene. Upstream of the structural region (S), there is a promoter (P) and possibly an operator (O) or activator (A) sequence. Downstream of the structural region, there is a termination region (T).

C. TRANSCRIPTION OF mRNA

The end product of transcription is a single-stranded mRNA molecule that is complementary to a sequence on one of the strands of DNA. This strand of DNA is the **coding** or **sense** strand, and the strand of DNA that is not transcribed is the **noncoding** or **nonsense** strand. The sense strand forms a **template** for the construction of a complementary strand of mRNA, in a similar fashion as strands of DNA act as templates for the construction of new DNA strands during DNA replication (see Figure 2.10).

The driving force behind transcription is RNA polymerase. This enzyme is similar to DNA polymerase in that it requires an RNA primer before it can build a complementary strand and it builds the new strand from the 5′ to the 3′ end. The direction that RNA polymerase moves is **downstream** and the opposite direction is **upstream**.

To understand transcription, it is useful to examine the structure of the coding strand of DNA (Figure 2.11). It consists of a sequence of DNA (the **structural** region) that codes for mRNA that will be translated into a protein. On either side of this region are regions of DNA that are not transcribed or translated into protein. The region directly upstream of the protein-coding region is essential for transcription. It is the promoter region. This sequence of bases is required for transcription; it allows RNA polymerase to recognize start points. The promoter in eubacteria is always upstream of the protein-coding region and consists of discrete DNA sequences that interact directly with the RNA polymerase protein. This is an example of protein–DNA interaction, which is a crucial element in many aspects of gene activity and regulation.

In eubacteria, one of the subunits (e.g., the sigma factor) of RNA polymerase is often required to assist in binding of RNA polymerase to the promoter. This gives bacteria a large-scale regulatory tool. Sigma factor binds to the promoter of certain genes, whereas other factors bind to the promoters of other genes. Therefore, the particular DNA-binding factor that is present will greatly influence the pattern of gene activity.

Directly downstream of the promoter, many eubacterial genes also have an **operator** region, which is important in regulation of transcription. **Repressor** proteins bind to the operator region (another example of DNA–protein interaction) and block the action of RNA polymerase. This **negative** regulation is relieved, in some cases, by the presence of an **inducer** molecule that binds to the repressor protein, preventing its binding to the operator.

Eubacterial genes sometimes use **positive** regulation acting on **activator**-binding sites that are upstream of the promoter region. Activator proteins bind to this region and enhance binding of RNA polymerase to the promoter, thus increasing the frequency of transcription of the gene.

The promoter region tells RNA polymerase where to start transcription, but other sequences are required to indicate the point when transcription should end. This is done by **terminator** sequences.

In eukaryotes, transcription is more complex. Numerous additional proteins, called **transcription factors**, are required to help RNA polymerase bind to the sense strand of DNA. Promoters are present upstream of the protein-coding region and generally have the same purpose as in eubacteria. However, eukaryotes do not use negative regulation. Instead, they tend to rely on positive regulation, through the use of **enhancers**, DNA sequences that are usually upstream from the promoter. The promoter in eukaryotic genes often consists of a number of short DNA sequences (**modules**) interspersed with less important sequences. One module, the **tata box**, tells RNA polymerase where to start transcription.

Many eukaryotes are multicellular organisms, with a complex hierarchical structure made up of cells, tissues, and organs. This hierarchy is often reflected by specificity of promoters; for example, many promoters in plant genes are tissue specific. This is important to biotechnologists; for example, when transferring a gene into a plant, it is often possible to restrict activity of the new gene to a specific tissue or organ, through the use of specific promoters.

D. EDITING OF RNA TRANSCRIPTS IN EUKARYOTES

As mentioned in Section III of this chapter, one of the fundamental differences between eubacteria and eukaryotes is that RNA transcripts of eukaryotes are edited to remove introns, before they can function as mRNA. This **splicing** process is usually (but not always!) absent in eubacteria and archaea. When it does occur in prokaryotes, it occurs through a fundamentally different mechanism than in eukaryotes. Eukaryotic RNA transcripts (**pre-mRNA**) are spliced by a complex of RNA and protein (ribonucleoprotein) called a **spliceosome**, whereas prokaryotic splicing is done autocatalytically (i.e., the RNA molecule itself has the ability to edit out introns and join the exons).

E. TRANSLATION OF MRNA INTO PROTEIN

Translation describes the process of protein synthesis by ribosomes, using a mRNA transcript. The combined processes of transcription and translation lead to the **expression** of proteins. How does translation proceed? In eubacteria, mRNA associates with ribosomes soon after transcription has begun. Recall that ribosomes are made up of several subunits of ribosomal RNA (rRNA) associated with specific proteins. In an orderly fashion, transfer RNA (tRNA) molecules charged with amino acids interact with the ribosome and the mRNA, resulting in a chain of amino acids whose sequence is determined by the base sequence of the mRNA. Examination of

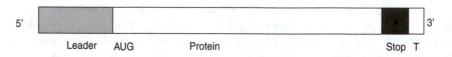

FIGURE 2.12 Structure of eubacterial mRNA. T = trailer.

a typical eubacterial mRNA molecule (Figure 2.12) reveals that the AUG **start** codon marks the beginning point of translation (i.e., the point where the ribosome begins to construct a polypeptide). To the left of this point (toward the 5′ end), there is a **leader** region that is not translated. It is essential for translation, though, because it contains the **Shine–Dalgarno** sequence, part of the **ribosome-binding site**, which is required for the initial binding and interaction of the mRNA with a ribosome. The ribosome-binding site also overlaps into the initial part of the protein-coding sequence of the mRNA. Near the 3′ end of the mRNA, a **stop** or **nonsense** (because it does not code for an amino acid) codon (e.g., UAA) signals to the ribosome the end of the polypeptide. The new polypeptide and the mRNA are then released from the ribosome. To the right of the stop codon (toward the 3′ end) there is a noncoding region referred to as the **trailer**.

In eukaryotes, the mechanism of translation at the ribosomes is similar, but the structure of the mRNA is different. Enzymes in the nucleus add methyl (CH_3) groups to specific points of several bases at the 5′ end of the leader region of the molecule; this is the **methylated cap**. This cap apparently helps the mRNA bind to ribosomes in the cytoplasm. Another processing event occurs in the nucleus — poly A polymerase adds 100 to 200 adenine nucleotides onto the 3′ end of the mRNA transcript. The function of this **poly A tail** is unclear, but may be related to transport of mRNA from the nucleus to the cytoplasm.

F.   Posttranslational Processing of Polypeptides

In eubacteria, polypeptides leave the ribosomes and achieve their final folded conformation, often with the help of small proteins called **molecular chaperones**. This is an essential process, because protein function depends on the three-dimensional conformation of the protein. Disulfide bond formation is a crucial part of this folding process and is partly dictated by the cytoplasmic environment — if the cytoplasm contains an abundance of compounds that have strong reducing power (i.e., a "reducing" environment), disulfide bond formation is less likely to occur. **Disulfide oxidoreductases**, oxidative enzymes that catalyze the formation of disulfide bonds, are also essential to this process in both bacteria and eukaryotes.

Eubacteria have simple cellular structure. When a protein is produced in the cytoplasm, it will function in one of three locations: in the cytoplasm, in the cell membrane, or outside of the cell (**secreted proteins**). Proteins that are destined for the membrane have a specific **signal** sequence of amino acids at their **amino-terminal** end (this is the end that is first synthesized by the ribosome; the opposite end is the **carboxy-terminal** end). As a secreted protein passes through the membrane, a protease removes the signal sequence. A protein that contains a signal sequence is called a **preprotein**.

In eukaryotes, proteins must be **sorted** and **targeted** to various locations in the cell (nucleus, endoplasmic reticulum [ER], vacuole, etc.). This complex process is also guided by amino-terminal (N-terminal) signal sequences. Biotechnologists are primarily interested in secreted proteins; these proteins are synthesized by ribosomes closely associated with the ER. Regions of ER with many associated ribosomes are known as rough ER. Secreted proteins have a signal sequence that leads to rapid interaction with **signal recognition particles** (SRPs). This occurs before the polypeptide has left the ribosome. The SRP binds to **docking** proteins in the membrane of the ER, forming a complex that directs the new polypeptide into the lumen of the ER.

When the polypeptide is released from the ribosome, it begins its journey from the ER to the Golgi apparatus and out of the cell. While it is in the ER, the important process of **glycosylation** begins. This involves the addition of oligosaccharides to specific points in the polypeptide. These carbohydrate chains are added either to the free amino group of asparagine (**N-glycosylation**) or to the OH group of serine, threonine, or lysine (**O-glycosylation**). N-glycosylation begins in the ER and is completed in the Golgi, whereas O-glycosylation occurs only in the Golgi (see Figure 2.6). The exact pattern of glycosylation (i.e., the length of each chain and the nature of the subunits of the chain) varies among eukaryotes; for example, the pattern in yeasts is quite different from that in mammalian cells.

Secreted proteins travel from the ER to the Golgi via **transport vesicles** that are budded off from the ER. From the Golgi apparatus, proteins are transported to the membrane via **secretory vesicles**. The membrane of a secretory vesicle fuses with the cell membrane, releasing its contents to the outside of the cell. For many proteins, though, this is not the end of posttranslational processing. Many proteins exist in a **pro** form (e.g., pro-insulin). Such proteins must be modified (i.e., some of the amino acid sequences must be removed) before the protein achieves its final, active configuration. In some cases (e.g., **prochymosin**), the amino acid sequence is able to catalyze this final processing step; in others, proteases are required. In some cases, environmental factors (e.g., strong acid in the stomach lumen) drive the final conversion to an active protein. Because of this final processing step, secreted proteins are associated with confusing jargon. For example, chymosin (rennet) is initially produced as **preprochymosin**. When the signal sequence has been removed, it is then called prochymosin, and when it has been cut into its final active form, it is called chymosin.

G. RELEVANCE OF DNA TO BIOTECHNOLOGY

The preceding sections may seem like overkill — why do food biotechnologists need to know so much about DNA? Many biotechnological processes involve the transfer of genes from one organism to another. It should now be clear that if this transfer occurs between organisms that are quite distantly related (e.g., from a mammalian cell to a eubacterial cell), problems are likely to occur. Will the promoters, termination signals, ribosome-binding sites, and signal sequences work in the new cell? Will the RNA transcript be correctly edited before translation? Will the new cell have the machinery for proper posttranslational processing? These are crucial questions, and one of the goals of Chapter 3 is to address them.

H. Working with DNA

One of the major barriers confronting newcomers to biotechnology is understanding the jargon associated with the day-to-day manipulation of DNA. Four essential techniques will be described here.

1. Purification of Nucleic Acids

It is often necessary to separate DNA from other cellular constituents. One important reason for this is that cells contain **nucleases** that cut nucleic acid polymers into small fragments or individual nucleotides. Fortunately, it is relatively easy to purify DNA from cells. The cell must first be lysed (broken). With animal cells, this is easily accomplished through the use of solutes such as sucrose (high levels of sucrose cause water flow out of cells, resulting in bursting of the cell), membrane solubilizing agents such as sodium dodecyl sulfate (SDS), and proteases. Cells with walls (e.g., fungal, plant, and bacterial cells) may require additional or different treatments. For example, yeast cells are usually treated with cell-wall-degrading enzymes to yield **sphaeroblasts** (cells bound by a membrane only), which lyse easily.

Once cellular contents have been released, a chemical such as phenol is added; this causes denaturation of any proteins present. Organic solvents such as chloroform are added to solubilize lipids. This mixture is then centrifuged, which results in three distinct layers: one made up of phenol, one aqueous layer, and a layer of chloroform. The aqueous layer is carefully removed. The next step is to add cold ethanol; this precipitates DNA and RNA, which can then be pelleted through centrifugation and resuspended in a small amount of water or buffer. Such DNA samples are usually treated with RNAse (a nuclease that degrades RNA) to eliminate contaminating RNA. RNA samples must be handled very carefully, because RNA is less stable than DNA, and RNAses are much more common in the environment. Gloves can prevent contamination of RNA samples with RNAses present on skin. Using water (in buffers, etc.) that contains chemical inhibitors of RNAses is another preventive measure.

2. Gel Electrophoresis

When working with DNA, it is often necessary to separate fragments of different lengths. Also, determining the length of DNA fragments is frequently useful. Both aims can be accomplished using gel electrophoresis. This technique involves the addition of a sample containing DNA to a gel (agarose or polyacrylamide) that is suspended in a buffer. DNA is loaded into small slots cut into the gel. An electrical current is then applied, and the charge that DNA carries (due largely to the phosphate groups of the backbone) will move the DNA fragments toward the positive pole (Figure 2.13). However, the fragments do not move through the pores of the gel at equal rates; small fragments flow rapidly, whereas large fragments are impeded by the gel structure, and move slowly. Thus, fragments of different sizes are separated. A number of samples can be added to each gel, creating discrete **lanes**. Fragments of known length can be added to certain lanes, allowing measurement of DNA fragment length in the other lanes. Once electrophoresis is completed, DNA in the

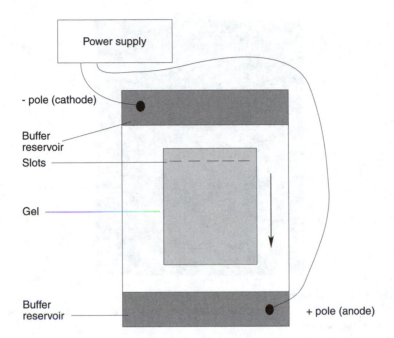

FIGURE 2.13 Gel electrophoresis. The gel is submerged in buffer, and DNA fragments migrate toward the anode (indicated by the arrow).

gel can be visualized by adding **ethidium bromide** to either the gel or the DNA sample. Ethidium bromide binds to DNA and fluoresces visibly when exposed to ultraviolet light (Figure 2.14). Molecular biologists measure DNA size in units of **base pairs** or **kilobase (kb) pairs** (thousands of base pairs). Gel electrophoresis can accurately separate and measure small fragments of DNA, but for segments larger than 40 kb, special techniques (e.g., **pulsed-field gel electrophoresis**) are required.

3.   Blotting and Hybridization

Blotting and hybridization techniques are commonly used to detect sequences of DNA. For example, **southern blotting** can determine if a specific DNA sequence is present in a sample of DNA. This is often done after gel electrophoresis. Blotting refers to the wicking of DNA from a gel or other substance onto a membranous filter. Water is encouraged to flow through the gel and the membrane filter by immersing the gel in a solution and placing a paper towel or similar blotting paper over the filter. As water flows to the blotting paper, DNA flows with it but is trapped on the membrane filter.

Once on the filter, the DNA can be denatured into single strands through immersion in a solution of NaOH. Then, a **probe** can be added (after removing the NaOH). A probe consists of a specific sequence of DNA or RNA that is **labeled** in some way. Two common methods of labeling are to introduce radioactive isotopes to the probe or to covalently link a fluorescent compound to the probe.

1   2   3   4   5   6

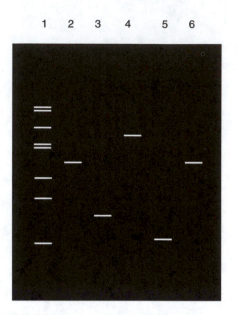

FIGURE 2.14 Separation of DNA fragments by gel electrophoresis. Lane 1 contains frag-ments of known molecular weight. This **ladder** can be used to measure the size of DNA fragments in lanes 2 through 6.

If a sequence in the DNA sample is complementary to the probe, **hybridization** will occur (Figure 2.15). The unhybridized probe is then washed from the filter, and hybridized probes can be detected using either radiation-sensitive film for radioactive probes or fluorescence (production of visible light upon exposure to ultraviolet light) for fluorescent probes. The above procedure can also be used to detect specific mRNA transcripts. This is referred to as **northern** hybridization. RNA or DNA probes can be used in both northern and southern hybridizations.

Hybridization can also be accomplished without gel electrophoresis. **Dot blots**, for example, are made by placing drops of a sample containing DNA onto a nitro-cellulose filter. Labeled probes can then be added. Unhybridized probes are washed away, and hybridization can then be detected. Dot blots offer a quick test for the presence of a particular sequence of DNA and are widely used.

4. DNA Sequencing

Since the advent of the human genome project, DNA sequencing has become part of everyday language. However, it is useful to gain a deeper understanding of the process, because of its central importance to understanding and using genes. If one knows the sequence of a gene, it can be compared to genes from other organisms, and its amino acid structure can be deduced through our understanding of the genetic code. This often leads to substantial insights into protein structure and function and may be highly relevant to biotechnological applications of the gene. Sequencing also has directly practical applications, in that it allows the design of gene probes

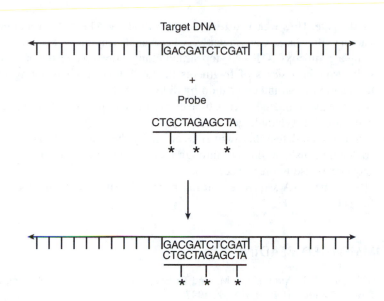

FIGURE 2.15 Hybridization of a DNA probe to a DNA sample. The DNA sample must first be converted to single strands. The probe then is able to hybridize with the complementary sequence in the sample. The probe is labeled by covalently linking it to an enzyme of fluorescent compound (*).

that can be used to detect the gene, as well as the design of primers to allow amplification of the gene using PCR.

DNA sequencing can occur through either the **Maxam–Gilbert** or the **Sanger dideoxy** method. We will describe the Sanger dideoxy method here because it is the most commonly used.

- Obtain a sample of DNA fragments containing the desired gene.
- Divide the sample into four tubes.
- To each tube, add nucleotides of the four bases plus a synthetic analogue of one of the bases. For example, the A tube would contain guanidine triphosphate (GTP), cytosine triphosphate (CTP), thymidine triphosphate (TTP), ATP, and **dideoxy ATP**. All of the nucleotides are labeled to allow detection of either radioactivity or fluorescence (see blotting section, above).
- Add DNA polymerase and DNA primers to each tube. DNA polymerase uses the triphosphate nucleotides ATP, CTP, TTP, and GTP to add nucleotides to the primer, creating a new strand of DNA. However, it is unable to add dideoxy ATP, CTP, TTP, or GTP onto a strand. Thus, when DNA polymerase is forced to use a dideoxy nucleotide, strand elongation stops.
- Consider the A tube; it contains ATP as well as dideoxy ATP. As DNA polymerase adds nucleotides to the new strand, it will add ATP whenever it "sees" thymine in the template DNA (i.e., the DNA fragment that is being sequenced). If it adds a normal ATP, then the strand will continue

to elongate. However, if it tries to add a dideoxy ATP, then strand elongation will cease. Consequently, wherever T (thymine) occurs in the template, dideoxy ATP will stop strand elongation *some of the time*. This will result in a series of fragments in the A tube, each resulting from termination of strand elongation by dideoxy ATP.

- Separate the fragments in each tube by high-resolution gel electrophoresis, using a polyacrylamide gel.
- You can then detect the fragments, either by the use of autoradiography (radiation-sensitive films) or through fluorescence (many automated gene sequencers use fluorescence).
- "Read" the DNA sequence, starting from either the top or the bottom of the gel.

RECOMMENDED READING

1. Madigan, M. T., Martinko, J. M., and Parker, J., *Brock Biology of Microorganisms*, 8th ed., Prentice Hall, New York, 1997.
2. Sahm, H., Prokaryotes in industrial production, in *Biology of the Prokaryotes*, Lengeler, J. W., Drews, G., and Schlegel, H. G., Eds., Verlag, Stuttgart, 1999.
3. Deacon, J. W., *Modern Mycology*, 3rd ed., Blackwell Science, Oxford, 1997.
4. Lodish, H., Berk, A., Zipursky, S. L., Matsudaira, P., Baltimore, D., and Darnell, J. *Molecular Cell Biology*, 4th ed., Freeman, New York, 2000.
5. Lewin, B., *Genes*, 7th ed., Oxford University Press, Oxford, 2000.
6. Wolff, R. and Gemmill, R., Purifying and analysing genomic DNA, in *Analyzing DNA: A Laboratory Manual*, Cold Spring Harbor Laboratory Press, Cold Spring Harbor, New York, 1997, chap. 1.

3 Gene Cloning and Production of Recombinant Proteins

I. WHAT IS A RECOMBINANT PROTEIN?

Consumers generally recognize the importance of protein to their diet. However, few people are aware of the important role of specific proteins as food modifiers or additives. Amylases, for example, are used to transform starch into sweeteners such as glucose. Other amylases are used by the brewing industry to help convert starch into fermentable carbohydrates. Proteases can be used to tenderize meat and to coagulate milk during cheese making.

Usually, such enzymes are derived from plant, animal, or, more commonly, microbial sources. These are "natural" sources of the proteins. In contrast, recombinant proteins are derived from organisms that do not naturally make the protein. The ability to make the protein is transferred into the organism. In molecular terms, the DNA that codes for the protein is transferred. As an example, consider a protease that is naturally found in pineapples. Using recombinant DNA technology, we can isolate the DNA coding for this enzyme from the rest of the pineapple DNA and introduce it into a bacterium such as *Escherichia coli*. Then, we can grow large amounts of the bacterium and purify the enzyme from the culture. The protease would then be a **recombinant** protein. In the parlance of molecular biology, it would also be called a **heterologous** protein, because it is produced by an organism different from the original source organism.

Food biotechnologists usually produce recombinant proteins in bacteria or fungi, because of the ease of large-scale culture of these organisms. In contrast, mammalian or human cell lines are often the source of recombinant proteins for human therapy; in the future, therapeutic proteins will also be produced in whole plant and animal systems.

A more common food-related application of recombinant proteins is in the development of transgenic (genetically modified) crops. The recombinant protein improves the crop, benefiting food producers, processors, or the consumer. This important and controversial application of recombinant DNA technology will be discussed in Chapter 4.

II. WHY BOTHER MAKING RECOMBINANT PROTEINS?

To make a recombinant protein, one has to transfer the specific gene from one organism to another. This is not always easy. Why, then, do people bother? Why not just obtain the protein from the organism that naturally produces it?

Sometimes it is not practical to use the natural producer. For example, human growth hormone is used to treat dwarfism, a condition that arises when children are congenitally deficient in human growth hormone. This specific protein is found only in humans; consequently, the only natural source of this hormone is from human cadavers. Unfortunately, only minute amounts can be obtained from cadavers. However, once the gene for human growth hormone was transferred to mouse cell lines, it could be produced relatively cheaply in large amounts.

Bovine growth hormone (BGH), also known as bovine somatotropin (BST), is found in cows. Similar to human growth hormone, BGH cannot easily be isolated from its natural source. However, in the 1980s, the gene for BGH was transferred from cows to *E. coli*, and the resultant recombinant bacteria were able to produce intact BGH (rBGH, or recombinant BGH, to differentiate it from natural BGH). This made it cost effective to sell rBGH to dairy farmers. When rBGH is injected into cows, milk production usually increases. We will discuss this controversial recombinant protein further in Section XII of this chapter.

Chymosin is another example of a protein that is available from either natural or recombinant sources. This enzyme has been used for thousands of years to increase the rate and extent of coagulation of milk during cheese production. Most of the cheese produced in North America is made using a recombinant form of chymosin. One advantage of recombinant chymosin is that it has more consistent effects on its substrate (casein in milk) than chymosin from rennet, the traditional source. Recombinant chymosin will also be discussed at the end of this chapter (Section XI), because the story of its development demonstrates many of the technical problems associated with making recombinant proteins in microbes.

In some cases an enzyme may be readily available from plant or microbial sources, but the enzyme may have undesirable qualities. For example, several microbial proteases can be used instead of chymosin, but they tend to produce off flavors in cheese, because of excessive proteolytic activity. The enzymatic activity could theoretically be directly modified by changing the amino acid sequence of the active site. This can be done through **site-directed mutagenesis**, and the resulting protein would be considered a recombinant protein, even if it was produced by the organism that produced the original protein.

The aim of this chapter is to explain the processes that are used to **clone genes** into bacteria and fungi and that allow the production of recombinant proteins from these cloned genes. These processes are important because of the increasing presence of recombinant proteins in the food supply and because many of the techniques described in this chapter are used in other kinds of food biotechnology. Understanding the basic methods of gene cloning is essential to many of the subsequent chapters of this book. Understanding the techniques of producing recombinant proteins is also crucial for a complete understanding of the controversies associated with recombinant food additives (e.g., recombinant chymosin) and recombinant plants (e.g., herbicide-resistant soybeans). One of the most controversial aspects of genetically modified foods is the inclusion of antibiotic resistant genes. Why are these genes necessary? A close look at gene cloning principles is necessary to answer this

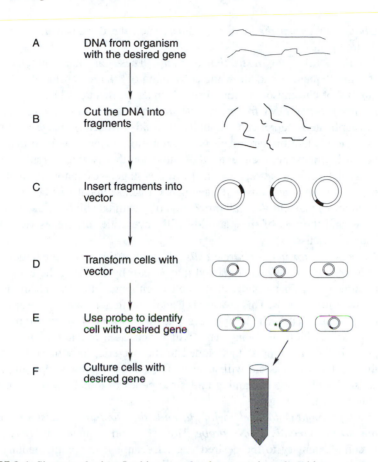

A — DNA from organism with the desired gene

B — Cut the DNA into fragments

C — Insert fragments into vector

D — Transform cells with vector

E — Use probe to identify cell with desired gene

F — Culture cells with desired gene

FIGURE 3.1 Shotgun cloning. In this example, the vector is a plasmid.

question. Gene cloning is the necessary first step toward production of a recombinant protein in a microbe or plant.

The transfer of a gene from one organism to another can be conceptually divided into two processes: (1) cloning the gene and (2) transferring the cloned gene into the desired organism.

III. HOW AND WHY ARE GENES CLONED? (THE BIG PICTURE)

To transfer a desired gene from one organism (the *source*) to another (the *target*), one must *separate* the gene from all other genes in the source organism. Otherwise, one loses the principal advantage of a recombinant organism — precise transfer of genes, rather than wholesale mixing of large numbers of genes. One way of achieving this goal is to use the following approach, commonly referred to as **shotgun cloning** (Figure 3.1):

1. *Obtain DNA from the source organism* (i.e., the organism that has the desired gene).
2. *Cut the DNA using restriction enzymes.* These enzymes cut DNA at specific sequences and allow the conversion of a large chromosome, or a number of chromosomes, into a set of linear segments of DNA.
3. *Incorporate the DNA fragments into a vector.* A vector is an agent that can replicate in target cells. Usually plasmids or bacteriophage are used as vectors when the target cell is a bacterium. For now, assume that the vector is a plasmid, a circular DNA molecule that is able to replicate in host cells (see Chapter 2, Section II). At this stage, we cannot differentiate between wanted and unwanted segments. Each segment will be incorporated into a plasmid. A small proportion of plasmids will have the desired gene, and the rest of the plasmids will have undesired genes from the source organism.
4. *Move the vector into the target cells.* Usually, at this stage, the target cell is *E. coli*. Plasmids can be moved into *E. coli* by exposing the cells to a treatment that opens pores in the cell membrane, allowing plasmids to diffuse into the cells. This process is usually inefficient, and it is necessary to eliminate cells that do not have plasmids. Plasmid vectors typically have an antibiotic-resistant gene, which acts as a **marker** or indicator gene. If the antibiotic is then added to the bacteria, only those bacteria that contain the plasmid will survive. The next step allows separation of cells with plasmids containing undesired genes from cells with plasmids containing desired genes.
5. *Use a probe to isolate cells with plasmids that contain the desired gene; the most common is a DNA probe.* This is a short sequence of DNA that is complementary to the desired gene. Because it is complementary, it will anneal to the desired gene, but *it will not anneal to any other fragment of DNA.* The probe must also be linked to a compound that can be easily detected; radioactive or fluorescent compounds are most popular.
6. *Grow large amounts of cells containing the desired gene.* These cells have the required genetic information for producing the desired recombinant protein. The protein can therefore be harvested from cultures of these cells. *The desired gene has been successfully cloned.*
7. The target cell should be able to efficiently synthesize large amounts of the protein coded for by the desired gene. Often, though, the target cell, for a variety of reasons, will not synthesize enough of the desired protein. You may need to repeat the procedure, transferring the desired gene to a cell that is better suited to the synthesis of large amounts of the protein. This procedure is referred to as **subcloning** and is conceptually similar to shotgun cloning. However, it is usually much easier than shotgun cloning, because the interfering "background" of the source organism's genome is absent.

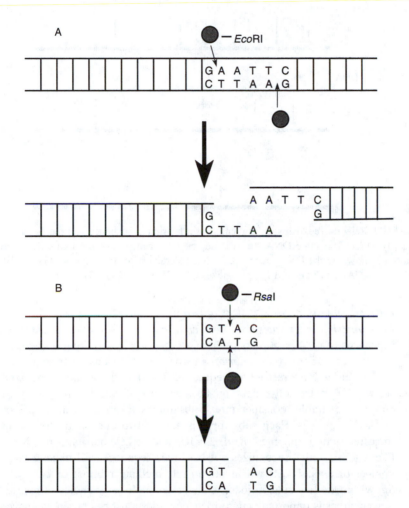

FIGURE 3.2 Effects of the restriction enzymes *Eco*RI (A) and *Rsa*I (B) on DNA.

IV. GENE CLONING: THE DETAILED PICTURE

A. RESTRICTION ENZYMES

Restriction enzymes are classified as endonucleases because they cut DNA at points within the molecule. The crucial advantage they have over other endonucleases is that they do not cut DNA at *random* points. Instead, they cut at *specific* DNA sequences. For example, EcoRI will cut DNA whenever the enzyme encounters the sequence GAATTC (Figure 3.2). Therefore, the catalytic action of the enzyme is *restricted* to certain sequences of DNA. The following points about restriction enzymes are important to their use in gene cloning:

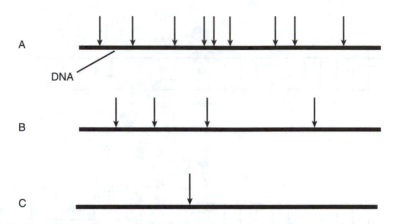

FIGURE 3.3 Relationships between the length of a restriction sequence and the frequency of cleavage of DNA. The same DNA strand is depicted in each part of the figure. Each arrow indicates a point where the DNA strand is cut. *Rsa*I (A) recognizes the sequence GTAC; *Eco*RI (B) recognizes GAATTC; and *Sfi*I (C) recognizes GGCCNNNNNGGCC (N = any base).

- *Each enzyme usually recognizes only one sequence.*
- *Many restriction enzymes have been characterized.* Some recognize short sequences (e.g., 4 base pairs), whereas others recognize long sequences. A strain of *Streptomyces fimbriatus* recognizes a 13 base pair sequence, an unusually long restriction sequence. The length of the recognized sequence is extremely important because that affects the frequency of cuts. For example, consider three subsamples taken from a sample of DNA (Figure 3.3). Each subsample is exposed to one of the following three restriction enzymes: *Rsa*I, *Eco*RI, and *Sfi*I. What happens? Most likely, *Rsa*I, which recognizes a 4 base pair sequence will make a large number of cuts. *Eco*RI will make an intermediate number of cuts, and *Sfi*I will make few cuts. Evidently, the biotechnologist's choice of restriction enzymes is important; appropriate enzymes must be chosen whenever restriction enzymes are used in a cloning procedure. For each type of vector, there is a limit to the amount of DNA that can be inserted. The correct restriction enzyme will result in the correct fragment size for a particular vector.
- *Many restriction enzymes create "sticky ends."* These are "loose" single strands of DNA protruding at the cleavage site (see Figure 3.2A). Sticky ends are useful because complementary sticky ends will anneal to each other. This allows the *joining* of DNA from different sources. If each of two samples of DNA has been treated with the same restriction enzyme, and if this enzyme generates sticky ends, then, when the samples are *mixed*, annealing will occur between DNA strands from the *two different samples*. However, sticky ends from the original sample may also re-anneal, because they are also complementary. Also, certain restriction enzymes create **blunt ends** (see Figure 3.2B) rather than sticky ends.

- *Usually, the DNA sequence recognized by a restriction enzyme is a* palindrome. The sequence of DNA is identical in both strands, when read from the same orientation (e.g., from 5′ to 3′; see Figure 3.2).

Restriction enzymes are isolated from bacteria. At first encounter, these enzymes appear to be unusual. Why would a bacterium make such unusual enzymes? It turns out that restriction enzymes are part of a bacterial defense system against bacteriophage. This system works because restriction enzymes degrade foreign DNA, such as phage DNA that has entered the cell. Of course, bacterial DNA always contains sites recognized by its own restriction enzymes. Specific enzymes recognize restriction sites and methylate them. This protects bacterial DNA from digestion by its own restriction enzymes.

B. PLASMID VECTORS

Plasmids are vital tools for biotechnologists. Over the past 20 years, these circular strands of DNA have been extensively used to clone genes. Plasmids are useful because they are small, easily manipulated, and can be "grown" in large quantities. Their use in cloning is best explained with the following scenario: you have isolated a fungus from soil that produces an enzyme (amylase X) that is able to efficiently break down starch into valuable food ingredients. Unfortunately, the fungus is difficult to grow in large-scale culture, so you decide to try to move the gene (*AmylX*) coding for this enzyme into *E. coli* (Figure 3.4). You use pBR322, a commonly used plasmid vector. Examination of pBR322 (Figure 3.5) reveals that it contains an *ori* sequence, necessary for replication in *E. coli,* and two antibiotic-resistant genes. These antibiotic-resistant genes are vital in the cloning process, which will now be explained in detail.

Step 1: Purify the DNA containing the desired gene. This is usually a straightforward procedure (see Chapter 2, Section V). One of the advantages of cloning procedures is that only small amounts of DNA are required. Therefore, if the organism that contains the desired gene is in short supply, or grows very slowly, this is not usually a debilitating problem. We will refer to this DNA as foreign DNA.

Step 2: Purify plasmid DNA. This is done by growing *E. coli* containing pBR322, and then breaking open the bacterial cells to release cellular contents. Plasmid DNA can then be easily purified and separated from chromosomal DNA, because the small size of plasmids gives them physicochemical characteristics different from chromosomal DNA.

Step 3: Cut both plasmid and foreign DNA with the same *restriction enzyme.* In this case, *Bam*HI is a good choice. This enzyme cuts DNA at the following sequence: GGATCC. There are three reasons for this step:

  a. The fungus contains several different chromosomes, which are all present in the purified DNA sample. However, only one chromosome contains *AmylX,* and you want to clone only the portion of the chromosome that contains *AmylX.* If the DNA is cut into small pieces, each

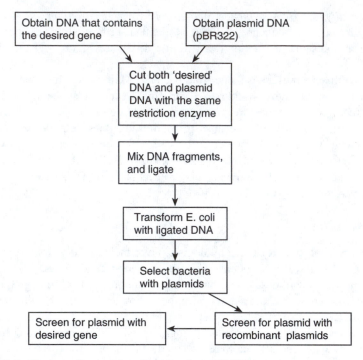

FIGURE 3.4 Cloning·of a gene into *E. coli* using pBR322.

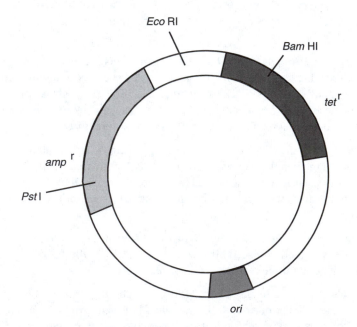

FIGURE 3.5 Structure of pBR322. There are two genes for antibiotic resistance: *tet*^{r} and *amp*^{r}.

piece can theoretically be integrated into separate plasmids. It is then relatively easy to separate the plasmid containing the desired gene from plasmids containing the "noise" (the rest of the fungal DNA).

b. It is necessary to cut the fungal DNA into small pieces because it is impractical to attempt to incorporate more than 4 kb (four thousand bases) of DNA into pBR322. This restriction applies to many plasmid vectors, but other vectors, such as lambda phage, cosmids, and yeast artificial chromosomes, can accommodate larger segments of DNA.

c. *Bam*HI creates sticky ends. This will allow the segments of foreign DNA to anneal to plasmid DNA. Note that there is only one *Bam*HI **restriction site** (G↓GATCC) in pBR322 (i.e., there is only one sequence in pBR322 that *Bam*HI recognizes and cuts). Therefore, the only effects of digestion with *Bam*HI will be to convert the circular plasmid to a linear plasmid and to generate sticky ends. If there was more than one restriction site, pBR322 would be cut into several pieces, and it would no longer be a useful vector because each piece would likely be missing an *ori,* or an antibiotic-resistant gene. It is also important to note that the *Bam*HI restriction site occurs *within the tetracycline-resistant gene.* This will become important in step 6.

Step 4: Mix the DNA containing the desired gene with the vector DNA (Figure 3.6). Random collisions between pieces of DNA with complementary sticky ends will result in annealing, or joining, of DNA segments. This annealing is not permanent, unless **DNA ligase** is added, because simple annealing does not create covalent links between the sugar phosphate backbones of the segments. DNA ligase covalently welds (**ligates**) annealed segments of DNA. The main goal here is to join segments of the foreign DNA with plasmid DNA, forming recombinant plasmids. If this step is successful, you will obtain a population of recombinant plasmids, some of which have the desired gene and many of which have other segments of foreign DNA. In addition to recombinant plasmids, the population usually contains re-annealed plasmids that lack any inserted DNA. These re-annealed plasmids can be separated from recombinant plasmids in step 7.

Step 5: Transform E. coli *with the mixture of annealed and ligated DNA.* The desired plasmids cannot be isolated unless they are introduced into a microorganism. The most common way to introduce plasmids and other DNA sequences into *E. coli,* and other bacteria, is via **transformation**.

Transformation is a confusing word, because it has been given diverse meanings by different groups of biologists. In the context of gene cloning, transformation refers to *the uptake of DNA from aqueous solution by microorganisms.* Most microorganisms cannot do this; one notable exception is *Streptococcus pneumoniae.* This bacterium has specific proteins that allow DNA from other cells of *S. pneumoniae* to enter the cell. The DNA may then be integrated into the cell's chromosome.

E. coli, however, does not naturally take up DNA in this manner. Bilayer lipid membranes will not allow the diffusion of DNA into cells, and *E. coli* lacks the proteins that allow *S. pneumoniae* to take up DNA. Fortunately for biotechnologists,

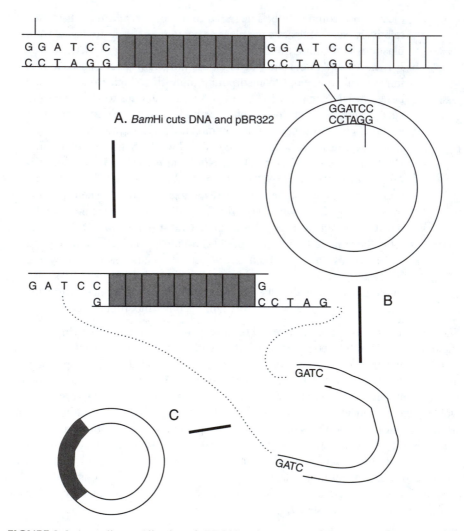

FIGURE 3.6 Annealing and ligation of pBR322 and a segment of DNA containing the desired gene. Both plasmid and DNA are cut with *Bam*HI (A). They are then mixed together, allowing annealing of plasmid and DNA (B). Finally, ligase is added, to covalently bond the plasmid to the DNA insert (C).

E. coli can be treated (i.e., be transformed) to increase their ability to take up DNA. The most common transformation method for *E. coli* is to expose it to a dilute solution of cold $CaCl_2$. This appears to transiently "freeze" lipid membranes, resulting in the formation of fissures or pores that are large enough to allow the diffusion of DNA into the cell. One disadvantage to this method is that it is extremely inefficient — only $1/10^6$ of the bacterial cells will be transformed. An additional problem is that many bacterial species cannot be transformed using this method.

Once the plasmid DNA is inside *E. coli*, it can be stored there indefinitely. DNA polymerase will bind to the *ori* region of the plasmid and will replicate it, preventing

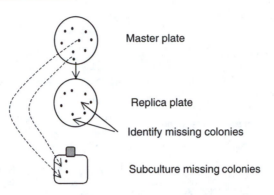

Master plate

Replica plate

Identify missing colonies

Subculture missing colonies

FIGURE 3.7 Separation of bacteria containing recombinant pBR322 from bacteria with nonrecombinant pBR322. Recombinant bacteria have a disrupted tetracycline gene and are sensitive to tetracycline, which is present in the replica plate but not the master plate. Once sensitive bacteria have been identified, they are subcultured from the master plate.

loss of the plasmid as the bacterium divides. The amount of replication that occurs varies widely among plasmids. Plasmids that replicate frequently are considered to be **high-copy-number** plasmids. PBR322 is a low-copy-number plasmid and is consequently not as popular as other, more frequently replicating plasmids, such as the pUC group of plasmids.

Transformation will yield a mixture of bacteria. Most will remain untransformed and will lack plasmids. A small proportion of the transformed bacteria will contain plasmids with the desired foreign gene, and the remainder will contain plasmids with undesired sequences of foreign DNA. Bacteria that contain plasmids with inserted segments (**inserts**) of foreign DNA are considered to be **recombinant**.

Step 6: Select bacteria transformed with plasmids. Untransformed bacteria can easily be removed by adding ampicillin. *E. coli* is susceptible to ampicillin, so the only bacteria that will survive in the presence of ampicillin are those that have been successfully transformed with pBR322. The *bla* gene, found in pBR322, codes for β-lactamase, which degrades ampicillin.

Step 7: Screen for recombinant clones. Application of ampicillin results in a mixture of bacteria that have been successfully transformed. Even if steps are taken to prevent the re-annealing of plasmids without inserts, it is usually necessary to separate re-annealed plasmids from plasmids containing inserts. Plate the mixture of transformed bacteria onto a solid agar medium containing ampicillin (the **master plate**; Figure 3.7). The bacteria must be diluted sufficiently, so that each colony grows from a single cell. Bacteria within each colony, then, are *clones*; they are genetically identical, because they share the same plasmid and the same insert within the plasmid.

A sterile velvet cloth is then carefully pressed against the master plate and pressed onto an agar plate that contains tetracycline. This will result in the transfer of bacteria from each colony onto the tetracycline-containing plate. It is essential to maintain the orientation of the velvet cloth, so that the bacteria are pressed onto the

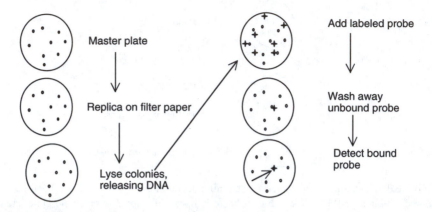

FIGURE 3.8 Screening of recombinant libraries using a DNA probe.

tetracycline-containing plate in the same pattern as the pattern of colonies on the ampicillin-containing plate. This plate is referred to as a **replica plate**.

Bacteria that have the original, recircularized plasmid are able to grow on media containing tetracycline, because of the presence of the gene giving tetracycline resistance (see Figure 3.5). However, bacteria that have recombinant plasmids are killed by tetracycline, because they do not have an intact tetracycline-resistant gene. Insertion of foreign DNA at the *Bam*HI restriction site (see Figure 3.5) breaks the tetracycline-resistant gene into two fragments, with the foreign DNA between them. This results in transcription of a gene that is not functional; i.e., it does not give the cell resistance to tetracycline. Therefore, cells with recombinant plasmids (recombinant clones) grow on the ampicillin-containing plate but not on the tetracycline-containing plate. Recombinant clones, once identified, are subcultured from the ampicillin-containing plate. The resulting mixture of recombinant clones is often referred to as a **DNA library**.

Step 8: Screen the DNA library. Few of the recombinant plasmids in a shotgun DNA library contain the desired gene; most have inserts of undesired foreign genes. It is now necessary to screen the recombinant clones to identify those that contain the desired gene. The most common strategy is to use a **DNA probe** (Figure 3.8). This is a short segment of single-stranded DNA that is complementary to a portion of the desired gene. It must be **labeled** in some way. Often, radioactive elements such as ^{32}P are incorporated into the probe, but nonradioactive alternatives are also widely used. Nonradioactive probes are either fluorescent or biotin labeled. Fluorescein and several proprietary compounds can be covalently linked to oligonucleotides; hybridization of these probes to DNA can be detected by applying ultraviolet light. Fluorescent compounds emit light of a longer wavelength (often in the visible spectrum) when exposed to ultraviolet light. Thus, hybridized probes will "light up." Biotin-labeled probes work on a different principle; biotin is a vitamin that binds specifically to avidin (a protein found in eggs) or streptavidin (a bacterial protein). Avidin can be covalently linked to enzymes such as horseradish peroxidase that, when given synthetic analogues of their normal substrate, form a colored end product. Thus, hybridization can be detected through the formation of a complex consisting of DNA bound to biotin-labeled probe, which is in turn bound to enzyme-linked avidin. When the synthetic substrate is added, a colored end product will form.

The first step in probing a library is to plate out the library on an agar medium (master plate). A filter paper is then pressed onto the master plate, removing bacteria from each colony while preserving the orientation of the master plate. Cells are then lysed by the application of alkaline chemicals (e.g., NaOH), which also denatures plasmid DNA, yielding single strands. This allows the single-stranded probe to anneal specifically to complementary single-stranded DNA on the filter paper. This is an example of **hybridization** (the annealed double-stranded DNA is a hybrid of one strand of probe DNA and one strand of plasmid DNA; see also Chapter 2, Section H).

The probe is then added to the filter paper and, after incubation, the filter paper is washed to remove unbound (not annealed) probe. Hybridization is then detected via radioactive or fluorescent emission, depending on the type of probe.

Once the desired recombinant colonies have been identified, the bacteria can be picked from the master plate and grown in large volumes. At this point, **the gene has been cloned.** Usually, biotechnologists will take further steps to confirm that the correct gene has been transferred to *E. coli*. Confirmation can be made by determining the DNA sequence of the insert or by testing for presence of the desired protein in cultures of the recombinant bacterium.

DNA probes cannot be used if nothing is known of the DNA sequence of the desired gene. However, this is rarely the case; for most genes, sequence information is available for similar genes in other organisms. If such information is not available, then it may be necessary to use another method of library screening. If the desired gene is an enzyme, then it may be possible to screen clones for the desired enzyme activity.

V. THE cDNA ALTERNATIVE

Shotgun cloning works well when the desired gene is from bacteria or other organisms that have a small genome. However, it is less successful when we try to clone genes from eukaryotes, especially eukaryotic plants and animals, which have particularly large and complex genomes. One reason for this lack of success lies in the large number of recombinant clones that must be screened (most screening processes are laborious and time consuming).

However, there is a more fundamental problem. As discussed in Chapter 2, Section II.B, DNA sequences in eukaryotes are not directly transcribed and translated into protein. The primary mRNA transcript typically contains several segments (introns) that are "edited out" before translation. The sequences that survive editing (exons) are spliced together into a mRNA molecule that leaves the nucleus and is translated by ribosomes.

This makes it difficult to use shotgun cloning for eukaryotic genes. If the entire DNA sequence for a eukaryotic gene is transferred to *E. coli*, the protein that is produced will be completely different from the native eukaryotic protein. *E. coli* lacks editing ability, and the entire primary mRNA transcript would be translated

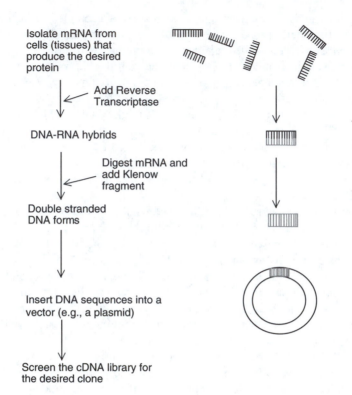

FIGURE 3.9 The use of reverse transcriptase to make a cDNA library. Reverse transcriptase uses mRNA as a template to produce a complementary strand of DNA. Other enzymes are then used to eliminate the mRNA and produce a double-stranded DNA molecule. The DNA segments can then be inserted into vectors, leading to construction of a cDNA library. This library can then be screened in the usual manner.

into protein. It is highly unlikely that such a protein would retain the properties of the original protein.

Fortunately, there is a solution, which comes from an unlikely source: **retroviruses**. Retroviruses use RNA to store their genetic information. As part of their infection cycle, they convert the RNA to DNA, which is then incorporated into the DNA of the host cell. **Reverse transcriptase** catalyzes this conversion. This enzyme is used by biotechnologists to generate DNA from edited mRNA (Figure 3.9). The DNA produced is double-stranded and does not contain introns. This strategy is attractive because mRNA can usually be isolated from the organism that naturally produces the desired protein. Reverse transcriptase then uses the mRNA strands as templates to generate a complementary strand of DNA. This results in a **hybrid** molecule consisting of one strand of RNA and one strand of DNA. The mRNA can then be digested away using specific nucleases, and the **Klenow** fragment of DNA polymerase I will use the single-stranded DNA as a template for a complementary DNA strand. The Klenow fragment of DNA polymerase I is able to synthesize complementary strands of DNA from a DNA template, but it lacks some of the nuclease activity of intact DNA polymerase. It is used for a variety of molecular biology procedures.

The combined action of reverse transcriptase and the Klenow fragment results in a double-stranded DNA molecule. The rest of the cloning procedure is similar to that described for shotgun cloning; the only difference is that the DNA used in step 3 is obtained from edited mRNA transcripts, not from the genomic DNA of the organism that naturally produces the desired gene.

DNA produced from mRNA is referred to as **cDNA** (c = copy), and libraries obtained by this method are **cDNA libraries**. This approach has allowed the cloning of a large number of eukaryotic genes that could not be cloned using shotgun cloning.

VI. POLYMERASE CHAIN REACTION: A REVOLUTION IN CLONING

A. OVERVIEW

The last decade has seen an extraordinary increase in the use of **polymerase chain reaction** (PCR) for cloning and other molecular biological procedures. This revolution has occurred because PCR allows repeated replication (amplification) of specific sequences of DNA. One drawback is that information about the sequence of the desired gene is required. More specifically, the sequence at both ends of the gene must be known; these sequences are used to construct **primers**. The only other requirements are a heat-stable DNA polymerase, a mixture of nucleotides, and a **thermal cycler**, which is simply a device that can rapidly heat and cool a solution in a programmed sequence. The heat stability of the DNA polymerase is essential to the process. Fortunately, such enzymes can be isolated from a variety of prokaryotes that inhabit water close to geothermal vents; many of these organisms can grow in water at or above 100°C.

PCR is often used to simplify gene cloning. If we know enough about a desired gene to construct primers, we can use PCR to locate the DNA coding for the gene and incorporate it into a plasmid vector. Cloning is completed by transformation of the recombinant plasmid into a suitable host. This works because PCR amplifies only the desired gene. Therefore, virtually all of the noise (i.e., the rest of the genome) is eliminated at an early stage of the process.

DNA amplified from eukaryotic genomic DNA contains exons and introns, and consequently will not be useful. However, DNA coding only for exons can be obtained by starting with mRNA, using reverse transcriptase to convert the mRNA to cDNA. We can then use PCR to amplify the cDNA.

B. THE MECHANISM OF AMPLIFICATION

First, it is important to understand primers. Each primer is a short sequence (usually ~20 base pairs) of single-stranded DNA, and the base sequence of the primers is complementary to the flanking ends of the gene (Figure 3.10A). The other requirement is that the primers must be complementary to different strands. If the primers are identical to flanking ends of the *same strand* of DNA, the reaction will not work.

A

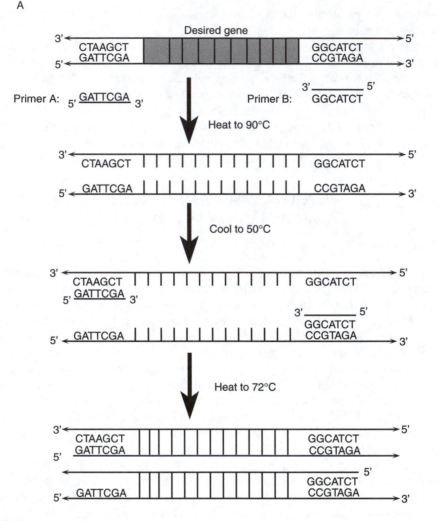

FIGURE 3.10 Amplification of DNA by polymerase chain reaction. In the first cycle (A), primers bind to each strand, allowing DNA polymerase to synthesize complementary strands. Note that each new strand extends past the desired region (shaded) in one direction. In the second cycle (B), primers bind again to each strand. However, note that one of the new strands *runs only between the primers*. The noise (undesired DNA) outside of the shaded region has not been amplified. In the third cycle (C), the amplified segments containing only the region between the primers are starting to dominate (numerically) segments containing regions outside the primers. This domination increases exponentially as the cycles continue.

Once we know the flanking sequences to a desired gene, we can simply construct (or buy) oligonucleotides of the correct sequence, which can then be used as primers. The synthesis of oligonucleotides (short segments of DNA or RNA with a defined base sequence) is an automated, relatively inexpensive procedure.

Primers are added to the DNA sample that contains the desired gene, a heat-stable DNA polymerase is added, and this **reaction mixture** is placed in a thermal

B

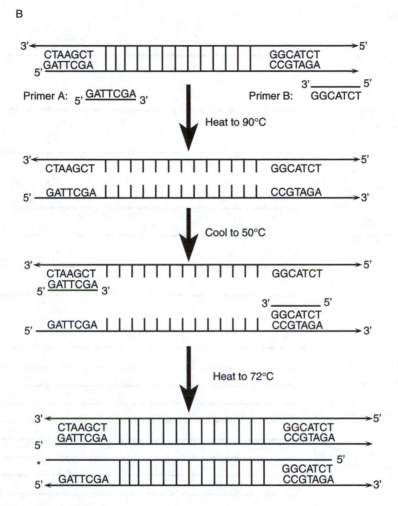

FIGURE 3.10 (continued).

cycler. The PCR consists of a number of repeated cycles. The following steps are part of a typical PCR cycle:

1. The reaction mixture is heated to 90°C. This results in separation of double-stranded DNA into single strands.
2. The mixture is cooled to 50°C. This allows annealing of the primers to the single-stranded DNA templates. Note that the primers anneal to *separate* strands of DNA.
3. The mixture is heated to 72°C. This allows DNA polymerase to initiate DNA synthesis, *starting at the primers*. DNA polymerase requires a primer for initiation of synthesis of a new DNA strand, and it synthesizes the new strand in the 5′ to 3′ direction.

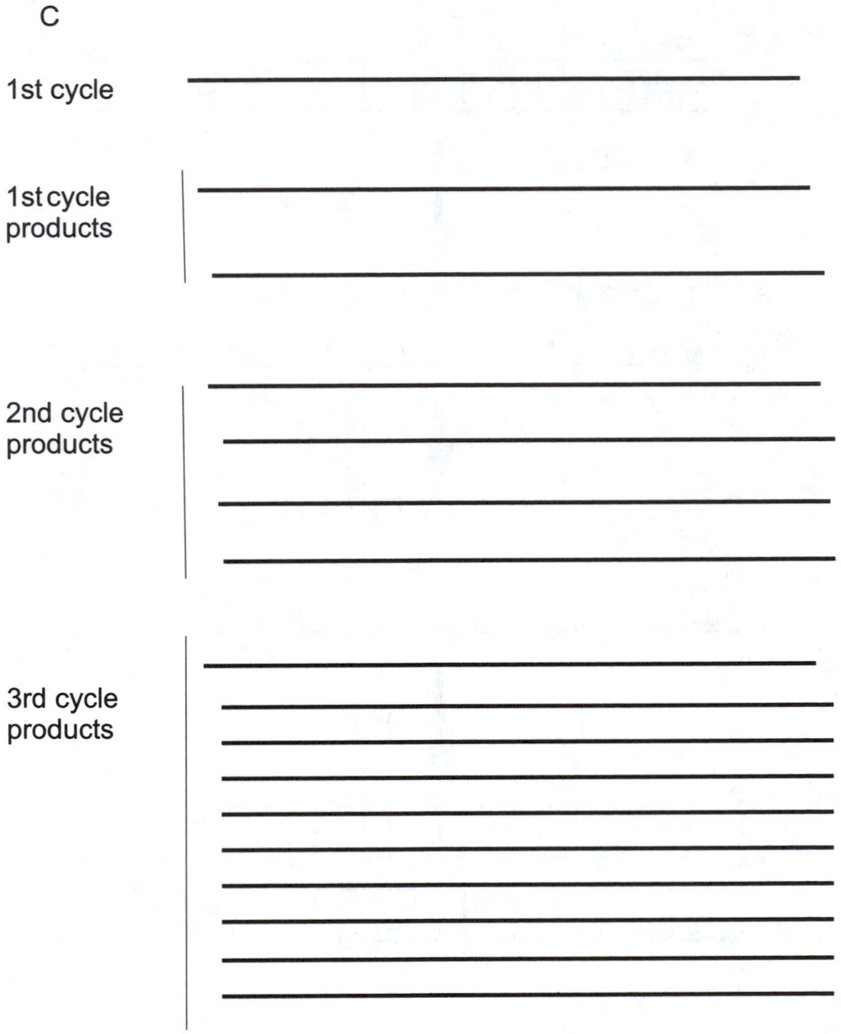

C

1st cycle

1st cycle
products

2nd cycle
products

3rd cycle
products

FIGURE 3.10 (continued).

The cycle is then repeated 20 to 30 times. This results in amplification of the DNA sequence *between the primers*. Close examination reveals why other DNA sequences are not amplified. The orientation of the primers always directs DNA polymerase *into the gene*. After several cycles of PCR, the sequence between the primers is present in much higher numbers than the rest of the DNA sample. Eventually, after 30 to 40 cycles, the amplified DNA has been copied 10^9 times, effectively drowning out the original sample of DNA.

Let's go through the first few cycles, to illustrate how this works. In the first cycle (Figure 3.10A), the DNA is heated to 90°C and the double-stranded DNA denatures into single strands. When the temperature decreases to 50°C, the primers anneal to complementary sequences. Note the orientation (5′ to 3′) of each DNA segment; remember that complementary sequences will anneal only if the two strands are antiparallel, and will not if they are parallel (same orientation). Note also that each of the template strands at this stage extends in both directions.

What happens when we allow DNA polymerase to replicate new strands of DNA? DNA polymerase *must have a primer*; it requires a sequence of DNA that can be used as the initial segment of the new strand. By using designed primers, we control the starting point for DNA replication. Also, DNA polymerase can synthesize the new strand of DNA in *only one direction*. The new strand is synthesized in the 5′ to 3′ direction. So, after one round of replication, each of the original template strands has been amplified.

The second cycle (Figure 3.10B) reveals the specific nature of the amplification. To keep things simple, consider what happens to only one of the newly synthesized strands. As in the first cycle the first step is to increase the temperature to 90°C, resulting in denaturation of the double-stranded DNA into single-stranded DNA. Then we decrease the temperature and allow annealing to occur. Note that only *one* of the primers has the correct orientation to anneal to this strand. DNA polymerase then synthesizes a new strand, starting from the primer. The new strand is (as always) synthesized in the 5′ to 3′ direction. The newly synthesized strand is a short segment that *includes only the two flanking primer regions and the DNA sequence between the primers*. The rest of the original segment of DNA is no longer present. This is the crucial feature of the PCR; it leads to *selective amplification* of the DNA between the primers. Subsequent cycles result in more and more of these short segments (Figure 3.10C), and eventually, after 20 or more cycles, they are present in large enough quantities that they can be visualized on an agarose gel. In other words, the DNA sequence between the primers has been *amplified*. After about 30 cycles, amplified DNA is present in about 1 million times as many copies as the original DNA.

C. PCR Is Not Perfect

PCR has revolutionized molecular biology and can shorten and simplify many processes, such as gene cloning. However, problems often arise when using PCR. The most serious complication is **DNA contamination**. If extra DNA is present, the PCR will be unsuccessful if the contaminating DNA contains sequences complementary to the primers. Part of the contaminant DNA will be amplified, resulting in a mixture of desired and undesired amplified products.

The **DNA polymerase** that is normally used in PCR is another source of problems, because it lacks proofreading ability. Most DNA polymerases can detect mismatched bases; for example, if the DNA template reads A, and DNA polymerase adds C to the new complementary strand, it recognizes the mistake, removes the incorrect base, and replaces it with the correct base. The enzyme usually used in PCR (TAQ™) is derived from the thermophilic eubacterium *Thermus aquaticus*. It lacks the normal proofreading ability of DNA polymerase. As a result, the wrong base is sometimes integrated into new strands. This happens in a random manner and is not usually a problem unless it occurs early in the cycle. Also, many applications of PCR aim to *detect* genes; in such applications, occasional mistakes in

DNA replication should not affect the amplification process nor reduce the ability of PCR to detect a gene. However, if the aim is to isolate a gene for the purpose of cloning it, even a rare mistake could eventually cause problems. For this reason, it is advisable to use a heat-stable DNA polymerase with proofreading ability when using PCR to obtain DNA for cloning. For example, the DNA polymerase of *Thermococcus litoralis*, a thermophilic Archaen, has proofreading ability.

In the context of gene cloning, the biggest drawback of PCR is that it is much less effective when the sequence to be amplified is longer than 5 kb. This is a serious problem because many genes are longer than this. However, there are ways to address this, and PCR remains an extremely useful tool in gene cloning programs.

D. VARIATIONS ON PCR

Primers do not always bind to only one target. If another gene in the DNA sample has a similar sequence, it will also be amplified, resulting in a mixture of products. This problem can be sidestepped by amplifying the mixture of amplified DNA using a second set of primers. The second set of primers amplifies *within* the DNA amplified by the first primers. It is highly unlikely that a sequence that is amplified by both sets of primers exists within a sample. This approach is known as **nested PCR**.

Hotstart PCR is another alternative method that is frequently used. It can be effective in solving **mispriming** problems, that is, binding of primers to nontarget DNA. Sequence similarity (discussed above) can cause mispriming, but sometimes primers will bind to DNA that lacks complementarity. This happens because most DNA samples contain a small amount of single-stranded DNA; at room temperature primers may bind nonspecifically to these fragments, and DNA polymerase may amplify them, even though it is relatively inactive at room temperature. You can prevent this by withholding a crucial part of the PCR mixture (e.g., magnesium) until the DNA sample has been heat denatured.

One advantage of PCR is that it can be used to *modify* the DNA sequences that are amplified. Changing the DNA sequence is possible because PCR will still amplify a sequence even if one of the primers is slightly mismatched. The new DNA sequence is included in one of the primers, and as the cycles proceed, it will be incorporated into the amplified fragments. The aim in such a maneuver could be to alter the active site of an enzyme (protein engineering) in order to improve enzyme or protein function. More prosaically, it is often useful to introduce a sequence recognized by a specific restriction enzyme, so that a specific sequence can be cloned. The restriction site is usually added at the 5′ end of the primer, rather than the 3′ end, because the 3′ end is most responsible for the specificity of primer annealing. The addition of restriction sites is often an important step in the cloning of a PCR product, because it allows the introduction of sites that can be used to generate sticky ends. If successful, insertion of the amplified fragment into a plasmid vector is then straight-forward. Longer sequences (up to 45 base pairs) can also be introduced at the 5′ end of a primer. Promoters and other regulatory elements are often added to an amplified segment in this way. Molecular biologists have also devised simple but

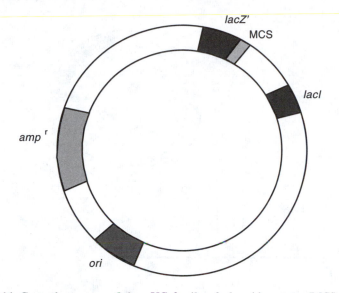

FIGURE 3.11 General structure of the pUC family of plasmid vectors. (MCS = Multiple cloning site.)

elegant methods to introduce deletions or base substitutions of single base pairs or short sequences within a gene.

VII. ALTERNATIVE VECTORS: pUC

The plasmid vector pBR322 is useful for explaining the basic strategy of gene cloning, but it is no longer a popular plasmid for most gene cloning procedures. The three major disadvantages to using pBR322 are (1) it is a low-copy-number plasmid; (2) the requirement of replica plating onto tetracycline plates is cumbersome and time consuming; and (3) most inserts in pBR322 are not strongly expressed (i.e., not much recombinant protein is produced).

The pUC family of plasmid vectors (Figure 3.11) is superior to pBR322 in all three respects. The *ori* sequence is different and binds more readily to DNA polymerase, resulting in more frequent replication. This makes it easier and faster to obtain plasmid DNA (fewer bacteria are required to obtain a given amount of DNA). The **promoter** sequence is also different in pUC vectors. In pBR322 the *tet* promoter controls the transcription of inserted genes; in pUC vectors, the *lacZ* promoter has this function. The *lacZ* promoter binds to RNA polymerase more efficiently than the *tet* promoter, resulting in higher levels of mRNA formation and higher levels of translated protein.

The final advantage of pUC vectors is in the method of selection and screening of transformed bacteria (Figure 3.12). Ampicillin treatment of bacteria transformed with pBR322 eliminates all bacteria that do not have the plasmid. Because pUC contains an ampicillin-resistant gene (*amp*), the same strategy works with pUC. In contrast to pBR322, though, identification of bacteria containing **recombinant pUC**

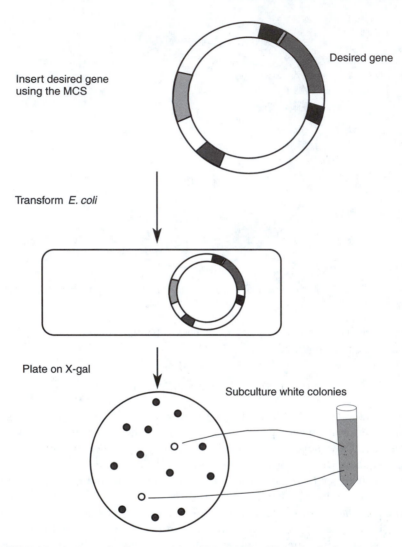

FIGURE 3.12 Cloning a foreign gene into pUC. (MCS = Multiple cloning site.)

does not require replica plating. In pUC, restriction sites are clustered in a **multiple cloning site (MCS)**, also known as a **polylinker**, that has been inserted at the end of the *lacZ'* gene. The MCS consists of a series of restriction sites, allowing the user to pick the most appropriate restriction enzyme for digestion of the plasmid DNA and the foreign DNA that contains the desired gene.

Successful recombination, then, results in incorporation of a foreign DNA sequence at the end of the *lacZ'* sequence. This sequence contains only part of the *lacZ* gene, which codes for **β-galactosidase**, the well-characterized enzyme that is part of the lac operon in *E. coli*. When using pUC vectors, one uses a strain of *E. coli* that lacks the *lacZ'* sequence but has the rest of the *lacZ* gene within its chromosome. When pUC is present in *E. coli*, the polypeptide expressed by the

left arm   MCS   *red+*   *gam+*   MCS   right arm

FIGURE 3.13 Structure of λEMBL3. Only the left and right arms are required for lytic infection of *E. coli*. A multiple cloning site (MCS) is present next to each arm.

lacZ' sequence in the plasmid interacts with the polypeptide produced by the chromosomal fragment in *E. coli*. This produces a functional β-galactosidase enzyme, which catalyzes the breakdown of lactose into glucose and galactose. Microbiologists have discovered that β-galactosidase also converts a synthetic carbohydrate (5-bromo-4-chloro-3-indolyl-β-D-galactopyranoside, or **X-gal**) into a **blue end product**. Therefore, if a bacterium is transformed with pUC that has an intact *lacZ'* gene, β-galactosidase will convert X-gal into a blue compound, and the colony will appear blue. However, if the bacterium is transformed with a recombinant pUC that has a foreign gene inserted onto the end of the *lacZ* gene, then the polypeptide expressed by *lacZ'* will no longer interact successfully with the polypeptide expressed by the chromosomal fragment. As a result, the cell will lack β-galactosidase activity. Therefore, it will not convert X-gal into a blue compound; the colonies will remain white. This new protein is referred to as a **fusion protein**, because it can be visualized as the β-galactosidase protein with a foreign protein fused to it.

Because the fusion protein lacks β-galactosidase activity, identification of recombinant clones is easy. The bacteria are diluted and plated on agar containing X-gal. White colonies are picked off and constitute a library of recombinant clones. Screening of a pUC library is similar to screening of a pBR322 library. The only advantage of using pUC is that it is easier to identify recombinant proteins because of their higher levels of expression in pUC. One final note about pUC vectors: they usually contain the *lacI* gene. This gene codes for the repressor protein that normally controls transcription of *lacZ* in *E. coli*. This protein prevents transcription of *lacZ* unless the inducer (lactose) is present. *lacI* allows pUC to regulate expression of the fusion protein. In practice, a synthetic inducer (isopropylthiogalactoside, or IPTG) is used instead of lactose.

VIII. ALTERNATIVE VECTORS: PHAGE LAMBDA

A. LAMBDA CLONING VECTORS

The Achilles heel of plasmid vectors such as pUC and pBR322 is their inability to efficiently incorporate sequences greater than 10 kb. For this reason, phage vectors, which can accommodate 20-kb segments of inserted DNA, are often used in the initial stages of gene cloning. λEMBL3 (Figure 3.13) is a commonly used example of a phage vector. (For a review of phage replication, see Chapter 2, Section IV.)

λEMBL3 is very different from plasmid vectors. It is linear and it does not have an *ori* or antibiotic-resistant genes. Instead, it is made up of three regions: a left arm, a right arm, and a "stuffer" region. The stuffer region contains the *red+* and *gam+* genes. The crucial characteristic of the genes in the stuffer region is that they

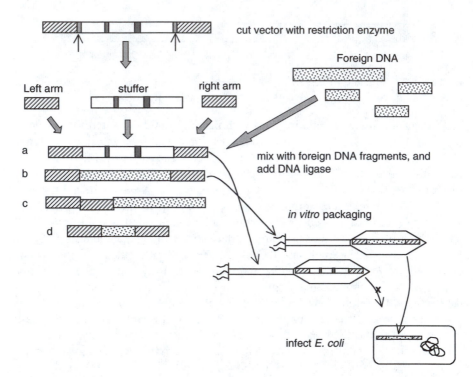

FIGURE 3.14 Gene cloning using λEMBL3. Many different products (a, b, c, and d) result from ligation of λEMBL3 and foreign DNA. Product d is not packaged into phage particles because it is too short. Product c is not packaged because the right and left arms lack an intervening stuffer region. Products a and b are both packaged because they are the correct length and they contain right and left arms in the correct positions. The phage-containing product b, however, is unable to infect the P2 lysogen strain of *E. coli* because it contains the stuffer region.

are not required for the lytic cycle. In contrast, the left and right arms are essential for successful completion of the lytic cycle.

Phage and plasmid vectors require a similar overall strategy. In both cases, the initial step is to cut the vector and DNA sample (containing the desired gene) with a restriction enzyme. λEMBL3 has been constructed so that it has MCSs at the border between the left arm and the stuffer region and between the right arm and the stuffer. Therefore, when restriction enzymes digest λEMBL3, the phage will be cleaved into three pieces: left arm, right arm, and stuffer. When the foreign DNA is cut with the same restriction enzyme and mixed with cut λEMBL3, a number of annealed products will appear (Figure 3.14). The only products of interest are strands made up of left arm ligated to foreign DNA ligated to right arm (product b in Figure 3.14). In these *recombinant* strands, the stuffer region has been replaced by a segment of foreign DNA. Our next challenge is to separate these recombinant DNA strands from the other annealed products.

In fact, this is a very straightforward procedure. First, an ***in vitro* packaging system** is added. This is a solution containing all the viral proteins required for assembly of infectious λEMBL3 particles. When this solution is added to the mixture of annealed DNA, DNA segments will be assembled into viral particles. However, not all DNA segments will assemble. Only DNA that has an intact left and right arm at either end and is approximately 47 kb long will be incorporated into new virus particles. Therefore, many of the annealed products will not be incorporated, either because they lack the left or right arms, or because they are the incorrect length (see Figure 3.14).

However, if the original phage DNA is still present (left arm + stuffer + right arm), then it will be repackaged into viral particles, because it is the correct length. These phage particles are eliminated in the next step: infection of a susceptible strain of *E. coli*. A specific strain of *E. coli* is used — **P2 lysogen**. λEMBL3 will complete its lytic cycle in a P2 lysogen *only if it lacks the* red *and* gam *genes of the stuffer region*. Therefore, any original phage particles will contain the stuffer, with its *red* and *gam* genes, and will not be able to replicate in the P2 lysogenic strain.

Infection of *E. coli* will result in a library of recombinant viral particles that must be screened to find the virus that contains the desired gene. In principle, the screening is similar to the screening used to find the correct recombinant plasmid. The only difference is that virus particles from the library are diluted and plated onto a "lawn" of *E. coli*. When the diluted phage infect bacterial cells, they will progress through lytic infections, resulting in small cleared areas in the lawn (**plaques**). If the initial phage dilution was correct, then the plaques will be well separated, and it can be assumed that each plaque resulted from the initial infection of one bacterial cell by one virus. Each infected cell would then produce numerous new, identical virus particles, which would then infect neighboring cells. Eventually, enough neighboring cells would be infected and killed to produce a hole (plaque) in the lawn of bacteria.

Each plaque represents a **clone of recombinant λEMBL3**. DNA probes can be used to screen these plaques for the correct clone. This is usually done by transferring DNA from the plaques onto a filter paper and then probing as described in Figure 3.8. Certain phage vectors (e.g., λgt11) lead to pronounced synthesis of introduced genes, which facilitates screening procedures based on detection of recombinant proteins. For example, antibodies specific to the desired protein can be used to identify plaques containing the protein.

B. COSMIDS

Sometimes we need to clone segments of DNA that are more than 20 kb long. One alternative is to use **cosmid** vectors that can accommodate 40-kb segments. These are specialized plasmid vectors (Figure 3.15) that contain the two cos regions from lambda phage, as well as a restriction enzyme site within the cos regions, another restriction site, an antibiotic-resistant gene, and an *ori* allowing replication in *E. coli*. The cos sequences are the part of the left and right arms of the lambda phage that are required for packaging of phage DNA into new viral particles, as long as the DNA is the correct length (40 to 50 kb). Typical cosmids have an *ori* for replication in *E. coli*, as well as an antibiotic-resistant gene (*amp*^r in Figure 3.15). This allows cosmids to replicate in host *E. coli* cells, similar to a plasmid vector,

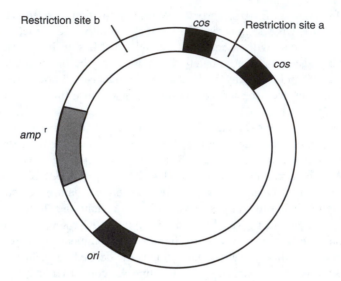

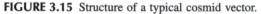

FIGURE 3.15 Structure of a typical cosmid vector.

and allows the use of antibiotics to eliminate cells that lack cosmids. Thus, cosmids have qualities of both lambda phage (*cos*) and plasmid (*ori*). Note also that our cosmid has two restriction sites that are cut by different enzymes (a and b).

Restriction enzymes a and b cut the cosmid into two fragments (Figure 3.16). Both fragments are mixed with DNA fragments (~50 kb long) from the organism that has the desired gene. It is important that these genomic fragments are cut with restriction enzyme b. The cosmid and genome fragments are mixed and allowed to anneal. Some of the annealed (and ligated) segments will contain genomic fragments between two cosmid fragments. These products are the correct length to be packaged into lambda particles using *in vitro* packaging systems (see preceding section). The resultant viral particles will then inject the DNA into a host cell. However, the DNA contains only the *cos* regions from the left and right arms of lambda phage; consequently, they are unable to replicate new viral particles. At this stage the DNA recircularizes and behaves as a plasmid vector. Normal methods (e.g., cold CaCl$_2$) of plasmid transformation of bacterial cells do not work with plasmids as large as cosmids. However, the phage injection system is able to introduce the large recombinant cosmids into host cells.

IX. ARTIFICIAL CHROMOSOMES

One can easily sequence small sections within the genome of a eukaryote. However, understanding the order of genes along each chromosome can be an arduous task. This **mapping** process can be done more efficiently using vectors that can accommodate large sequences of DNA. Phage vectors and cosmid vectors are usually used, but artificial chromosomes are now available as an alternative. Artificial chromosomes were originally used in yeast. They contain centromere sequences that allow

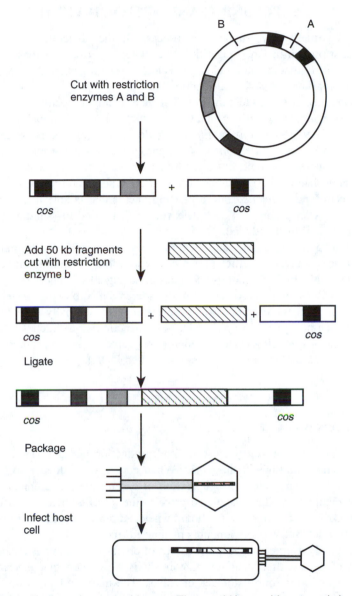

FIGURE 3.16 Cloning using a cosmid vector. The cosmid is cut with two restriction enzymes, generating two fragments. After ligation, the only segments that will be packaged into viral particles are those that are the correct length (because of the inclusion of the foreign insert) and have *cos* regions at either end.

a large segment of foreign DNA to function as a chromosome within a eukaryotic cell. Now they are also available for mammalian cells. Their latest application is as vectors for introducing DNA into animals, in addition to mapping genomes and locating specific genes within a set of chromosomes.

X. SUBCLONING AND VECTOR DIVERSITY

Cloning is virtually never a one-step process. If the objective is to produce a recombinant protein in large quantities while keeping costs at a reasonable level, then it is unlikely that the initial cloning vector, and the organism "cloned into" will both be appropriate for large-scale production.

In many cases we need to **subclone**. This is basically a repetition of the cloning process. For example, consider a situation in which we want to transfer a gene from *Lactococcus lactis* to *Bacillus subtilis*. It would theoretically be possible to set up a DNA library directly in *B. subtilis*, if an appropriate plasmid or phage vector were available. However, it would be easier to set up the library in *E. coli*, screen the library, and isolate the desired clone. The desired gene could then be cut out of the cloning vector and *subcloned* into a plasmid vector that is able to replicate in *B. subtilis*. This is an example of a **shuttle vector** (it is used to "shuttle" the cloned gene from *E. coli* to *B. subtilis*).

Subcloning can have other aims. Often, the vector used in the primary cloning procedure does not result in efficient **protein expression**; subcloning into an **expression vector**, with strong promoters and appropriate signal sequences (see Chapter 2, Section V), will often increase the amount of secreted heterologous protein. Vectors with signal sequences are also known as **secretion vectors**. Finally, before the advent of PCR, which has greatly simplified DNA sequencing, cloned genes were usually subcloned into **sequencing vectors**, which have characteristics appropriate for DNA sequencing (e.g., the generation of single-stranded DNA).

XI. RECOMBINANT CHYMOSIN

A. CHYMOSIN AND CHEESE MAKING

Chymosin has a long history of use in food production and is essential to cheese making practices throughout the world. When added to acidified milk, chymosin increases the rate and extent of coagulation of casein, the predominant protein in milk (see Figure 1.3). This coagulated protein (the curd) is then treated in various ways to make cheese. Typically, the liquid whey is drained or pressed from the curd, which is then left for variable amounts of time to ripen.

Traditionally, **rennet** has been the source of chymosin. Rennet is taken from the stomachs of calves after slaughter. Chymosin is a protease that is produced by epithelial cells in the fourth stomach of calves. The main problem with the use of chymosin from calves is that the supply is unstable and is affected by global demand for veal. Purity is an additional problem; different batches and chymosin from different suppliers are not always of similar purity and activity. Purity is important because animal rennet is contaminated with variable levels of **bovine pepsin**, which causes undesirable proteolysis and unwanted flavors if present during curd formation.

Chymosin is a complex protein. The protein that is translated from the mRNA transcript in calf epithelial cells is initially inactive (Figure 3.17). This form is *preprochymosin*. The "pre" part of the protein is a **signal sequence** of amino acids

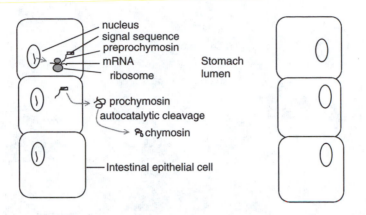

FIGURE 3.17 Formation of chymosin in the lumen of the fourth stomach of bovine calves. The mRNA is exported from the nucleus and is translated into **preprochymosin**. The signal sequence is removed and **prochymosin** is secreted from the cell. The low pH of the stomach lumen causes an autocatalytic cleavage of prochymosin, yielding chymosin, the active form of the enzyme.

that informs bovine intestinal epithelial cells that preprochymosin is a **secreted** protein. This signal sequence is cleaved from the protein as it passes across the plasma membrane into the intestinal lumen. Therefore, **prochymosin** is the form of chymosin that is secreted into the lumen. Interestingly, it has *autocatalytic activity* and can "cut itself up." However, this cleavage occurs only in acidic conditions. When prochymosin is secreted into the acidic environment of the fourth stomach, part of the protein is cleaved and the remaining protein (**chymosin**) is an active protease.

In cheese production, chymosin is added to milk shortly after the starter culture. Lactic acid bacteria in the starter ferment carbohydrates in the milk and produce lactic acid as an end product. This results in increased acidification of the milk. In the absence of chymosin, decreased pH causes coagulation of casein, the predominant protein in milk. However, a soft curd results, similar to soft cottage cheese. To get a firm curd, chymosin must be added. Chymosin causes increased coagulation because it cleaves casein at points in the amino acid chain that are covalently attached to carbohydrate chains (Figure 3.18). Intact casein exists in aqueous solution as a micelle, and aggregation of micelles is prevented by electrostatic repulsion between carbohydrate chains in adjacent micelles. However, when chymosin removes these carbohydrates from the micelles, nothing prevents aggregation, and a firm curd results.

The mechanism of action of chymosin is very specific, so it is essential that the recombinant protein be identical in structure to the protein produced in calf stomachs. Various microbial proteases can be substituted for chymosin, but they usually yield inferior cheese because of excessive proteolysis. A better alternative would be to transfer the bovine gene for chymosin to a microorganism that can be grown on a large scale at low cost. Several research groups attempted to do this in the early 1980s.

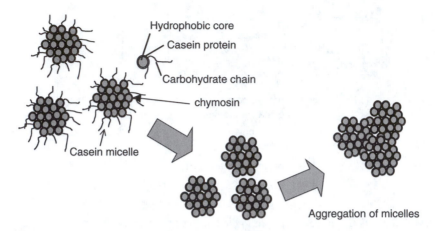

FIGURE 3.18 Mode of action of chymosin on casein micelles.

B. Inclusion Bodies

Initial efforts to clone chymosin into *E. coli* were partially successful. mRNA was isolated from calf intestinal epithelial cells, and cDNA libraries were constructed using plasmid vectors such as pBR322. These libraries were screened using a variety of strategies (one research group used DNA probes designed from part of the chymosin amino acid sequence). Recombinant clones of *E. coli* were isolated, and the presence of the prochymosin gene was confirmed by sequencing of the inserted DNA.

However, when the chymosin gene was subcloned into an expression vector with strong promoters, problems arose. The protein was produced in large quantities (5% of total cellular protein), but it was not secreted from cells; instead, it accumulated within cells in clumps of denatured protein. Such clumps, referred to as **inclusion bodies**, commonly occur when foreign proteins are expressed at high levels in *E. coli* and other bacteria. This is frequently a problem when transferring genes from eukaryotes into *E. coli*. The intracellular environment (e.g., pH, redox potential) is very different in bacteria than in eukaryotes. This often causes aggregation of eukaryotic proteins. Also, many eukaryotic proteins undergo processing in the endoplasmic reticulum and the Golgi apparatus, which form the internal membrane system of eukaryotic cells (Chapter 2, Section V). These membranous compartments are chemically different from the cytoplasm. For example, the **oxidizing environment** of the endoplasmic reticulum allows disulfide bond formation in proteins that would be unlikely to occur in the more reducing conditions of the cytoplasm. Furthermore, eukaryotes often have molecular chaperones that help ensure that proteins are folded correctly. In prokaryotes, the chaperones have different structure, sometimes leading to incorrect folding of proteins, which may lead to exposure of **hydrophobic** residues to the aqueous solution. The interaction of such exposed regions among different protein molecules results in extensive aggregation and denaturation of the protein.

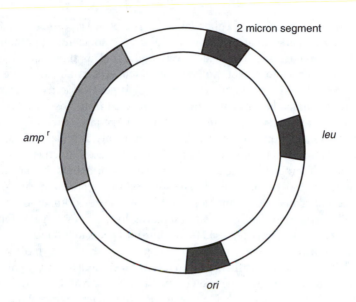

FIGURE 3.19 Structure of YEp (yeast episomal plasmid).

Inclusion bodies can be purified from *E. coli* and **renatured**, but this process is usually expensive. There are better alternatives. In the case of chymosin, the most successful approach has been to subclone the prochymosin gene into fungal hosts. Because fungi are eukaryotes, inclusion bodies are much less of a problem.

C. RECOMBINANT PRODUCTION BY YEAST

One of the advantages of using yeast (unicellular fungi) in cloning programs is that there are plasmids that are able to replicate in yeast and can be used as vectors. One such plasmid is **yeast episomal plasmid** (YEp). Episomes are segments of DNA that are able to integrate into host chromosomes or remain extrachromosomal. Under certain conditions YEp is maintained at a high copy number (30 to 50 copies per cell) in yeast cells and is therefore not diluted out as the cells divide by mitosis.

In comparison to bacteria, it is more difficult to transform yeast with DNA. However, several methods work; **electroporation** is particularly effective with yeast sphaeroplasts (cells whose walls have been enzymatically removed). This method requires the application of high voltage to a suspension of yeast cells. Small pores transiently form in yeast membranes, allowing the uptake of DNA into the yeast cells.

Biotechnologists have constructed YEp shuttle vectors that can replicate in both *E. coli* and the yeast *Saccharomyces cereviseae* (Figure 3.19). After cloning the prochymosin gene in *E. coli*, they subcloned the gene into a YEp vector and incorporated it via transformation into *S. cereviseae*. Large amounts of protein were produced, but unfortunately, it was in an insoluble, aggregated form.

This occurred because prochymosin was not secreted from the yeast cells. The original form of the gene in cows never accumulates intracellularly, because it is rapidly secreted from intestinal epithelial cells. Therefore, it seemed sensible to try to mimic this, by persuading recombinant yeast cells to secrete recombinant pro-chymosin. So, another subcloning step was required. This time, the prochymosin gene was subcloned into a **secretion vector**. Such vectors already have a **signal sequence** recognized by *S. cereviseae*. Any foreign genes inserted in front of this sequence have a much better chance of being secreted from host yeast cells.

The secretion vector was partially effective. A large proportion of the translated protein was secreted from the yeast cells. Unfortunately, the yield of recombinant protein was too low; prochymosin production was not cost effective.

The low yield problem is intrinsic to *S. cereviseae,* a fungus that normally does not secrete large amounts of proteins. The solution to the prochymosin conundrum finally came with a final subcloning step that allowed the transformation of a different yeast — *Kluyveromyces lactis*. This yeast normally secretes large amounts of a variety of enzymes, and has been used for a number of years to produce commercially sold lactase. When a recombinant plasmid containing the prochymosin gene was introduced into *K. lactis*, sufficient amounts of recombinant prochymosin were synthesized and secreted into the growth medium. Chymosin could then be easily collected, purified, and sold to cheese makers. Recombinant chymosin has also been produced successfully in filamentous fungi (e.g., *Aspergillus nidulans*).

The chymosin story is typical of many attempts to produce recombinant proteins in large quantities. Initial cloning efforts are rarely completely successful, and usually a number of subcloning steps and genetic modifications are necessary to produce recombinant proteins at a reasonable cost.

What about safety assessment of chymosin? The U.S. Food and Drug Admin-istration (FDA) considers recombinant chymosin to be sufficiently similar to natural chymosin. Consequently, toxicity assessment was not required, and permission was granted to use recombinant chymosin in the production of cheese. Several other factors influenced the FDA's decision; for example, purification of chymosin resulted in destruction of the microbe that produced the chymosin and efficiently removed impurities associated with large-scale culture of the fungus. Chymosin can also be used for cheese production in the European Union, Canada, and many other coun-tries.

XII. RECOMBINANT BOVINE GROWTH HORMONE

Before the advent of rBGH, it was not possible to use BGH to boost milk production in cows. This is not because of a lack of understanding of the potent biological effects of injected BGH on cows. As early as the 1940s, animal scientists knew that injections of this hormone would increase the rate of milk production in cows. However, at that time the only source of BGH was from dead cows, and the cost of obtaining useful amounts was prohibitive. Consequently, research on the potential uses of BGH stalled, until the development of recombinant BGH in the early 1980s. This illustrates again the major advantage of recombinant proteins — they are inexpensive. Once the gene for BGH was transferred to *E. coli*, it became econom-ically feasible to grow large amounts of the recombinant bacterium, purify the hormone, and sell it at a price that is cost effective for dairy farmers.

rBGH is an excellent example of the nontechnical problems that a biotechnological company can face when attempting to introduce a biotechnological product into the food supply. Attempts to introduce rBGH to markets throughout the world have met with extensive opposition. This is surprising, because of the lack of controversy surrounding many other recombinant proteins. For example, both recombinant insulin and recombinant chymosin have been successfully sold throughout the world, with very little controversy.

Opponents of rBGH fear its effects on cows and on humans who drink cow's milk. In the U.S., the FDA was principally responsible for assessing the safety of rBGH, based on studies by the company (Monsanto) that developed rBGH, as well as independent studies. When there is a question of potential toxic effects of a recombinant protein, it is necessary to test the toxicity of the protein using animal models. Animals are exposed to the protein and closely studied for adverse changes. Toxicological studies did not detect any adverse changes in animal health after administration of milk from rBGH-treated cows. However, many opponents to rBGH have little faith in standard toxicological studies, because of the inability of such studies to detect adverse changes over a long period of consumption of the product or compound.

Because of these problems in assessing the safety of long-term consumption of new compounds, it is important to carefully consider differences between the new product and analogous, currently used products. In the case of rBGH, then, the central question is whether or not milk from rBGH-treated cows is *different* from milk from untreated cows. It is virtually impossible to state unequivocally that rBGH milk is identical to normal milk; milk is a complex suspension and solution that contains a myriad of different biochemicals. For this reason, the FDA felt that it was feasible to examine only specific compounds, to see if levels were affected by rBGH treatment. They found that levels of BGH were similar in both kinds of milk, but there was a statistically significant increase in the amount of insulin-like growth factor (IGF-1) in milk from rBGH-treated cows. IGF-1 is a protein hormone that is influenced by growth hormones such as BGH; typically, increased levels of growth hormones lead to increased levels of IGF-1.

The controversy surrounding IGF-1 is largely related to its potential role in tumor formation. One of the principal actions of IGF-1 is to increase the rate of cell proliferation, which also occurs during tumor formation. To determine if a by-product such as IGF is potentially dangerous to human consumers of milk, the following questions must be considered:

1. Would IGF survive its voyage from the oral cavity to the intestine? To do so, a compound must not be affected by transit through the stomach, with its low pH.
2. Is IGF absorbed by the intestine into the body, and into the circulation?
3. Does bovine IGF exert any effects on cells or processes of the human body? Would a bovine hormone such as IGF bind to receptors in human cells?

If the answer to any of these questions is "yes," the potential risk to humans increases.

The FDA decided that the available scientific evidence indicated that increased levels of IGF-1 in rBGH-treated milk were not dangerous. The primary reason for this decision was that it appeared unlikely that ingested IGF-1 would remain intact after passage through the stomach. It also determined that the amount of IGF-1 ingested would be insignificant, partly because of the relatively large amount of IGF-1 present in normal human saliva. The ability of human intestinal epithelial cells to absorb IGF-1 and the ability of bovine IGF-1 to exert effects on human cells were considered irrelevant, because of the stomach and saliva considerations.

The FDA's approval of the sale of rBGH-treated milk has not eliminated controversy over this product. One of the principal points of controversy surrounds the ability of many pH-sensitive microbes (e.g., *Salmonella typhimurium*) to survive passage through the stomach under certain conditions (e.g., protection by lipids). Is it possible for IGF-1 to survive passage through the stomach?

The level of controversy surrounding IGF-1 and rBGH has now abated in the U.S., but most countries throughout the world (including the European Union) still do not allow the sale of rBGH-treated milk. In some countries (e.g., Canada) rBGH did not receive approval because of studies indicating adverse effects on cow health. Several studies have found that administration of rBGH results in increased rates of mastitis (infection of the udders) and other health problems. These problems are probably linked to increased milk production, and not directly to rBGH treatment.

It may seem odd that BGH has generated so much controversy. Part of the problem may lie in the basic nature of BGH; many people are offended by the idea of "supercharging" cows with growth hormones to increase milk supply. Thus, the fact that rBGH is a product of recombinant DNA technology may be less important than the fact that it is a potent hormone. The use of "natural" BGH derived from carcasses would likely be considered equally objectionable to many people. However, a significant number of consumers appear to be offended by the "unnatural" aspects of rBGH (i.e., it is a mammalian protein produced by a bacterium). Interestingly, rBGH is safer than natural BGH, because natural BGH could be contaminated with viruses or the causative agent of bovine spongiform encephalopathy (mad cow disease).

Consumer benefit is another important issue. The use of rBGH does not directly benefit the consumer; it does not increase the quality of milk and it is unlikely to result in decreased milk prices. This adds to consumer resentment. In contrast, the use of recombinant chymosin is simply a change in the source of an enzyme that has always been used to make cheese. Thus it is less likely to be seen as a technological intrusion into a traditional food. Also, the use of recombinant chymosin may make cheese more palatable to lacto-vegetarians, who are likely to be offended by the use of chymosin derived from slaughtered calves. Also, Muslims and Jews may prefer to use cheese made from recombinant chymosin to ensure that they are not consuming cheese made using porcine (pig-derived) chymosin.

There is an important lesson here for food biotechnologists — the general public is particularly sensitive to products that contain elements that are central to human metabolism, such as growth hormones. Other heterologous proteins, such as recombinant chymosin or recombinant amylases (used in starch processing), have largely been ignored by antibiotechnology activist groups, and generate very little controversy. Another lesson: if a biotechnologically derived food or process can be seen to benefit the consumer, it is much more likely to have a seamless introduction into the marketplace. Evidently, the potential public reaction to a biotechnological product should be a primary point for biotechnologists to ponder in the *early* stages of new product development.

RECOMMENDED READINGS

1. Sambrook, J. and Russell, D., *Molecular Cloning: A Laboratory Manual,* 3rd ed., Cold Spring Harbor Laboratory Press, Cold Spring Harbor, New York, 2001.
2. Rapley, R., Molecular analysis and amplification, in *Molecular Biology and Biotechnology,* 4th ed., Walker, J. M. and Rapley, R., Eds., Royal Society of Chemistry, Cambridge, 2000, chap. 2.
3. Rapley, R., Recombinant DNA technology, in *Molecular Biology and Biotechnology,* 4th ed., Walker, J. M. and Rapley, R., Eds., Royal Society of Chemistry, Cambridge, 2000, chap. 3.
4. Hengen, P. N., Methods and reagents — fidelity of DNA polymerases for PCR, *Trends Biochem. Sci.,* 20, 324, 1995.
5. McPherson, M. J., Møller, S. G., Beynon, R., and Howe, C., *Pcr (Basics: From Background to Bench),* Springer-Verlag, Heidelberg, 2000.
6. Newton, C. R. and Graham, A., *PCR,* 2nd ed., Springer-Verlag, New York, 1997.
7. Harris, E. and Kadir, N., *A Low-Cost Approach to PCR: Appropriate Transfer of Biomolecular Techniques*, Oxford University Press, Oxford, 1998.
8. Glick, B. R. and Pasternak, J. J., *Molecular Biotechnology: Principles and Applications of Recombinant DNA,* ASM Press, Washington, D.C., 1994.
9. Glazer, A.N. and Nikaido, H., *Microbial Biotechnology: Fundamentals of Applied Microbiology,* Freeman, New York, 1995.
10. Ward, M., Chymosin production in *Aspergillus,* in *Molecular Industrial Mycology: Systems and Applications for Filamentous Fungi,* Leong, S. A. and Berka, R. M., Eds., Marcel Dekker, New York, 1991, chap. 4.
11. Beppu, T., Production of chymosin (rennin) by recombinant DNA technology, in *Recombinant DNA and Bacterial Fermentation*, CRC Press, Boca Raton, FL, 1988, chap. 2
12. Juskevich, J. C. and Guyer, C. G., Bovine growth hormone: human food safety evaluation, *Science,* 249, 875, 1990.
13. Bonneau, M. and Laarveld, B., Biotechnology in animal nutrition, physiology and health, *Livestock Prod. Sci.,* 59, 223, 1999.

4 Plant Biotechnology

I. OVERVIEW

The plant biotechnology field is of prime importance to anyone studying food production, food science, or nutrition. Some new plant biotechnologies (e.g., the use of plant cell culture) have gained public acceptance and are used throughout the world; others, particularly those surrounding the development of transgenic plants, are controversial and, in many parts of the world, have not been accepted by the public.

The central aim of this chapter is to explore the kinds of biotechnologies that affect crop production, the processing of food plants, and the composition of food plants. An emphasis is placed on transgenic plants because they have great potential to improve a diverse array of problems associated with food production and processing as well as health-promoting aspects of food. Also, transgenic plants are clearly the most controversial application of food biotechnology, and consequently have the potential to disrupt matters ranging from the global (international trade of food commodities) to the local (changed patterns of food consumption due to consumer concerns about the safety of transgenic crops).

This chapter explores plant cell and tissue culture techniques because of their central importance in traditional plant breeding and the production of transgenic plants. Plant cell culture allows the rapid production of genetically identical **clones** of valuable plants; this is exploited on a day-to-day basis in the cultivation of certain crops (e.g., potatoes) and is a valuable tool for plant breeders. Techniques such as embryo rescue and protoplast fusion allow plant breeders to expand the range of plants that can participate in breeding programs to plants that belong to different species or genera. Normally, breeders must choose parents from varieties of the crop species, restricting the range of available genetic material. With *in vitro* (culture of cells or tissues outside the plant environment) techniques, though, breeders can transfer valuable genes between species and sometimes between genera.

Plant cell culture is also a potentially useful technology in the context of food processing; a number of compounds used to influence food flavor and aroma (e.g., saffron) are expensive to obtain from intact plants. An alternative is to grow cells in culture and isolate the desired compound from the cells or the culture medium. To date, though, this technology is not commercially popular because of the relatively high cost of growing plant cells.

Cell culture techniques are also required for the production of transgenic plants. Most transgenic plants that have been released to growers have been aimed directly at helping the grower, by making it cheaper or easier to grow commodities such as corn, soybeans, and canola (low-erucic-acid rapeseed). However, a number of transgenic tomato, canola, and rice varieties have been developed with food processing or nutritional objectives. Most available tomato transgenics have slower ripening, or less softening accompanying ripening, than conventional tomatoes. One of the transgenic canola cultivars that has been released has an altered pattern of production of polyunsaturated fatty acids (PUFAs). In this case, the new PUFA has nonfood industrial applications, but the same technology is currently being tested in the context of altering PUFA production to increase levels of health-promoting PUFAs (e.g., linolenic acid). Published studies have also demonstrated that rice and other crops can be modified to have increased levels of provitamin A or vitamin E. Although these transgenic cultivars have not yet been cleared for commercial release, they demonstrate the power of transgenic technology to alter important properties of food.

The reaction to transgenic plants is not always enthusiastic, though. Transgenic plants have received an enormous amount of attention from the media and public. This attention has been triggered by vigorous action of mainstream and radical activist groups, both at international and grassroots levels. Because of this public interest, it is vital that plant scientists, food scientists, nutritionists, and biotechnologists understand the technology that is used to develop transgenic plants.

Because of the public backlash against transgenic plants, many biotechnology companies have scaled down their transgenic plant research and development programs. However, several large biotechnology companies have continued to release transgenic cultivars, perhaps because they have the financial resources to risk public rejection of transgenic food. Commercialization of transgenic crops requires many years of basic research and testing, followed by exhaustive (from the point of view of biotechnology companies) submissions to regulatory agencies such as (in the U.S., for example) the Food and Drug Administration (FDA) and the Environmental Protection Agency (EPA). To minimize risk, companies attempt to commercialize transgenic crops that will be attractive to a large number of growers, usually because of decreased pesticide costs. Thus there has been little direct benefit to consumers, which is one of the reasons for the low level of public acceptance of transgenic crops in many parts of the world. It is predicted that the next generation of transgenic cultivars will offer consumers great benefits (e.g., increased levels of nutraceuticals in food), making transgenic technology more palatable to the public. However, the financial risks associated with transgenic plant commercialization have made it difficult for biotechnology companies to take the risk of developing cultivars aimed at the consumer market. This has led to a great deal of dissatisfaction among biotechnologists involved in the development of transgenic plants. They feel that antibiotechnology activists have unfairly targeted their technology, and that the risks of transgenic plant technology have been greatly overstated. On the other hand, antibiotechnology activist groups, as well as a number of consumer lobby groups, feel that transgenic crops have been inadequately assessed for health and environmental risks, and that the release of transgenic seeds to the farming community, especially in the U.S. and Canada, constitutes an unacceptable lapse in governmental regulation. It is unlikely that this issue will be resolved soon to the satisfaction of either camp. It is unlikely that transgenic technology will disappear, and it is unlikely that antibiotechnology groups will ever be satisfied with the processes used to regulate commercialization of transgenic crops. Some of the environmental concerns

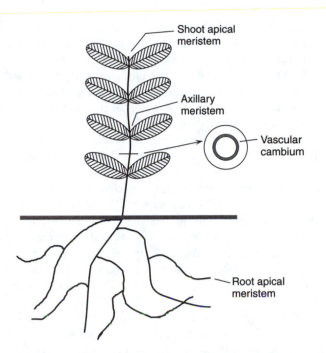

FIGURE 4.1 Points of growth in plants. Axillary meristems represent potential branching points of the stem (shoot), and the vascular cambium is responsible for increasing the girth of a stem or root. The cork cambium is not shown in this diagram, but it is required for the production of bark in woody plants. Vascular and cork cambia are lateral meristems.

associated with transgenic plants will be discussed in this chapter (Sections IV.A and IV.C). Health concerns are discussed in Chapter 9.

In this chapter, plant cell culture and several of its applications will be discussed first. This will be followed by an explanation of the techniques that are used to directly transfer genes into crop plants, then examination of the range of realized and potential applications of transgenic plant technology.

II. PLANT CELL AND TISSUE CULTURE

A. CONTROL OF PLANT GROWTH

To understand how plant cells are manipulated *in vitro*, we must understand how plants grow and how such growth is controlled. Plants, like all multicellular organisms, grow through the combined action of two processes: cell division and the expansion of individual cells. In most animals, a well-defined pattern of cell growth and division occurs early in life, leading to an adult organism that no longer increases in size. However, in many plants (especially trees), the full extent of growth is more variable, and growth occurs throughout the plant's lifetime. This growth is carefully controlled by the plant, and it only occurs in **meristems** (Figure 4.1), which are most easily seen at the tips of roots and stems (**apical meristems**). Less obvious are **axillary meristems**, dormant buds that are located in the axils of leaves. An axillary meristem, when given the appropriate hormonal stimulation, develops into a new growing apical meristem; this is how branches form. **Lateral meristems**, consisting of the vascular cambium and the cork cambium, are found in the interior of woody stems and roots. These meristems produce woody tissue and bark, acting to increase the girth of a plant to support continued upward growth of the apical meristems. If this did not occur, the stem would be unable to support the weight of the new growth occurring at the tips of branches.

So, how is meristem growth controlled? Evidently, control is essential; uncontrolled growth and subsequent tumor formation is damaging and wasteful for plants, just as it is for animals. In plants, control of growth occurs via **plant hormones**. The most well-characterized hormones are the **auxins**, **gibberellins**, and **cytokinins**. They are a source of great confusion because of the myriad of effects they have on plant cells. Effects vary within plants; for example, auxin will likely inhibit cells in an axillary meristem (i.e., the cell will not divide or expand), whereas it would stimulate cells to divide in the vascular cambium. To complicate matters further, cells from different plant species have different patterns of response to plant hormones.

Fortunately, we can make some general comments on these classes of hormones:

- Auxins tend to stimulate the expansion of cells, by triggering the loosening of cell walls. This allows turgor pressure to increase the size of cells (similar to the increase in size of a balloon when water flows into the balloon).
- Gibberellins tend to stimulate both cell division and cell expansion. This is particularly evident when gibberellins are applied to germinating seedlings.
- Cytokinins tend to stimulate cell division. These hormones are often produced in apical meristems and cause cells in the meristem to divide.

B. The *In Vitro* Life Cycle

These hormones are obviously interesting to plant scientists, but why are they important to food biotechnologists? The answer lies in the use of plant hormones to stimulate the growth of plant cells *in vitro*. To illustrate this idea (Figure 4.2), consider a tobacco plant growing in soil in a pot. If you remove a tobacco leaf and place it in soil in another pot, it will almost certainly die and quickly decompose. However, if you use a hole-puncher to cut out pieces of a leaf (**explants**) and place the pieces on an agar medium containing mineral nutrients, a source of carbohydrates (usually sucrose), and plant hormones, then growth will occur. The *type* of growth can be manipulated by changing the types of hormones present and the concentration of each hormone. For example, **callus** formation is sometimes useful. A callus is a disorganized mass of dividing, undifferentiated plant cells. Often, a relatively high concentration of auxin in the agar medium stimulates callus formation.

Once a callus has formed, various paths are possible. The clumps of cells can be broken up in a liquid medium, forming a cell suspension. Such cell suspensions can be grown indefinitely, if nutrients and the correct mixture of hormones are

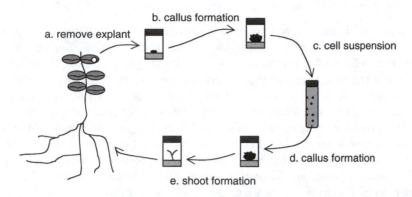

FIGURE 4.2 Tissue culture of a tobacco plant. In (a), a leaf explant is transferred to agar media containing a specific combination of plant hormones. This combination of hormones induces the formation of callus tissue (b). Callus tissue can be converted into cell suspensions in liquid media (c). Transfer of callus or cell suspensions to agar media with another specific combination of plant hormones results in induction of the growth of intact plantlets (d). Plantlets can be transferred into soil (e), and then into the field, if desired.

supplied. As long as the suspension is gently shaken, the cells will remain suspended and callus tissue will not form. Cell suspensions can in turn be returned to solid agar media, leading again to callus formation. Pieces of callus can then be transferred to fresh media, allowing the culture of callus indefinitely. Often, though, it is desirable to manipulate callus cultures so that intact plants ("plantlets") grow from the callus. Specific changes in the concentrations of hormones in the agar medium stimulate plantlet formation. For example, transfer of calli to agar containing a high concentration of cytokinins and a low concentration of auxin may result in the **regeneration of shoots**. Once shoots have formed, further manipulation of the hormones may be necessary to induce root development. Once shoots and roots are present, the plantlets can be transferred to soil. This transfer is often very traumatic, and plantlets usually have to be carefully nurtured in soil to prevent an early death. This conversion from cells to callus to intact plants can start with **single plant cells**, helping to ensure that the resulting plant is genetically uniform.

In comparison to animals, manipulation of plant cells is relatively easy. The main reason for this is that **totipotent** cells are easier to obtain in plants. A totipotent cell has the potential to differentiate into any of the cells present in the adult plant. In animals, totipotency is found only in the early stage of embryo formation. Plants, in contrast, always have totipotent cells in their meristems, especially in shoot apical and axillary meristems. Thus, adult plants retain small populations of embryonic cells that continue to divide and differentiate throughout the life of the plant. In some plants, (e.g., tobacco), totipotent cells can also be obtained from nonmeristematic cells (e.g., photosynthetic cells in the leaf). In other plants (e.g., many cereals), it is difficult to isolate totipotent cells; they may be found only in embryonic tissue. This is the major reason for the lag in development of transgenic barley, wheat, and rice, despite the great economic importance of these crops.

C. MICROPROPAGATION

This ability to circumvent the normal life cycle of plants is useful. It is often desirable to produce large numbers of plants that are genetically identical. Such plants may have characteristics that lead to faster growth, higher yield, or other useful agricultural characteristics. Alternatively, they may have superior processing characteristics; for example, a particular potato variety may have low levels of sucrose in the tubers and thus be better suited for the production of fried, frozen potato chips. This variety of potato could be propagated in the usual way, by growing plants in the field, harvesting tubers, and selling them to farmers as seed potatoes. Alternatively, the variety could be **micropropagated** on agar media, potentially producing hundreds of thousands of genetically identical plantlets. These plantlets could then be sold to farmers as an alternative to seed potatoes.

One advantage of micropropagation is that it is easier to ensure that the propagated plants are free of viruses and other pathogens. This can be a problem for plants that are propagated through vegetative means (i.e., not by seeds), such as potatoes. It is difficult to ensure that potato tubers grown conventionally are virus free, but virus-free plants can readily be obtained using tissue culture techniques. One of the best strategies for producing virus-free plants is to establish cell cultures from cells taken from apical meristems. These cells are not fed directly by the plant's vascular system, and, because many viruses spread through the vascular tissues, meristems remain virus free.

Micropropagation is particularly useful for crops such as potato and banana, which are not propagated by seed, and is less popular for seed-propagated crops such as cereals and legumes. However, micropropagation is frequently an essential tool in plant breeding programs aimed at improving seed-propagated crops.

Micropropagation is not risk free. Biotechnologists try to avoid callus formation when micropropagating plants. Cells in callus culture are genetically unstable; mutations frequently occur that may damage the cell or result in the loss of useful characteristics. This is due to **somaclonal variation** (see Section II.D.3 of this chapter).

Somaclonal variation can be devastating if desirable characteristics are lost or if undesirable characteristics are gained. For this reason, micropropagation systems are designed so that they minimize the time spent in callus culture. Large-scale micropropagation requires the monitoring of propagated plants, to ensure that somaclonal variation does not gradually change their genetic constitution.

D. PLANT CELL CULTURE AND TRADITIONAL PLANT BREEDING

1. Plant Breeding Basics

The improvement of crop plants has traditionally been the realm of the plant breeder. Most plant improvement is still achieved through traditional plant breeding techniques; cell culture is often an important component of plant breeding programs.

What is traditional plant breeding? It is important for food biotechnologists to understand the answer to this question, because plant breeding is the method used to develop most new food crops. Traditional breeding is preferred to a transgenic approach when numerous genes control the target for improvement. Yield is a good example — the amount of harvested product yielded by a plant is affected by (1) genes that affect distribution of photosynthesis-derived carbohydrates, (2) genes controlling plant architecture (e.g., how tall the plant is or how many branches it has), and (3) genes relating to the ability of the plant to tolerate stresses such as dry soil. Recombinant DNA technology is very effective at inserting single genes into plant cells, but as the number of inserted genes increases, the difficulty quickly increases. Consequently, the transgenic plants that have been released to date consist mainly of one- or two-gene insertions.

As transgenic technology evolves, and as we learn more about the genetics of plant development and function, people will develop transgenic plants with complex characters. Actually, this is already in progress; biotechnologists have successfully developed transgenic rice varieties with increased yield, as well as wheat varieties with increased tolerance to drought stress. In the future, transgenic technology will be used more often to improve consumer-important characteristics such as appearance, flavor, and functionality. However, traditional plant breeding is likely to continue to be the method of choice for many plant improvement programs. For this reason, it is essential to understand the primary tool for improvement of plant foods.

Traditional plant breeding (Figure 4.3) requires first and foremost the identification of **desired characteristics**. For example, an agronomist (one who is concerned with the behavior of field-grown plants) may wish to develop a variety of wheat that stays compact and flowers quickly, so that the plant invests most of its energy in reproduction (i.e., harvestable grains). Alternatively, a company that sells frozen vegetables may wish to have a variety of carrot that does not deteriorate in quality during storage. A competent plant breeder can grant both of these wishes, although it might take 5 to 10 years.

Once the desired characteristic has been identified, the breeder starts with a cultivar (varieties are called **cultivars** if they are being actively cultivated by producers) that is currently successful with growers and that will be useful to the intended market. For example, if a breeder wants to develop a variety that will be used by greenhouse tomato growers in the American midwest, then s/he will select a cultivar that is currently popular with these growers. Plant breeders usually select cultivars that will appeal to a large number of producers. The next task is to identify cultivars or varieties, or, in some cases, related wild plants that have the desired new character. For example, a common target of plant breeding programs is enhanced resistance to fungal pathogens. Cultivar X is the chosen cultivar that performs well but lacks resistance to a particular fungus. Cultivar Y lacks many of the valuable characteristics of cultivar X, but it is resistant to the fungus. The plant breeder will then cross cultivars X and Y, by taking pollen from one cultivar to pollinate the other. Some of the offspring (the **F1** generation) from this cross should have enhanced resistance to the fungus, while retaining most of the good characteristics of cultivar X. In terms of classical genetics, the phenotype (resistance) of cultivar X has changed, because of alterations in the genotype (genetic structure) of cultivar X. In the second round of crosses, the breeder selects plants from the F1 generation that have enhanced levels of fungal resistance.

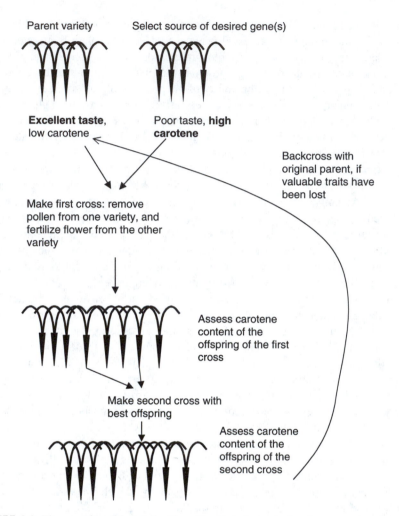

FIGURE 4.3 The use of traditional plant breeding techniques to develop a carrot variety with increased levels of β-carotene. The parent variety has a number of valuable traits (e.g., sweet taste), but has low levels of β-carotene. The breeder introduces genes from a carrot variety that has high levels of β-carotene, through careful pollination. Seeds are grown from this cross, and the plants are tested for β-carotene and other traits. The second cross is made between plants with enhanced levels of β-carotene. This process is repeated until the desired level of β-carotene is achieved. However, the resulting plants may lack some of the original valuable characteristics of the parent variety (e.g., they may have low levels of sucrose). Backcrossing with the parent variety is then required to restore the missing traits. Once this is achieved, the breeder attempts to breed a "true-breeding" line (variety) that does not vary in its important traits. Finally, registration is sought; this requires a submission to a regulatory agency of test results demonstrating that the new variety performs acceptably under a variety of growing conditions in a broad geographic range, and that the new variety offers a substantial improvement over previous varieties. If successful, the breeder will then have "plant breeder's rights," which in most countries offer a degree of intellectual property protection that is less powerful than patent protection but still valuable to the breeder.

The biggest problem with this approach is that after several generations the plants will likely have lost many of the valuable characteristics of cultivar X. This occurs because of the nature of meiosis, the process that leads to pollen and ovule formation. Each pollen or ovule has only one half of the chromosomal complement of the parent plant. So in a sense, the breeder has created an equal mixture of good genes from cultivar X and not-so-good genes from cultivar Y (with the exception of the desired genes for fungal resistance). **Backcrossing** (see Figure 4.3) is a common approach to restore valuable characteristics. Several rounds of backcrossing are usually necessary before obtaining a plant that has all of the characteristics of the parent variety, but also the additional desired characteristic. This requirement for backcrossing is a major reason for the protracted length of time (and amount of labor) required for traditional breeding programs.

Despite these problems, traditional plant breeding programs have been dramatically successful throughout the world. Virtually all of the food that we eat is from plants and animals that have been bred in this way, and newer techniques, such as transgenic development, likely will not substantially decrease the need for traditional breeding programs in the future.

2. Protoplast Fusion

The major disadvantage of traditional breeding programs is that the range of sources of new characteristics is narrow; only closely related plants (usually within the same species) can be used. This is frustrating for plant breeders, who often see highly desirable characteristics in unrelated plants. Protoplast fusion partially solves this problem by allowing the recombination of chromosomes from a wider range of parents into a new cell, and ultimately, into a new plant.

Protoplast fusion (Figure 4.4) often begins with mesophyll cells (mesophyll is the main photosynthetic tissue in leaves). The cells are treated with enzymes that break down the cell wall, resulting in **protoplasts**, plant cells that have an intact membrane but no cell wall. These cells can then be persuaded, through electroporation or the addition of polyethylene glycol (PEG), to **fuse** with protoplasts derived from other *species* of plants. When treated in this way, cells that collide undergo **cell fusion**. The membranes of two colliding cells fuse, and the contents of the cells are mixed. Nuclear fusion also occurs, and, if the plant breeder is fortunate, the cell will be able to cope with the mixture of chromosomes and divide mitotically. This usually occurs via the loss of chromosomes, but the resulting cell usually has some of the characteristics of the two original cells.

Fused protoplasts can then be induced to re-form their cell walls and are then plated on agar media. If all goes well, fused cells will grow into a callus. Manipulation of hormones then leads to the formation of embryos (these are referred to as **somatic embryos**) that grow into plantlets. The resulting plants can then be backcrossed to restore valuable characteristics that were lost during the cell fusion process.

Protoplast fusion has been less successful and less widely practiced than originally envisioned, primarily because of the difficulties in obtaining fused cells that will divide normally. However, this approach is useful if a desirable characteristic

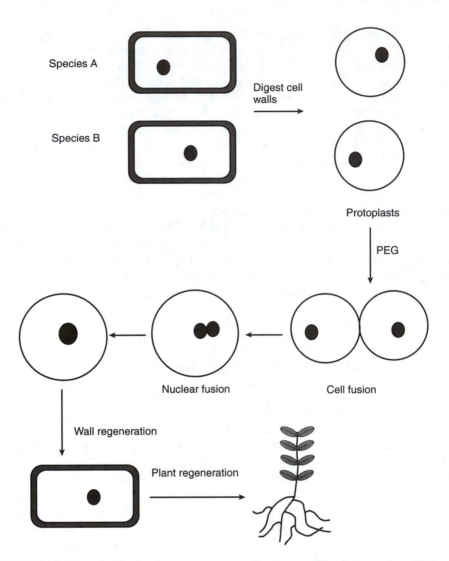

FIGURE 4.4 Protoplast fusion between two sexually incompatible plant species. (PEG = Polyethylene glycol.)

resides in a species that is too distantly related to be crossed using conventional techniques, but is related closely enough to allow chromosomal re-assortment and successful division of cells fused with the parent variety. In plant breeding terms, then, protoplast fusion sometimes allows the use of **wide crosses**.

3. Somaclonal Variation: Problem or Opportunity?

The manipulation of plant cells from plant to callus to plantlet was described as a relatively straightforward process. For some plants (e.g., tobacco) it is straightforward, but, unfortunately, the tissue culture of many plant species poses problems.

One of the problems shared by virtually all tissue culture systems is **somaclonal variation**. This is genetic variation that arises during callus culture. For reasons that are largely unknown (but much speculated upon), such variation frequently arises spontaneously, especially if callus culture occurs for a long time.

In some cases, somaclonal variation can be useful. Plant breeders sometimes are frustrated by *lack of variation*. If there are no sources of a desired characteristic, then the breeder has nothing to start with. The parent crop, however, can be grown as a callus, broken up into individual cells, and diluted out onto an agar medium. Each of the resulting calli will then (if the dilution was successful) grow out from *one original cell*, just as bacterial colonies on a streaked plate grow from single cells. These calli can then be assessed for the presence of new, desirable characteristics. Although not widely practiced, this approach has resulted in the selection of improved plants (in one case, a crunchier carrot was obtained). Somaclonal variation is particularly useful if the desired characteristic can be detected during the callus stage. Otherwise, plantlets must be induced and transferred to soil before screening for the desired characteristic can proceed.

Plant tissue culture is a complex art that has applications that are useful to food biotechnologists. However, there are many associated problems. Many important plants are difficult or impossible to grow in tissue culture. Furthermore, tissue culture is expensive in terms of materials and equipment, especially when compared to soil-based methods of plant culture. Despite these limitations, plant tissue culture is an important tool for plant biotechnologists. This is especially true for people interested in developing transgenic plants; tissue culture techniques are usually required at one or more stages in the development of transgenic plants.

III. TRANSGENIC PLANTS

A. OVERVIEW

Few subjects are as dear to the heart of a food biotechnologist (or as controversial to the public) as transgenic plants. These are plants that have received *specific, well-defined genetic modifications*. How is this different from traditional plant breeding? In a traditional breeding program, the changes that occur are usually undefined at the molecular level; the resultant plant might taste better, grow faster, and be more drought tolerant, but the breeder usually does not know exactly *which genes have been gained or lost*. In a transgenic approach, *specific genes or groups of genes* are transferred from a *source organism* into a *target plant*. Therefore, transgenic plants are conceptually similar to bacteria or fungi that have been given new genes using the techniques described in Chapter 3. Indeed, the techniques used in the development of transgenic plants are similar to the techniques used for the transfer of genes to bacteria and fungi.

The major advantage of the transgenic approach over traditional approaches is that theoretically any organism can be a source of transferred genetic material. Genes can be transferred from distantly related plants, from bacteria, fungi, or viruses, and even from animals. Furthermore, the potential exists for exquisite control over the activity of transferred genes, in terms of the amount and the timing of gene expression.

This control is achieved through the use of specific promoter elements. Some promoters will allow expression only in certain tissues or organs (e.g., leaves, seeds, pollen, fruits) or in response to specific stimuli (e.g., wounding of the plant). This level of control can be useful. For example, consider a gene that codes for a protein that triggers the synthesis by plant cells of compounds that are toxic to fungi. Rather than have this gene continually expressed, it would be preferable to have expression only after wounding of the plant, and only in wounded tissues. Such tissues are vulnerable to attack by fungal pathogens, so the production of antifungal compounds would be beneficial. In contrast, the production of antifungal compounds in unwounded plants would be unnecessary and would waste energy and carbon. It might also be desirable to limit gene expression to parts of the plant that are not eaten, if there are any human toxicity concerns related to the antifungal compound.

Most transgenic plants are developed using *Agrobacterium tumefaciens*, an unusual bacterial plant pathogen that has evolved the ability to transfer some of its genes to plant cells, where they are inserted into chromosomes to become a permanent part of the genome. Molecular biologists have modified this process to allow transfer of specific genes via *A. tumefaciens*, without the transfer of any genes of *A. tumefaciens*. Bacteria are the source of many of the genes that have been transferred to plants using *A. tumefaciens*. Bacterial genes conferring herbicide resistance and genes from *Bacillus thuringiensis* (Bt) have been transferred to numerous plants. In contrast, some transgenic plants are created through the insertion of genes native to the plant that have been altered in some way. These may be **antisense** versions of genes, which act to decrease expression of specific genes. For example, this is the usual approach taken to delay ripening in tomatoes and other fruits. Antisense genes decrease the levels of enzymes involved in ripening, such as enzymes that lead to ethylene production, and degradative enzymes such as polygalacturonase (PG), which lead to fruit softening.

In some cases, the gene that is transferred originates from another plant. Transgenic canola varieties with high levels of laurate are an example. Laurate is a fatty acid that has many industrial (e.g., detergents) and food applications. Normally, canola oil does not contain this lipid; using *A. tumefaciens*, biotechnologists introduced a thioesterase from the California bay laurel tree into the canola genome, resulting in high levels of laurate in canola seeds.

Most transgenic plants that have been released have been aimed at producers, especially toward the goal of easier and cheaper control of insect or weed pests. Biotechnology companies have concentrated on these applications because of the large, receptive market (agrochemicals are a major cost for most farmers, so they tend to be attracted to technology that allows decreased reliance on chemical control of pests) and because herbicide and insect resistance can often be induced through the addition of single genes. These types of transgenics will be more fully explored in Sections IV.A and C of this chapter, but several comments are appropriate here. Herbicides harm plants through disruption of crucial metabolic pathways (e.g., glyphosate, which disrupts synthesis of aromatic amino acids) or through a breakdown in hormonal control of plant growth (e.g., 2,4-D, which mimics the action of auxin hormones and results in uncontrolled growth that kills the plant). There are several ways to induce resistance to herbicides, but arguably the most common approach is to find a bacterium that is able to degrade the herbicide, and then clone a gene for the enzyme that is responsible for this degradation. Then, transfer this gene to the plant using *A. tumefaciens* or an alternative method. The transgenic plant will then have the ability to degrade the herbicide before it can exert its toxic effects. This makes weed control easier and more efficient; fields can be sprayed when and if weed control is necessary, without having to worry about adverse effects on the crop.

Insect control can also be given to plants through the addition of a single gene. There is a proviso, though: the specific insect pest must be affected by a strain of Bt. This soil bacterium produces proteins that are toxic to specific groups of insects; organic farmers have long exploited this property by spraying Bt spores onto crops. The transgenic approach is to clone the gene for a toxic protein from Bt and then to transfer the gene into the crop genome. Vegetative tissue of the crop will then be toxic to the insect pest, resulting in reduced use of chemical insecticides and fewer agrochemical costs to the farmer.

At first glance, it may seem that herbicide- and insect-resistant transgenics are relevant to agriculture but not to food science and nutrition. However, the wide acreage of insect- or herbicide-resistant corn and soybeans, two commodities that are widely used in the food industry, makes these transgenics quite important. For example, the use of Bt transgenics should reduce the amount of pesticide residues on fruits and vegetables, thereby increasing food safety. Insecticide levels are generally low and considered to be safe, but in North America (and likely globally as well) there is a sporadic incidence of high levels of pesticide residues on fruits and vegetables that are consumed with minimal processing. Any factor that decreases reliance on pesticides is likely to decrease the incidence of such contamination.

In contrast, the use of herbicide-resistant crops may not decrease the amount of herbicides used and in some cases leads to increases. However, the herbicides that are targeted by transgenics are typically less toxic and more rapidly degraded in the environment than many other herbicides. Therefore, the use of these transgenics may improve environmental quality and decrease the amount of toxic herbicide contamination of produce.

However, this is a highly contentious issue, particularly in Europe, where there is a perception that farmers are much less reliant on agrochemicals than North Americans. Many people in Britain, for example, fear that the use of herbicide-resistant crops will result in zero tolerance for weeds. This could result in a dramatic loss of plant biodiversity in British fields, which could have serious ramifications on wildlife. It is difficult to assess this risk, because it is difficult to assess the current "weediness" of British agriculture, and it is difficult to assess the importance of weeds in agricultural fields to wildlife. However, at least one study has shown that use of herbicide-resistant crops could theoretically reduce food availability (weed seeds) to skylarks.

Similarly, the potential of Bt transgenics to reduce the use of chemical insecticides is controversial. With Bt corn that is toxic to the European corn borer, nontarget effects on monarch butterflies are an additional source of contention. Many scientists and antibiotechnology activists fear that pollen released by Bt corn will adversely affect monarch larvae. This issue will be further explored in Section IV.A of this chapter.

Consumers throughout the world are also concerned about safety issues associated with transgenic plants. Concern is particularly strong in Europe, where fear of bovine spongiform encephalopathy (BSE) and distrust of food regulatory agencies lingers. In 1998, mandatory labeling legislation was passed in Britain, making it illegal to sell food containing transgenic organisms (popularly known as genetically modified organisms, or GMOs) without labels. This led to the virtual disappearance of such food from British supermarkets, primarily because of fears by grocery chains that consumers would shun foods carrying GMO labels.

Consumer groups and antibiotechnology activists are also pressing for labeling of meats and dairy products produced by cows fed feed containing transgenic crops. Many farmers in the U.K. are against such labeling requirements because of anticipated difficulty in obtaining sufficient transgenic-free feed.

The situation in Europe and Britain illustrates the need for food companies and anyone that counsels people about food consumption to understand the techniques used to create transgenic crops, as well as the environmental safety issues that surround them. For example, a clinical dietitian may be interviewing someone who has recently been found to have high levels of cholesterol. The dietitian tells the client that s/he should increase consumption of fruits and vegetables. However, the client responds that he does not want to do that, because of "all those GMOs out there." To respond effectively, the dietitian needs to understand how transgenic plants are made, as well as the risks and benefits associated with transgenic crops.

Although most of the currently released transgenic plants have been aimed at producers, this will likely change in the next 10 years. Transgenics will be increasingly targeted to solve nutrition-related problems such as low vitamin content, high levels of antinutrients, amino acid imbalances, and allergenicity. The food industry will also benefit from transgenics with improved processing or storage qualities (see Chapter 1, Table 1.3). For example, the glutenin alleles present in a wheat variety strongly affect bread-making qualities of the variety. Transgenic technology allows fine control over the expression of these genes. These and other applications of transgenic technology to processing and nutritional aspects of food will be further discussed in Section V of this chapter. The main objective of the remainder of this chapter is to explain the techniques required for the development of transgenic plants and the range of opportunities associated with this technology.

B. The Process of Making Transgenic Plants

Consider the following scenario: you work for a biotechnology company that specializes in the production of transgenic food plants. You have been told that the company would like to consider developing an oilseed plant with high levels of a specific polyunsaturated fatty acid (PUFAx). What would be the overall plan of action? Although many variables should be considered, we can divide the process into **pretransformation** and **posttransformation** steps. In some cases, the posttransformation process may be more time consuming than the pretransformation process.

Pretransformation

1. Identify the **target** plant. Soybeans and canola are popular oilseed crops, as well as olive, corn, palm, and several other plants. Canola (low-erucic-acid rapeseed) is a popular choice for transgenic oilseed production because of the high oil content (40%) of the seeds and good agronomic qualities (e.g., cultivability over a wide geographic range). Canola is also relatively easy to grow in tissue culture and is easily transformed.

2. Identify the **source** plant. You need to find a plant that produces high levels of PUFAx. In many cases it will not be feasible to use this plant directly as a source of the oil, perhaps because it is not consumed as food, and thus has uncertain safety or edibility. However, it may be possible to transfer the trait to canola or another oilseed producer.

3. Isolate the **enzyme** that is responsible for the production of PUFAx. This will allow characterization of the enzyme's properties and sequencing of the protein. Nucleic acid probes (Chapter 2, Section V.H.3) can then be designed, based on the amino acid sequence of the protein.

4. Isolate mRNA from the source plant and establish a **cDNA library** (Chapter 3, Section V). You will then be able to use the probe developed in step 3 to screen the cDNA library for the *PUFAx* gene. This will allow cloning of the *PUFAx* gene.

5. Confirm that the cloned gene is the correct gene by analyzing proteins expressed by the gene. This is normally done in a bacterial or fungal host cell.

6. Isolate genomic DNA from the source plant and set up a **DNA library** (Chapter 3, Section IV). Use the probe to screen this library for the desired gene. This is done so that the **promoter** and other genetic elements can be characterized. This is important, because transgenic oilseeds usually use promoters that restrict gene expression to the seed. Expression of fatty acid synthetic genes in other organs and tissues usually negatively affects plant growth. The promoter cannot be cloned from the cDNA library because the promoter region is not transcribed into mRNA. Sometimes, the researcher will already know of a promoter that will restrict gene expression to the appropriate tissue or organ. If so, this step may be less crucial. In most cases, though, it is advisable to know as much as possible about the gene and how it is regulated before proceeding to the next step.

7. Construct a plasmid vector (probably in *Escherichia coli*) that has the correct gene (from the cDNA library) and promoter (from the DNA library).

8. Insert this vector into the target genome, using *A. tumefaciens*, bolistics, electroporation, or micro-injection. These procedures virtually always require *in vitro* culture of the target plant. Following successful transformation, you will need to grow intact transgenic plants from these cultures.

Posttransformation

9. Reject transgenic plants that are clearly unsuitable, because of stunted growth or other abnormalities. Current methods of creating transgenic plants involve insertion of foreign genes into random sites within the plant cell's chromosomes. This sometimes disrupts important genes, resulting in abnormal plants. The transgenic events that survive this screening process can then be assessed in subsequent steps.

10. Confirm that the desired gene is present in the transgenic plants. This can be done by southern blotting (Chapter 2, Section V.H.3). At this stage it is useful to determine how many copies of the gene have been inserted into the plant's genome. In most cases, it is most desirable to have single-gene insertions, because this should lead to greater stability (see step 12).

11. Assess the **phenotype** of the transgenic lines. In other words, find out if the desired trait has been introduced. In our example, this would require analysis of the lipid composition of the seeds, to determine if PUFAx is produced.

12. Suitable lines are then tested for stability of gene inheritance and expression. This involves assessment of gene expression after crossing a transgenic line with nontransformed lines. It is essential that the recombinant gene be retained by at least some of the offspring of these crosses.

13. Use traditional breeding techniques to develop true breeding lines, so that all offspring of a transgenic plant will have the recombinant trait. This is impossible in some crops, but in such cases, alternative approaches such as hybrid crops are possible. Hybrid seed is made by crossing two varieties and selling the resulting seed (F1 generation) to farmers. Usually valuable traits are lost in subsequent generations, so every year the seed grower must repeat the cross, and every year the farmer must buy a new lot of seed.

14. Ensure that the resulting transgenic line has the necessary agronomic qualities (e.g., is able to grow in a specific range of soil types and climates) and processing characteristics (e.g., release of the oil through seed crushing).

15. Assess the environmental and food safety risks of the transgene (see Chapter 9 for a discussion of this process). This is required for the regulatory process in most countries. This should begin early in the process. It would be foolhardy to go to all the trouble and expense required to attain this stage without careful considerations of human and environmental safety issues.

16. Register the new transgenic variety ("line" and "variety" are similar terms, but variety is usually used at a later stage of the process). This process is similar to that undergone by varieties derived through traditional plant breeding methods. Registration normally involves demonstration to regulators that the variety is different from other varieties and that this difference is potentially useful to producers or processors.

Registration confers on the breeder certain intellectual property rights, but companies that develop transgenic plants usually attempt to gain patent protection as well. This gives them greater control over the use and sale of their transgenic varieties.

17. Obtain regulatory approval for release of the transgenic cultivar (a "cultivar" is a variety that is being actively cultivated). This normally involves submission of a document that assesses the characteristics of the transgenic protein and its potential toxicity and allergenicity to humans, as well as an assessment of potential environmental risks. In our example (PUFAx), the environmental assessment would likely be straightforward, but assessment of toxicity and allergenicity may be more difficult (see Chapter 9). In some regions (e.g., Europe), the company would have to obtain approval from two regulatory paths; one is set up to allow release of transgenic propagules (usually seeds) to farmers, and one is to allow use of the transgenic in food.

18. In many cases, it may also be necessary to ensure that the current path of the crop from the farm gate to processors can cope with the proposed transgenic variety. In our example, it would be necessary to **segregate** transgenic harvests from harvests of other cultivars (one would not want to mix PUFAx seeds with non-PUFAx seeds because it would result in dilution of the new valuable lipids). Segregation is often desirable for exporting purposes, as well; for example, shipments to Europe that are destined for use as food must be segregated into transgenic (GMO) and nontransgenic (non-GMO) lots, so that foods can be appropriately labeled. Also, most of the transgenic crops that have been approved in the U.S. and Canada have not been approved in Europe; in these cases, they must be segregated from European shipments, because it is illegal to sell them in the European Union.

The development process for a transgenic plant is not fast, although in many cases it may be quicker than an equivalent process using traditional plant breeding, because of the precision of the genetic transfer (i.e., only specific genes are transferred using a transgenic approach). We will now concentrate on step 8, and develop an understanding of the methods used to transform plant cells with foreign genes.

C. *AGROBACTERIUM TUMEFACIENS*: A NATURAL DNA VECTOR

We now have a problem. Assuming that we have cloned our *PUFAx* gene into *E. coli*, how do we get it from the bacterial cell, through the plant cell wall, into the nucleus, and permanently integrated into the plant's genome? Theoretically, this could be done in two ways: (1) by integration of *PUFAx* into a plasmid that is able to replicate in plant nuclei, or (2) by integration of *PUFAx* into a plant chromosome, which would allow *PUFAx* to be transferred into daughter cells through the normal segregation of chromosomes that occurs during mitosis. The second option is feasible for plant cells; unfortunately, plant cells rarely have plasmids, so the first option is not useful.

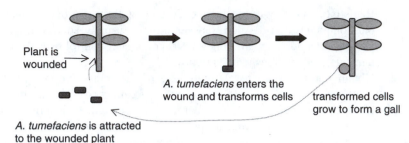

Plant is wounded

A. tumefaciens enters the wound and transforms cells

transformed cells grow to form a gall

A. tumefaciens is attracted to the wounded plant

FIGURE 4.5 Production of a gall by *A. tumefaciens*.

Getting foreign sequences of DNA into plant cells is not easy; obtaining a stable integration of foreign DNA into plant chromosomes is even more difficult. However, plant biotechnologists have been able to harness the ability of *A. tumefaciens* to act as a vector. This bacterium has been used to develop many of the transgenic plants that have been released to growers.

D. THE NATURAL LIFE CYCLE OF *AGROBACTERIUM TUMEFACIENS*

A. tumefaciens is a Gram-negative aerobic bacterium that, until the 1970s, was known primarily as an unusual plant pathogen that induces tumor formation in plants. When it was discovered that it is able to transfer bacterial DNA into plant cells and plant chromosomes, plant biotechnologists and molecular biologists quickly saw the potential for harnessing the unique abilities of this bacterium.

The life cycle of *A. tumefaciens* moves from a dormant to an active state when a plant is wounded (Figure 4.5). *A. tumefaciens* is able to survive free-living in soil, primarily in a dormant state. A number of compounds are released from plant cells after wounding, and *A. tumefaciens* is chemo-attracted to these compounds. It then enters the plant via the wound. Once inside the plant, *A. tumefaciens* induces tumor (gall) formation, usually in stem tissues. The bacteria reproduce extensively within the gall, and when the plant dies, bacteria are released into the soil, where the life cycle continues.

A. tumefaciens is able to induce galls because of its large Ti (tumor inducing) plasmid (Figure 4.6). The *vir* and *T-DNA* regions of this plasmid are particularly interesting. *vir* contains genes that code for proteins that are able to direct and control the transfer of plasmid DNA from *A. tumefaciens* to plant chromosomes. However, the entire plasmid is not transferred; only the *T-DNA* is nicked out of the plasmid, transferred into the plant nucleus, and integrated into a chromosome. This is similar to **conjugation**, the process that allows bacteria to transfer plasmids from one bacterium to another. The transfer of DNA from bacterium to plant is complex and requires the coordinated activity of numerous genes in the *vir* region. Briefly, specific proteins recognize the DNA sequences of the left and right borders and excise a single strand of DNA from the region between the borders (the *T-DNA* region). This single-stranded DNA is coated with a DNA-binding protein and becomes part of the **T-transporter complex** (made up of 12 *vir* proteins), which moves the DNA from the bacterial cell into the plant cell. This apparatus probably moves through

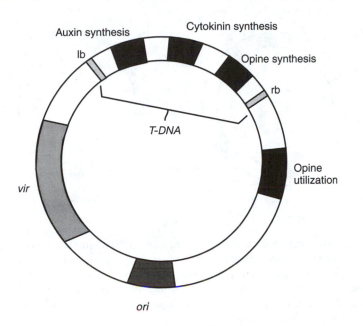

FIGURE 4.6 Structure of the Ti plasmid of *A. tumefaciens*.

membrane channels out of the bacterial cell and into the plant cell. The complex moves from cell to cell via a **pilus** (T-pilus), a flexible, proteinaceous tube. Assembly of the T-pilus is mediated by *vir* proteins. Some of the proteins in the complex then function to allow transport of the complex into the nucleus. Once in the nucleus, the *T-DNA* is integrated (spliced) into a plant chromosome. The mechanism for this is unknown, but it probably involves host enzymes that normally function in DNA replication and repair. The transfer of *T-DNA* from bacterium to plant is an elegant, choreographed process requiring the participation of genes of the *vir* region, as well as several bacterial chromosomal genes and host plant proteins. Initially, scientists believed that the *agrobacterium* system of gene transfer is unique; however, it is becoming increasingly clear that many bacterial pathogens can transfer DNA and proteins into host plant, animal, or fungal cells.

The *T-DNA* region of the *Ti* plasmid contains a number of genes that are necessary for the life cycle of *A. tumefaciens*. Several genes are involved in the synthesis of **auxins** and **cytokinins**, which trigger uncontrolled growth of plant cells (similar to callus formation in tissue culture systems). Other genes code for enzymes that synthesize **opines**, low-molecular-weight compounds that are an important source of carbon and energy for *A. tumefaciens* growing within galls. Opine production is a clever trick by the bacterium — each strain of *A. tumefaciens* produces a specific form of opine that can be catabolized only by that strain. Because most soil bacteria and fungi cannot utilize opines as a source of carbon and nitrogen, *A. tumefaciens* has exclusive use of these compounds. The production of a gall of plant cells producing opines, then, is in a sense a "private party" that only *A. tumefaciens* can attend.

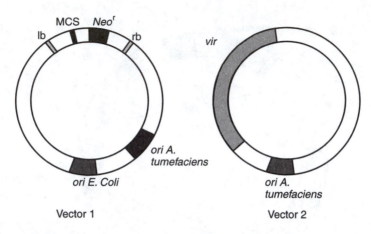

FIGURE 4.7 The binary vector system, derived from the *Ti* plasmid of *A. tumefaciens*. Most of the genes of the *T-DNA* have been removed, leaving the left and right borders. The vir region has been moved into another plasmid (vector 2). Note that plasmid 1 has an *ori* region that allows it to replicate in *E. coli* and an *ori* that allows it to replicate in *A. tumefaciens* (therefore, it is a shuttle vector). (MCS = Multiple cloning site, lb = left border, rb = right border, Neo^r = neomycin resistance.)

E. The Use of A. *tumefaciens* to Create Transgenic Plants

It would clearly be unwise to use unmodified *A. tumefaciens* to develop transgenic plants. Such plants would have extensive galls and would not grow well. Consequently, many modifications have been made, particularly to the *Ti* plasmid. Most biotechnologists use a **binary vector** system (Figure 4.7), which uses two plasmids — one that contains the *vir* region (vector 2) and one that contains the **left and right border regions** of the *T-DNA* (vector 1). The major reason for constructing two smaller plasmids is that smaller plasmids are easier to manipulate; they replicate to higher copy numbers in bacteria, and it is much easier to get them into bacterial cells via transformation.

Vector 1 contains the left and right borders of the T-DNA but none of the genes between these regions that are present in the *Ti* plasmid. All of the genes between the left and right borders of the *T-DNA* have been removed, so that tumors will not form. However, if a segment of foreign DNA is placed between the left and right borders, then the *vir* proteins will recognize the borders, excise the intervening DNA, and incorporate it into a plant chromosome.

The following steps (Figure 4.8) are required for the transfer of a foreign gene (*PUFAx*, in our example) to a plant using an *A. tumefaciens*–derived binary vector system:

1. *PUFAx* is cloned into *E. coli* using an appropriate vector, as described in Chapter 3. The cloned gene is then subcloned into vector 1. Because the multiple cloning site (MCS) is between the left and right borders, *PUFAx* will be inserted between the borders. *E. coli* is used at this stage because it is easier to grow and manipulate than *A. tumefaciens*.

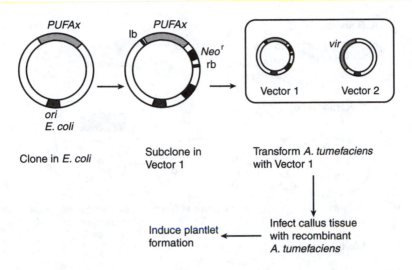

FIGURE 4.8 The use of the binary vector system of *A. tumefaciens* to transform plants.

2. Vector 1 is purified from *E. coli* and incorporated into *A. tumefaciens* via transformation. Plasmid 1 will be able to replicate in *A. tumefaciens* because it has an additional *ori* that allows replication in *A. tumefaciens*. The strain of *A. tumefaciens* used must contain vector 2, which has the *vir* genes required for transfer of *PUFAx* to plant cells. Such strains are readily available from a number of biotechnology companies.

3. *A. tumefaciens* now has vectors 1 and 2. The next step is to inoculate the bacteria onto cultures of callus derived from the target plant. *A. tumefaciens* will infect these cells, and plasmid 2 will produce proteins that will direct the transfer of *PUFAx* from plasmid 1 into plant cells. *PUFAx* will then be inserted into a plant chromosome.

4. The callus is transferred to a medium that contains **neomycin (kanamycin).** Plant cells are susceptible to neomycin, so any plant cells that lack the introduced foreign DNA will be killed. Vector 1 has a **neomycin-resistant gene** between the left and right borders; if the transfer of *PUFAx* is successful, the kanamycin-resistance gene will also be transferred, and will be integrated into the plant genome.

5. The callus is then transferred to a medium that allows regeneration of plantlets.

Many variations of the above scheme are possible. The inclusion of antibiotic-resistant genes in transgenics is controversial, so it may be preferable to use other marker genes. The **GUS** system is one alternative — this method requires that vector 1 contain a **β-glucuronidase** gene. Plant cells normally lack this gene, which codes for an enzyme that is able to convert X-glucuronide (a synthetic carbohydrate) to a blue end product. This color change allows identification of recombinant plant cells. Such cells can then be induced to grow into intact plants. GUS is called a **reporter gene**, because its presence indicates the successful transformation of a plant cell.

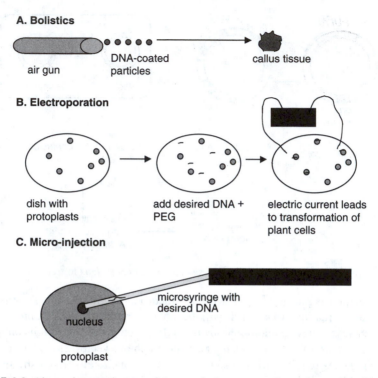

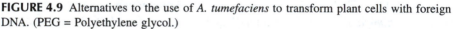

FIGURE 4.9 Alternatives to the use of *A. tumefaciens* to transform plant cells with foreign DNA. (PEG = Polyethylene glycol.)

Another common modification is to avoid callus culture and directly wound and infect plantlets. This avoids the genetic instability that often occurs during callus culture. This is particularly useful with plants that are difficult to manipulate in tissue culture. In some plants (e.g., rice), *A. tumefaciens* is able to infect only embryo cultures. These are cultures of cells that have been hormonally induced to form embryos.

F. ALTERNATIVES TO *A. TUMEFACIENS*

The *A. tumefaciens* system works well with many dicots, but is less effective with monocots (e.g., cereals). For this reason, extensive research has been aimed at developing alternative systems for the transformation of monocots and other plants (Figure 4.9). One relatively successful system has been **bolistics**. By this method, DNA containing the desired foreign gene is attached to tiny beads, or microprojectiles. An air gun is then used to bombard plant cells or tissues with the microprojectiles. A small percentage of cells are successfully transformed in this way.

Plants can also be transformed by electroporation of protoplasts and adding DNA after pretreatment of protoplasts with PEG. Finally, micro-injection of DNA into intact plant cells has been successful in some cases. One of the major disadvantages of all of these systems is that integration of foreign DNA into the plant genome rarely occurs; integration is much more frequent when *A. tumefaciens* is used as a vector. Another disadvantage is that foreign DNA is inserted at random points in plant chromosomes. Consequently, disruption of essential genes sometimes occurs.

IV. APPLICATIONS OF TRANSGENIC PLANTS TO FOOD PRODUCTION

A. TRANSGENICS RESISTANT TO INSECT PESTS

Many of the early transgenic crops were aimed at reducing reliance on chemical insecticides through the introduction of genes coding for insect-toxic proteins. This would have obvious benefits — farmers would be able to save large amounts of money through reduced purchase of pesticides, and the public would benefit through reduced consumption of pesticides in crops and decreased levels of pesticide residues in soil and water.

The selection of a suitable gene for such transgenics is not easy. Three key criteria must be met: (1) the gene must code for a protein that is produced in plant tissues at a high enough concentration to be toxic to insect pests; (2) the protein must be toxic to insects but not to humans or other mammals; and (3) the protein should be highly specific, only killing targeted pest insects rather than a wide range of insects, because many insects are beneficial.

Many proteins meet these criteria but proteins produced by the spore-forming bacterium Bt were the first to be exploited using transgenic technology. One reason for this is that Bt has a long history of safe and effective use in agriculture and forestry. As mentioned before, organic farmers rely heavily on Bt pesticides, applied to crops in the form of spore suspensions. These spores are toxic to insects because they contain crystalline proteins (both the proteins and the genes responsible are designated by *cry*). When certain species of insects consume the bacteria, the crystals become soluble in the alkaline environment of the insect midgut. After dissolving, they are converted from a **protoxin** to a **toxin** through the action of insect proteases in the midgut (the active toxin is designated by **δ-endotoxin**). *cry* proteins then interact with specific receptors on the surface of gut epithelial cells, leading to the formation of pores in the membrane. The epithelial cells then die, because of cytoplasm leakage. Extensive loss of gut epithelia eventually kills the insect.

There are at least 135 different *cry* genes; they are classified into three groups (*cry*I, *cry*II, and *cry*III), based partly on the size of the polypeptides and partly on the effective host range. The bacteria are also classified into groups (pathotypes) based on the effective host range of their toxins. For example, *B. thuringiensis* var. *berliner* produces *cry*I toxins, which are effective against Lepidopterans (butterflies and moths); *B. thuringiensis* var. *israelensis* produces *cry*II toxins effective against Lepidopterans and Dipterans (flies and mosquitoes); and *B. thuringiensis* var. *tenebrionis* produces *cry*III toxins effective against Coleopterans (beetles). Fortunately, a number of important insect pests are Lepidopteran (e.g., the European corn borer), Dipteran (e.g., houseflies and mosquitoes), or Coleopteran (e.g., the Colorado potato beetle) and can potentially be controlled either with insecticidal sprays of Bt spores or through integration of *cry* genes into crop genomes.

δ-Endotoxins do not exert toxic effects on mammalian digestive systems. The safety of Bt toxins for humans is well established; large-scale spraying programs over large tracts of forest (aimed at control of spruce budworm) have shown no impact on humans, with the exception of uncommon allergic reactions to whole-spore preparations. This is mainly considered to be a hazard to people spraying crops, and not to people consuming crops that have been sprayed with Bt spores.

In comparison to chemical insecticides, Bt toxins are highly specific. Broad-spectrum chemical insecticides kill beneficial predatory insects as well as insect pests. This often leads to problems due to **secondary pests**, insects that are normally controlled by predatory insects. Consequently, Bt insecticides are considered to be less damaging to the environment. The main factor limiting the implementation of Bt insecticides in conventional agricultural systems has been cost; chemical insecticides are generally cheaper.

Biotechnologists quickly realized that this cost problem could be averted by creating transgenic crops that had *cry* genes integrated into their genomes. By the early 1980s, such transgenics had been successfully developed in the laboratory, and in 1996, the U.S. FDA allowed the release of transgenic insect-resistant corn carrying *cry* genes. In 2000, nearly one third of the American corn crop was Bt corn, a phenomenal success story, at least in terms of its rapid adoption by farmers. However, early estimates of the 2001 crop suggest that use of Bt corn fell slightly in 2001, possibly because of fears that major world markets (e.g., Europe) would not buy very much Bt corn..

Technically, the development of Bt transgenic plants is not easy. It is relatively easy to clone *cry* genes and transfer them to plants (both bolistics and *A. tumefaciens* have been used, depending on the target plant). However, early attempts resulted in plants with only a small amount of expression of the toxic protein — not enough to control the pest. Scientists soon realized that this low expression was partly due to **codon usage**. Remember that more than one base triplet (codon) can code for each amino acid. Unfortunately, bacteria tend to prefer certain codons and plants tend to prefer other codons for certain amino acids. Consequently, biotechnologists had to modify the DNA sequence of the *cry* genes to reflect this difference in codon usage. This, combined with the use of effective promoters, led to large increases in the amount of δ-endotoxin in plant leaves. The promoter from cauliflower mosaic virus (CaMV) has been used in some Bt transgenics, leading to high rates of expression in all tissues of the plant. Other transgenics have used promoters that restrict expression to either leaves or pollen. The introduction of enhancer elements (see Chapter 2, Section V.C) has also been effective in increasing levels of δ-endotoxin expression.

Transgenic Bt crops have been successful from a technical perspective (i.e., they effectively control target pests) and from a commercial perspective. However, Bt transgenic plants (especially Bt corn and Bt cotton) have also been attacked by a number of environmental and antibiotechnology activist groups. One controversy centers on the potential of target insects to quickly develop resistance to *B. thuringiensis* toxins. Organic farmers would lose one of their most valuable weapons against insect pests (organic farmers usually restrict their use of Bt sprays, partly for economic reasons and partly to reduce the risk of encountering resistant insects). The potential to develop resistance is a concern because: (1) resistance to *B. thuringiensis* toxins has already been observed in several instances, following long-term, intensive spraying programs, and (2) large-scale growth of Bt transgenics creates an environment that should efficiently select resistant insects. The response of the biotechnology industry to these criticisms is that resistance may occur, but it can be countered by the introduction of new transgenics, using different toxin genes, or **stacked** transgenics that contain more than one *cry* gene. Also, companies and regulators currently rely on the **high-dose/refugia** approach to delay resistance development. Refugia are areas with non-Bt plants within or adjacent to Bt plantings. Theoretically, the use of high-dose insecticides combined with refugia should delay resistance development. The reasons for this (and the use of refuges) are complex and involve the heritability of resistance genes. Recall that classical mendelian genetics describes genes as either **dominant** or **recessive**. If resistance genes are recessive, then refugia will allow Bt-sensitive (ss) insects to persist. They should breed with resistant (SS) insects in the Bt part of the field, leading to heterozygous (Ss) offspring. If the phenotype (resistance to Bt) is recessive, Ss insects should be susceptible to Bt. Refugia act as reservoirs of sensitive insects. However, if resistance is dominant, then Ss insects will be resistant to Bt. This is where the high dose becomes important (Figure 4.10). If the level of Bt in plant tissues is very high, then Ss insects will be susceptible, *even if resistance is a dominant trait*. This is an important point, because it is currently impossible to predict whether insect resistance will be dominant or recessive; both have been observed after intensive spraying of Bt spores. This illustrates the practical importance of the development of Bt transgenics that have high levels of expression of the Bt toxin.

In the next 5 years or so it should be possible to assess the success of the high-dose/refugia strategy. As an example of refugia, in 2000, the EPA recommended that U.S. corn growers establish refugia constituting 20% of their Bt corn fields and that they monitor their Bt corn closely, to establish an early warning system to detect resistant populations.

Resistance has appeared in some regions; for example, surveys of resistance to Bt cotton in populations of cotton bollworm in Australia revealed that resistant insects are present and are slowly increasing in number. The slow increase may be partly due to adverse aspects of resistance on the insects (e.g., reduced number of eggs produced) that weigh against the favorable aspects of resistance (i.e., ability to feed on Bt cotton).

Because of insect resistance, each Bt cultivar may have a relatively short life span in agricultural use. However, if resistant insects are less fit than nonresistant insects, it is possible that temporary removal of a particular Bt cultivar could result in a loss of resistant insects. If this happens, the cultivar could then be reintroduced. It should be stressed, though, that resistance to δ-endotoxins is relatively poorly understood, making it difficult to predict when and where resistance will occur, and how stable it will be.

The second controversy involving Bt corn concerns its effects on nontarget insects, particularly spectacular and popular insects such as monarch and swallowtail butterflies. These insects have several reproductive cycles every summer in the American midwest, leading to the occurrence of larvae feeding in and around corn fields. They do not feed on corn leaves, but they may accidentally ingest corn pollen

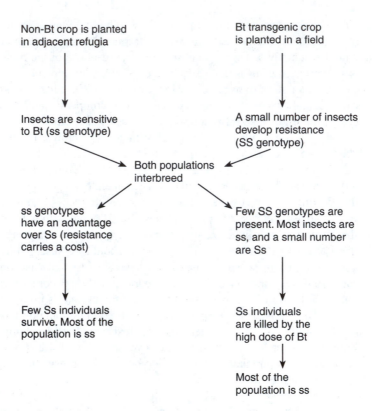

FIGURE 4.10 The theory behind the use of refugia and high doses of Bt to discourage the spread of resistant insects. This system should work even if resistance is a dominant trait in insects, as shown here.

containing δ-endotoxin by eating leaves of milkweed plants (main food of the larvae) that have corn pollen on their surfaces. Indeed, several laboratory studies have demonstrated toxic effects of Bt corn pollen when fed to monarch butterfly larvae. The investigators placed milkweed plants within and on the edge of Bt corn plots during the period of pollen release. Then they fed samples of the milkweed leaves to monarch larvae reared in the laboratory. Twenty percent of the larvae subsequently died, whereas none of the insects died that were fed leaves exposed to non-Bt corn pollen.

This raised a number of questions regarding the environmental safety of Bt corn because of the following points:

- Monarch butterfly larvae feed on milkweeds, which are commonly found within and adjacent to corn fields.
- Some Bt corn cultivars produce high levels of toxic proteins in the pollen; in some cases this is done deliberately through the use of pollen-specific promoters, because the European corn borer tends to consume large amounts of pollen.

The monarch butterfly is arguably the most popular insect on earth, and any threat to it raises the hackles of conservation groups. It has become the rallying cry of antibiotechnology activists throughout the world, and it is unlikely that hostility to Bt corn in Europe and elsewhere will abate without clear evidence that Bt corn does not harm the monarch butterfly. Biotechnologists have not been silent on this issue; many have argued that the evidence against Bt corn is inconclusive, for the following reasons:

- To date, no field study has demonstrated harm to monarch butterflies from ingestion of pollen from Bt corn. The environment in the field is much more complex than the controlled environment of the laboratory, and scientists often observe that effects observed in the laboratory do not occur in the field. In the case of the monarch butterfly, factors such as predation and lack of milkweed plants may be much more important than toxicity of Bt pollen.
- Large amounts of pollen are required to exert toxic effects on larvae; such large amounts are unlikely to occur anywhere except the interior of Bt corn fields, where the density of milkweeds is usually low.
- Toxicity to larvae is highest in transgenic cultivars with pollen-specific promoters; these tend to be less popular than transgenics with less specific (e.g., CaMV) promoters.
- The alternative to using Bt corn is to use chemical insecticides to control the European corn borer; these insecticides kill *all* lepidopteran larvae, including the monarch butterfly.

This is an area of active research, and it is anticipated that studies will soon be published that will help resolve this controversy. Until then, it is unwise to claim that Bt corn is either good or bad for monarch butterflies. Pro-Bt forces were recently encouraged by the publication of a study that demonstrated that Bt corn was not toxic to larvae of the swallowtail butterfly, another beautiful and popular butterfly in North America.

The third controversy is widely known as the **Starlink** scandal. In October 1999, it became clear that food in the U.S. had become contaminated with corn from a Bt cultivar (Starlink corn) that the FDA had allowed to be used as animal feed but not as food for humans. The FDA had made this decision based on the potential allergenicity of this corn, because the insect-toxic protein was less degradable under simulated stomach conditions (this is a characteristic of many food allergens).

The main company involved (Kraft) lost millions of dollars because of a recall of contaminated products such as taco shells. Trade to Japan and other countries has also been disrupted after the detection of Starlink corn in exported corn. The Starlink scandal, and its ramifications, will be more fully explored in Chapter 9.

B. Pathogen Resistance

Plants suffer from many diseases; some are physiological, caused by drought stress, mineral deprivation, and other environmental causes, but infectious agents (pathogens) also cause disease, reducing the amount of harvestable food by about 15%

globally. Post-harvest disease (spoilage) is also significant, resulting in the loss of about 25% of harvested food. In declining order of importance, fungi, viruses, and bacteria are the pathogens responsible for infectious plant disease, and fungi are responsible for most post-harvest food spoilage. Besides the economic losses, field and post-harvest diseases are a safety issue for the food industry and for consumers. Many of the fungi commonly involved produce **mycotoxins**. For example, *Aspergillus flavus*, which produces potent mutagens known as **aflatoxins**, is commonly found on cereal grains and peanuts grown in tropical or subtropical countries. Most countries have defined limits of aflatoxin contamination permitted in food and feed. For example, the U.S. has set limits of aflatoxin contamination in nuts destined for human consumption at 20 µg/kg.

Much of the fungal contamination that leads to mycotoxin contamination or food spoilage arises in the field; therefore, efforts to control fungal attack in the field may lead to improved food quality after storage. Interestingly, this is one of the points raised by proponents of Bt transgenic corn — when the European stem borer attacks corn plants, it creates a route of entry for fungi. It has been shown that levels of mycotoxins are lower on grains of Bt corn than on nontransgenic corn. Other transgenic crops are being developed that are better able to resist the growth of **mycotoxigenic** (capable of mycotoxin production) fungi such as *Aspergillus* and *Fusarium* spp. (the latter is a fungus commonly found in cereals grown in temperate countries).

Plant improvement is also the main strategy in the fight against plant disease in the field. The wheat stem rust epidemic that devastated wheat production in the mid-1950s in North America is an example of both the destructive ability of field plant pathogens and the potential of plant breeding to decrease the damage. The fungus that causes wheat stem rust (*Puccinia graminis*) produces small spores that can be transported via wind currents over wide distances. This allowed stem rust to spread over wide regions of North America, causing great loss of wheat yield. A concerted effort by Canadian and U.S. plant breeders led to the development of wheat cultivars resistant to the major pathovars of *P. graminis*.

Traditional plant breeders have successfully incorporated resistance genes into most crops. This resistance works through the incorporation into a plant of a gene that allows recognition of a fungus within plant tissues. The plant then reacts quickly (e.g., through the production of phyto-alexins that are toxic to the fungus) to eliminate the pathogen.

It is possible to transfer resistance genes from plant to plant using transgenic technology. This is an attractive option for many plant–pathogen combinations, because resistance genes are often present in wild plants that normally cannot be bred with crop plants. With the advent of gene cloning, and interplant transfer using *A. tumefaciens* and other methods, resistance genes can now be transferred from one plant to any other plant.

As an example of transgenic pathogen resistance, we will examine transgenic plants that are resistant to specific plant viruses. Viruses pose unique problems to farmers. They cannot be attacked directly using pesticides, and antiviral drugs are not available for use on plants. Furthermore, the antiviral drugs used for human and animal therapy are prohibitively expensive, and therapy of individual plants is impractical, due to the huge numbers of individual plants within a field.

This is an important problem, because of the economic losses suffered worldwide as a consequence of viral infection of crop plants. Current strategies in the fight against plant viruses are based on prevention. The most effective way to prevent costly virus infections is to use plant cultivars that are resistant to viral infection; unfortunately, resistant cultivars are not available for some crops. In these cases, the only alternative is to ensure that crop propagules (seed, in most cases) are virus free, and to control insects such as aphids that spread viruses from infected fields to noninfected fields. Consequently, current virus-control strategies rely heavily on chemical insecticides, with their associated risks, including occupational exposure of farmers to concentrated toxic solutions, insecticide residue on food, damage to nontarget insects, and accumulation of toxic residues in soil and water.

For these reasons, there is much interest in the development of virus-resistant cultivars. Several transgenic cultivars belonging to the Cucurbitaceae family (e.g., squash, watermelon) have been released in the U.S. Transgenic virus resistance is usually based on transfer of genes coding for either viral **coat** proteins or viral **replicase** genes. Replicase genes work through the phenomenom of **gene silencing** whereby the addition of a specific viral gene to the plant genome results in loss of expression of that viral gene when the virus infects the plant. The mechanism behind this silencing is unclear.

More commonly, viral coat protein genes are introduced into transgenic virus-resistant plants. Why are these genes effective? For a virus to replicate, it must take over the host cell and direct the cell's metabolic machinery toward the production of new viral particles (Chapter 2). Before they can exert this control, viruses that infect eukaryotic cells must first be **uncoated,** so that the viral genome can be released into the cytoplasm. This uncoating (see Figure 2.8) is essential, because viral control of the cell requires the expression of viral genes. Transcription of viral nucleic acids cannot occur unless they are liberated from their protective coat of capsid protein. Therefore, if uncoating is blocked, the virus will be unable to replicate.

For reasons that are unclear, large amounts of coat protein in a plant cell inhibit uncoating, and therefore give a plant a high level of resistance against viral infection. This resistance is highly specific; for example, a transgenic tomato that expresses the coat protein gene of tomato mosaic virus (ToMV) will be resistant against ToMV, but not *any other virus*. Therefore, if a particular crop is affected by a number of viruses, complete virus protection requires the addition of a large number of coat protein genes. Because this kind of protection requires high levels of expression of the transgene, the presence of numerous different coat protein genes may lead to unacceptably high levels of viral proteins within plant cells.

The major controversy surrounding virus-resistant transgenic plants lies in the presence of viral genes in crop plants. Certain scientists and public groups are concerned that such genes could mutate into more virulent forms and be incorporated into viruses, leading to more virulent forms of the virus. This has not occurred yet and does not appear likely.

C. Herbicide Resistance

Not surprisingly, the research leading to these transgenics has for the most part been funded by companies that manufacture herbicides. The driving force behind this research is the cross reaction that often occurs when herbicides are applied to growing crops — herbicides often damage crops as well as weeds. Because of this, farmers are forced to alter their herbicide application strategies. They may apply herbicides *pre-emergence* (before crop seedlings have germinated and emerged from the soil), or they may use herbicides that are less powerful but do less damage to crop plants. From the point of view of both farmers and herbicide companies, it would be useful to be able to grow plants that are resistant to herbicides; the herbicides can then be applied when the crop is actively growing, when weeds are most likely to interfere with crop growth.

How are herbicide-resistant plants developed? A variety of strategies are possible, but the most popular approach has been to manipulate herbicide targets. Herbicides often inhibit specific enzymes; for example, the broad-spectrum (affects most plants) herbicide glyphosate inhibits 5-*enol*pyruvylshikimate phosphate (EPSP) synthase, an enzyme that plants require for the synthesis of aromatic amino acids. Glyphosate (known commercially as Roundup™) is structurally similar to the normal substrate (phosphoenol pyruvate) of this enzyme. This similarity allows glyphosate to act as a competitive inhibitor of EPSP synthase; the plant is then unable to synthesize aromatic amino acids, and it dies.

There are several ways of producing transgenic plants that are resistant to glyphosate. One strategy is to add additional copies of the gene coding for EPSP synthase behind strong promoters; the level of EPSP synthase within plant cells may then become high enough to overcome the inhibition caused by the herbicide. Another successful approach has been to introduce genes that code for proteins that degrade specific herbicides. Such genes are often found in soil bacteria and fungi. Finally, some glyphosate-resistant transgenics have genes coding for bacterial versions of EPSP synthase that are unaffected by glyphosate. All three strategies have been successful. A small number of herbicide-resistant crops have also been developed through traditional plant breeding.

Opponents of herbicide-resistant transgenics believe that the use of such transgenics will lead to increased herbicide use, resulting in increased contamination of food and the environment with pesticide residues. The rejoinder from the herbicide companies is that it is not their intention to increase herbicide use, but only to increase the market share of their company. They also maintain that herbicide use will not increase, but will become more efficient, and that herbicide-resistant transgenics allow farmers to use less toxic, more biodegradable herbicides such as glyphosate. In 2001 the Centre for Agriculture and Environment (CLM) published an analysis, based partly on data provided by the U.S. Department of Agriculture (USDA), of herbicide use following the introduction of herbicide-resistant soybeans to the U.S. Farmer experiences varied, but the CLM concluded that these transgenics result in a mild decrease (0 to 10%) of herbicide use.

The possibility of creating "**superweeds**" has also received much attention, particularly in the popular media. Antibiotechnology activists fear that the cultivation of herbicide-resistant transgenics will result in weeds that cannot be controlled which will cause crop loss and invasion of natural habitats.

Regulatory agencies in the U.S. and Canada considered this risk when assessing transgenic herbicide-resistant soy and canola. In Canada, soy represents less of a risk than canola, because soy lacks weedy relatives that it could successfully pollinate (hybridize). Hybridization in this context refers to the formation of viable offspring from the union of gametes from a transgenic plant and a related species. In contrast, the type of canola used to develop transgenics (*Brassica napus*) can hybridize with seven related Canadian species, including *Brassica rapa* (the "other" canola), which is a weed in the Canadian prairies.

The Canadian Food Inspection Agency (CFIA) allowed the release of transgenic herbicide-resistant canola, despite the potential for spread of the gene to other species, for two major reasons: (1) acquisition of the resistance gene by weedy relatives would only benefit these relatives if they were exposed to glyphosate, and (2) glyphosate is not normally used to control these weeds. In other words, farmers that do not grow herbicide-resistant crops would not normally use these herbicides, so the only risk would be to growers using the transgenics. Presumably, then, if glyphosate-resistant weeds become a nuisance, farmers can use methods of weed control appropriate for nonglyphosate-resistant cultivars.

Increased weediness of the crop is another risk that must be assessed. Crops acting as weeds (volunteers) can sometimes be a problem. Weediness is controlled by a complex number of traits, including the time it takes for seed production, the number of seeds produced, the ability of the plant to grow vegetatively (e.g., by horizontal stems), and the ability of seeds to overwinter. It is unlikely that the transfer of herbicide-resistant genes will affect such developmental traits. However, regulators usually require documentation that some of the weed-related traits (e.g., seed production) are not different in the transgenic line. Plant biotechnologists were encouraged by a recent study of the overall weediness of a number of insect- and herbicide-resistant cultivars in 10-year-old experimental plots in the U.K. The transgenic crops had disappeared from all but one of the plots, allaying fears that transgenic plants would increase weed tendencies.

D. DELAYED RIPENING

In 1994, Calgene received permission from the FDA to release the first transgenic food, the Flavr Savr™ tomato. Although this transgenic has not been very successful commercially, many biotechnology companies remain convinced that controlling ripening through transgenic technology is a viable idea.

Fruit ripening is an essential but problematic process. Ripening is essential to food production because unripened fruit is usually inedible. Ripening is associated with changes in the texture, structure, and flavor of fruit. From the point of view of the consumer, ripened fruits are softer, taste much better, and are easier to digest than unripened fruit. These positive changes are also advantageous for the plant. Many plants depend on animals for seed dispersal — animals eat fruits and scatter seeds around, either because of their messy eating habits or through the safe passage of seeds through the animal's digestive tract. The best way to encourage animals to

TABLE 4.1
**Changes that May Occur during Fruit Ripening
and Their Relevance to Food Quality**

Change	Relevance to Food
Loss of chlorophyll	Less green appearance
Pigment accumulation	Fruit color development
Change in cell wall structure	Softening
Changes in organic acids	Flavor changes
Increased production of volatile chemicals	Flavor changes

eat fruit is to make fruit tasty and nutritious. Evolution, therefore, has favored fruits that taste good and are soft and easy to chew and swallow. People have expanded this trend through the selective breeding of plants.

This combination of soft and tasty fruit is great for the consumer, but it causes problems for food producers and processors. These problems are particularly acute in North America. Most of North America's fruit is produced in Mexico, California, or Florida and transported by rail or truck to points throughout North America. Many bumps and shocks are inevitable with this transportation, and the stress is exacerbated when fruit is transported overseas, as is the case with tropical fruits such as mangoes and bananas. Because of this transport stress, fruit producers have two choices: they must grow fruit cultivars that produce hardy fruit that can withstand transport, or they must transport fruit in an unripened state, when it is much more resistant to physical shocks and stresses. The transport of unripened fruit is a popular strategy. Some fruits, such as oranges, ripen quite well after transport; others, such as tomatoes, ripen much more efficiently if ripening is induced by the application of the plant hormone **ethylene**. Unfortunately, most tropical fruits (bananas are an important exception) cannot be induced to ripen and cannot be transported in the ripe state, making it virtually impossible to transport these fruits to distant markets. Storage of fruits is also a problem. Some fruits (e.g., apples) can be stored for long periods in **controlled atmospheres** (e.g., reduced O_2, increased CO_2, 4°C).

What happens during fruit ripening? In **climacteric** fruits, ripening is associated with a burst of respiratory activity triggered by increased production of ethylene in the fruit. In contrast, **nonclimacteric** fruits do not have a burst of respiration at the onset of ripening, and ethylene does not trigger or accelerate the ripening process.

In both fruit types, ripening is associated with a number of biochemical and structural changes (Table 4.1) that strongly affect fruit quality and perishability. In relation to perishability, softening is the most important factor. It is caused principally by degradation of pectin by **polygalacturonase** (PG), with the help of a number of other hydrolytic enzymes. Pectin is a heterogeneous polysaccharide composed mainly of a backbone of 1,4-linked galacturonan, with side chains of rhamnose residues. It acts as a cement in which cellulose microfibrils are embedded, producing a rigid, strong wall. When pectin is broken down, this strength and rigidity dramatically decline, resulting in fruit softening (Figure 4.11). Individual cells within fruit

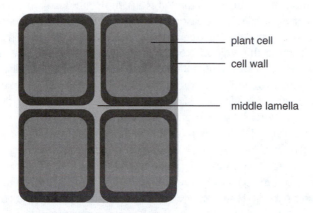

FIGURE 4.11 Polygalacturonate in plant cell walls. The cell wall contains pectic acid, and the middle lamella is particularly rich in pectic acid.

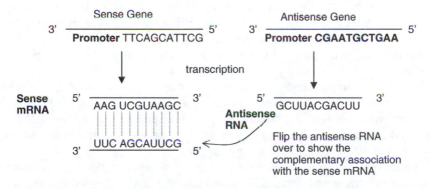

FIGURE 4.12 The relationship between a sense gene and an antisense gene. An antisense gene is constructed by reversing the sequence of the gene. This results in a mRNA product that is complementary to the sense mRNA. Association between the sense and antisense mRNA results in decreased translation of the sense mRNA and lower levels of the protein in the cell.

also lose adhesiveness because the pectin-rich middle lamella between cells is responsible for cell–cell adhesion. This leads to further fruit softening.

Transgenic fruits with delayed ripening have been made using two strategies: (1) reducing the amount of PG present in ripening fruits, and (2) reducing the amount of ethylene produced (ethylene triggers PG production in climacteric fruits). **Antisense** technology is usually the method of choice for both purposes. An antisense gene (Figure 4.12) is constructed by reversing the sequence of the gene. Transcription of an antisense gene produces an RNA molecule that is complementary to the sense mRNA sequence. Note from Figure 4.12 that the sense and antisense RNA strands are antiparallel and thus can form a double-stranded RNA molecule similar to double-stranded DNA.

The formation of this double-stranded RNA molecule leads to decreased translation of the sense mRNA. The reasons for this are not fully understood but may involve host cell defense systems against viruses. Double-stranded RNA is a red flag for cells, because this type of RNA is usually only present in virally infected cells. Therefore, it is not surprising that enzymes that recognize and degrade double-stranded RNA are common in eukaryotic cells. Antisense DNA may also work through **gene silencing**; under this scenario, DNA–RNA hybrids of a gene trigger methylation of that gene. Once a gene is methylated, it will not be expressed. Alternatively, antisense DNA may work through impaired translation of the endogenous (normal) gene. Formation of sense–antisense double-stranded RNA molecules may prevent translation by the ribosome.

The Flavr Savr tomato has lower levels of PG and consequently a decreased rate of softening after harvest. However, the difference in softening appears to be modest, probably because of the role of other wall-softening enzymes. Inhibition of ethylene production is an alternative approach to inhibition of softening. Plants produce ethylene from a methionine precursor; one of the enzymes involved (1-aminocyclopropane-1-carboxylic acid [ACC] oxidase) has been targeted through antisense technology. Transgenic tomatoes with antisense ACC oxidase genes have much reduced rates of softening and do not noticeably ripen. Fortunately, these transgenics will ripen after the application of exogenous (from outside of the plant) ethylene, allowing transport of the plants in an unripened (preclimacteric) state. After transport, the tomatoes are exposed to ethylene and ripening is triggered. These tomatoes are unlikely to offer improved tomato flavor or color to consumers, but the main application is to reduce the amount of fruit loss that normally occurs during transport. In conventional tomatoes, a substantial proportion of preclimacteric fruits will begin to ripen during transport, resulting in damage. With transgenic tomatoes that produce low levels of ethylene, this should not happen.

Tomato processors can also benefit from transgenic technology. Transgenic tomatoes with decreased levels of PG have increased Bostwick viscosity, a variable positively correlated with paste yield (the amount of paste obtained from a given number of tomatoes). Transgenic tomatoes with reduced levels of both PG and pectinesterase (another wall-softening enzyme) have increased levels of Bostwick viscosity, serum viscosity (this gives a more glossy paste), and soluble solids. All of these traits are valuable to tomato processors and may also be relevant to consumers. In the mid-1990s, one of the most popular brands of tomato paste in the U.K. was derived from transgenic tomatoes with decreased levels of PG. This tomato paste was perceived by consumers to be thicker and more flavorful than other pastes. Some chefs were also enthused by its ability to stick to pasta. Unfortunately, because of grocery chains' fears of consumer reactions to transgenic foods, this brand is no longer available in the U.K.

Because the Flavr Savr tomato was the first transgenic food, much attention focused on its safety assessment. It was the first transgenic crop to undergo an assessment of **substantial equivalence** (see Chapter 9 for a discussion of this concept). Under this strategy, we compare the levels of nutrients, toxins, and vitamins in the fruit to the parent variety. Calgene demonstrated that the only detectable differences between the transgenic tomato and the original tomato were in the levels of polygalacturonase (lower in the transgenic) and the rate of softening (slower in the transgenic). In other respects the tomato was unchanged; it was nutritionally similar and had similar levels of toxins (tomatoes normally contain low levels of tomatine, a toxic alkaloid).

This is perhaps an opportune point to discuss the presence of antibiotic-resistant genes, because this is the major criticism raised against the Flavr Savr tomato. As discussed previously in this chapter, antibiotic-resistant genes are included with the transgene so that selection of recombinant plants can be achieved simply by placing the plant cells on an agar medium containing the antibiotic. However, the use of antibiotic-resistant genes in this way leads to potential risks that need to be carefully assessed. Ingestion of plants containing antibiotic-resistant genes could theoretically lead to increased frequency of antibiotic resistance among bacteria in the gastrointestinal tract. This is certainly conceptually possible; DNA in plant cells could be released from lysed plant cells in the gut, and bacteria could then gain this DNA through transformation. Fortunately, successful transformation is unlikely. Relatively few bacteria are naturally **competent** (i.e., able to be transformed without treatment with electric currents or chemicals such as $CaCl_2$). The human colon is dominated by enterococci that do not appear to be capable of natural transformation. Finally, even if a bacterium in the gut became transformed with an antibiotic-resistant gene, this bacterium is unlikely to cause problems unless the person is taking the specific antibiotic that the resistance gene is effective against. If so, then resistant bacteria would be heavily favored, as compared to bacteria lacking the resistance gene. Kanamycin, the aminoglycoside antibiotic used most commonly in the selection of successful transgenic plant cells, is not widely used for therapy of microbial infections in humans.

For these reasons, the use of antibiotic-resistant genes is not generally viewed by the scientific community as a dangerous practice. However, because of public concerns, and because any dissemination of antibiotic-resistant genes is unwise, many biotechnologists are using or developing alternatives to antibiotic-resistant genes. The GUS system, which allows the detection of recombinant plant cells through a visual change in recombinant cells, is a popular alternative. Unfortunately, because this is not a *selective* process (i.e., it does not eliminate nonrecombinant cells), it is technically less efficient than the use of an antibiotic resistance strategy.

There are several other alternative strategies that can eliminate antibiotic-resistant genes in transgenic plants. For example, sequences can be introduced around antibiotic-resistant genes that are recognized by enzymes involved in the excision of transposons from DNA. The antibiotic-resistant gene can be used in the normal way to select transformed plant cells, and the transposon machinery can then be used to remove the antibiotic-resistant gene. This and other strategies are increasingly being used in the development of transgenic crops.

E. GOLDEN RICE

In 1994, the World Health Organization (WHO) estimated that 2.8 million children ages 0 to 4 suffer clinically from vitamin A deficiency disorder (VADD). Clinical deficiency is evident through several types of vision problems (xerophthalmias), such as night blindness and scars on the cornea. In severe cases, vitamin A deficiency leads to blindness. Subclinical deficiency is also globally widespread — in such cases, children and adults (particularly pregnant or lactating women) have decreased levels of retinal (a form of vitamin A) in their serum. They may not exhibit symptoms of vitamin A deficiency, but they are at high risk to develop symptoms. The WHO estimated in 1994 that 251 million children are subclinically affected.

The United Nations and others have been aware of this ongoing global tragedy since the 1960s, and concerted action has resulted in decreased incidence of clinical VADD, particularly in India, Bangladesh, and Indonesia. This decrease has been primarily driven by the widespread dispersal of vitamin A tablets. Unfortunately, VADD is still prevalent in many areas of the world and is most acute in subsaharan Africa and southeast Asia.

VADD is difficult to solve globally because it can be caused by a number of different factors. However, people suffering from VADD often fall into one or more of the following categories:

- *Communities that use rice as the staple food.* In many parts of the world, especially southeast Asia, rice is considered to be a satisfying and adequate food for young children. Unfortunately, milled rice has very low levels of β-carotene, one of the most important provitamin A compounds. Communities that use wheat as a food staple suffer lower levels of VADD than those using rice.
- *Low socioeconomic status.* In several countries, there is a strong negative correlation between income and incidence of VADD. Poverty results in an overreliance on cheap staple foods such as rice.
- *Periodic food shortages.* In drought seasons, or drought years, food shortages may lead to dramatically increased levels of VADD. For example, in the desert region of Rajasthan, India, a single drought year in 1987 led to an increase in xerophthalmia from 10 to 35% of the population.

Humans and other vertebrate animals produce vitamin A mainly through the conversion of dietary sources of carotenoids. The most common carotenoid thus used is **β-carotene**. This yellow-orange plant pigment is converted by the human body into all-*trans* retinal (Figure 4.13), a fat-soluble vitamin that is required for vision because of its role in the development of the cornea and its function in transmission of electrical signals from the retina to the brain. 11-*cis* Retinal (a compound derived from retinol) binds to the pigment opsin to form **rhodopsin** in the plasma membrane of rod cells of the retina. These cells are responsible for vision in low light. Similar complexes occur in the cone cells of the retina, which are responsible for color vision and vision in bright light. When light of the correct wavelength hits a rod or cone cell, rhodopsin undergoes a change in conformation, which results in an action potential (change in voltage across the cell membrane). This action potential is then transmitted via neurons to the brain, giving rise to the perception of sight. Vitamin A has a number of other roles in human metabolism, but clinical symptoms of VADD are usually related to vision; this is not surprising, considering the central importance of vitamin A to vision.

Beta-carotene

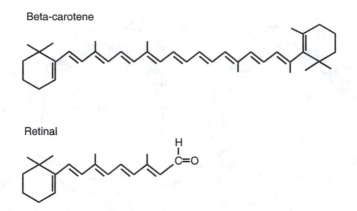

Retinal

FIGURE 4.13 Structure of β-carotene and retinal.

The continuing and widespread nature of VADD has led people to investigate a number of potential solutions. One option is to develop staple foods that have increased levels of β-carotene. The close link between rice dependency and VADD led researchers in Switzerland and Germany to develop **golden rice**. The research group, led by Ingo Potrykus and Peter Beyer, with funding from the Swiss government and the Rockefeller Foundation, developed a transgenic line of rice with increased levels of β-carotene. Because of the resulting yellow color in the endosperm, these seeds were dubbed "golden rice."

It was necessary to use a transgenic approach rather than classical plant breeding techniques because rice does not normally have β-carotene in the inner endosperm (the area of the cereal seed that stores starches and other nutrients). It is essential that the provitamin be stored in the endosperm because this constitutes the bulk of the rice grain after milling. Rice actually contains substantial amounts of β-carotene in the aleurone layer of the endosperm, but this oil-rich layer is removed by milling. If it is not removed, the rice has a greatly increased risk of developing rancidity, particularly in tropical regions.

The development of golden rice posed numerous technical difficulties. β-Carotene in rice is produced from a pool of geranyl geranyl diphosphate (GGPP), an isoprenoid that is used to form a variety of compounds. Three enzymes are necessary to convert GGPP to β-carotene; none of these enzymes are active in the rice endosperm. Introducing three genes simultaneously is much more difficult than introducing a single gene into a plant. Potrykus et al. used several *agrobacterium*-based approaches; we will focus on one of these. They inserted genes for two of the enzymes (***psy***, coding for phytoene synthase, and ***crt***, coding for phytoene desaturase) into one plasmid vector (pZPsC), and the other gene (*lcy*, coding for lycopene β-cyclase) into a separate plasmid (pZLcyH) that also had a selectable marker gene (*aph IV*, coding for hygromycin resistance). In each of these plasmids (Figure 4.14), left and right borders were present, allowing transfer of the genes by *vir* genes of *A. tumefaciens*. Two of the genes (*psy* and *lcy*) were preceded by endosperm-specific promoters (i.e., the genes would be expressed only in the

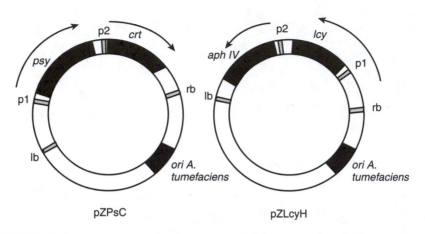

FIGURE 4.14 Structure of pZPsC and pZLcyH, the vectors used in the development of β-carotene-enriched rice (golden rice). Promoters are indicated by p1 (endosperm-specific promoter) and p2 (CaMV 35S promoter). Arrows indicate the direction of transcription from the promoters. (Modified from Ye X., et al., *Science,* 287, 303, 2000.)

endosperm); *crt* and *aph IV* were preceded by the CaMV 35S promoter, which induces constitutive (all tissues, all times) gene expression.

The researchers introduced the two vectors separately into *A. tumefaciens* by electroporation. The two resulting transformed bacterial strains were then co-incubated with immature rice embryos (these are produced through plant tissue culture methods). Transformed plant cells were then selected using hygromycin, which killed nontransformed plant cells. Plants were then regenerated from callus cultures of cells that survived the hygromycin treatment.

This co-transformation was a clever approach, because it turned out that lycopene cyclase is not required for the production of β-carotene in rice endosperm (it is probably constitutively produced). Because the vector that contained the *lcy* gene also contained the selectable marker gene for hygromycin resistance, it should be possible to remove the marker gene through traditional breeding methods, without losing β-carotene production. Genes from the two plasmids were probably inserted at widely different locations in the plant cell's chromosomes.

Transformed plants were grown to maturity and seeds were collected. Not all of the seeds were β-carotene enriched, but in a typical line, the average β-carotene content was 1.6 µg/g of endosperm tissue. Considering that this β-carotene measure was based on a mixture of β-carotene and non-β-carotene-enriched seeds, Potrykus et al. postulated that pure breeding lines (i.e., lines that produced 100% β-carotene-enriched seeds) would contain at least 2 µg/g β-carotene. This could correspond to 100 µg of retinol equivalents in a daily serving of 300 g of golden rice, possibly enough provitamin A to alleviate vitamin A deficiency.

The publication of the golden rice manuscript provoked a flurry of commentary from both pro- and antibiotechnology forces. The beleaguered corporate food biotechnology community, beset by widespread criticisms of arrogance (forcing transgenic crops onto the world), soon began to use golden rice as a poster child to promote transgenic technology. The antibiotechnology forces, led by Greenpeace, denounced golden rice, maintaining that more than 7 lb of dry golden rice would have to be consumed daily for an individual to gain enough provitamin A.

At this time, it is difficult to assess the utility of golden rice in decreasing the incidence of VADD, primarily because of uncertainties about the bioavailability of β-carotene in golden rice. It is well known that simply measuring β-carotene content of a fruit or vegetable does not give an accurate picture of the ability to deliver β-carotene after ingestion. Many factors affect this; the matrix surrounding β-carotene is particularly important. For example, many leafy green vegetables have high levels of β-carotene, but it is associated with protein complexes in chloroplasts. These complexes are poorly digested in the human gut, resulting in poor intake of β-carotene.

Several other factors add to the uncertainty. Required levels of provitamin A vary widely, depending on age and numerous other factors. Thus, the optimal level of intake is difficult to assess. Also, human absorption of β-carotene is linked to fat consumption; people with extremely low levels of fat intake (<5 g/d) are unable to absorb β-carotene. It is uncertain how fat intake is linked to VADD, but this could reduce the efficacy of golden rice in some situations. Severe VADD has decreased markedly in recent years; subclinical VADD is considered to be the larger problem, and golden rice may be more effective in alleviating this less severe form of VADD.

Despite the technological elegance of golden rice and its potential to help fight VADD, it is clear that golden rice is not a panacea that will end all incidence of VADD. Vitamin A deficiency is often tightly linked to cultural practices; in many regions, VADD sufferers potentially have access to excellent sources of provitamin A, but they do not use them for cultural reasons. In many cases, there are a number of approaches that can be used to fight VADD, and it would be unwise to ignore these alternatives because of the potential use of golden rice. That said, golden rice is a tremendous technological achievement and an excellent example of the power of transgenic technology to improve human health.

V. TRANSGENIC FOOD PLANTS UNDER DEVELOPMENT

A. AGRONOMIC TRAITS

Most transgenic crops grown today are aimed at helping farmers. This has been one of the driving forces behind consumer dissatisfaction with transgenic foods — they do not benefit the consumer. The future will bring more transgenics that are aimed at consumers, but farmers will not be neglected. Numerous transgenic crops currently in development aim to improve crop performance in the field. Many of these target problems that were difficult to tackle with the first generation of transgenic crops, which contained only one or two foreign genes. Numerous transgenic approaches to solve agronomic problems are possible, but we will focus on two promising examples: **disease resistance** and **stress tolerance**.

1. Disease Resistance

Biotechnologists have successfully created transgenic plants that are resistant to viral infection, but to date there have not been any successful introductions of transgenic crops that are resistant to fungal or bacterial pests. This will likely change in the future, as we acquire a better understanding of plant defense against these pathogens. One of the problems encountered has been the relatively short-lived nature of pathogen resistance; resistance genes can be transferred into a crop, but it usually does not take long for the selection of pathogen strains (races) that can overcome the resistance gene. Fortunately, there are other approaches; for example, the transfer of **elicitor** genes to a plant sometimes provides broad-spectrum protection from a range of fungal pathogens. Elicitors are compounds produced by pathogens that elicit a protective response (e.g., localized cell death that kills an invading pathogen). The key to this strategy is the use of pathogen-inducible promoters; these prevent expression of the elicitor gene, except in the presence of a fungal pathogen.

2. Stress Tolerance

Plants are nonmotile. They cannot use legs or wings to escape environmental stress; their only choices are stress tolerance (e.g., the ability to grow in dry soil) and stress avoidance (e.g., dormancy during times of drought). In the past, plant breeders mainly concentrated on increasing yield, and rarely ventured to increase crop stress tolerance. However, plant scientists have become increasingly aware that abiotic (arising from nonliving forces, such as water availability) stresses have strong effects on yield. This, coupled with increased global urbanization, which has forced many farmers onto land relatively unsuited for agriculture, has heightened awareness of the need for cultivars that are more tolerant to abiotic stress. The global outlook for agriculture is not promising — it is anticipated that increased population pressure and an increased proportion of unsuitable agricultural land will make it difficult to produce enough food to feed the world by 2020. Increasing stress tolerance of staple food crops is an important goal for both traditional plant breeders and biotechnologists.

The most serious abiotic stress in most parts of the world is water availability. Dry or saline soil seriously affects the growth of most crops. Dry soil is linked to climate, but saline soil is often exacerbated by agricultural practices. Excessive irrigation, for example, can lead to saline soil, because irrigation water always contains a certain level of ions; when the soil dries, these ions become more con-centrated and interfere with crop water uptake.

A number of compounds found in plant cells have been linked to drought and salinity tolerance. One strategy for crop improvement is to increase levels of such compounds in crop plants. This has been done at the laboratory level with transgenic tobacco and potato lines that produce **fructans**. These are polymers of fructose that are produced by many plants that tolerate seasonally dry climates, but they are absent in most crop plants. The exact role of fructans is unclear, but it may be related to hardening of plant cell walls, which reduces plant susceptibility to drought-induced wilting. In tobacco plants, the gene transfer was successful; transgenic fructan-producing tobacco plants acquired greater tolerance to drought stress.

B. Storage Proteins

Cereals and legumes are globally the most important sources of protein. Legumes are particularly rich in protein, and seeds of peas, chickpeas, common beans, and other legumes are usually 17 to 30% protein (expressed as a percentage of dry matter). Cereals have more modest protein content, typically having 7 to 15% protein. However, cereals such as wheat form a large part of the diet in many countries and are consequently the major source of protein. Unfortunately, cereals or legumes are not ideal sources of protein, especially when compared to meat, egg, or dairy sources. Eggs and dairy foods contain a large proportion of essential amino acids (those that humans must obtain in their diet), whereas plant foods tend to have relatively low levels of essential amino acids. Many cereals are particularly deficient in lysine and tryptophan, and many legumes are deficient in methionine and cysteine. Thus it is revealing that many indigenous cultures use a combination of legumes and cereals as a staple food — the southwestern U.S. combination of beans with corn tortillas is an effective way to gain adequate amounts of essential amino acids.

Biotechnologists have begun to test the use of transgenic technology to modify the amino acid composition of seed proteins, to increase their nutritional value. Seeds store amino acids in the form of **storage proteins**. Modifying amino acid content is technically easy — all that is required is to introduce DNA sequences that code for the desired amino acid into the DNA sequence of the seed storage protein. Unfortunately, this approach may not be successful, possibly because the new sequences decrease the stability of the storage protein. The structure of storage proteins is crucial to their function. In legumes, storage proteins are concentrated in membrane-bound protein bodies, where they must survive extreme desiccation and be mobilized (hydrolyzed enzymatically) when the seed germinates. Transgenic modification of storage proteins must not interfere with these basic duties.

Another option is to transfer a storage protein from one plant to another. This has been done with the methionine-rich albumin gene found in Brazil nuts. This gene was transferred to soybeans. However, the research team realized part way through the research program that the albumin protein that they transferred was the allergen responsible for allergic reactions to Brazil nuts! Development of this transgenic ceased, but attempts to transfer storage proteins continue.

Transgenic modification of storage proteins also offers potential benefits to food processors. Soybean protein, for example, has numerous uses in the food industry. Characteristics such as gel hardness, gel-forming rates, and emulsifying properties of soybean protein can theoretically be modified by changing the DNA sequence of soybean storage proteins, or by introducing subunits from other plants (storage proteins are usually composed of multiple subunits, and changing or adding subunits will change characteristics of the protein).

C. Antinutrients and Other Undesirable Compounds

Primary metabolites are compounds produced by plants that are directly used for growth, development, and function of the plant. **Secondary** compounds fill roles that are peripheral to the main activity of plants (photosynthesizing, taking up water,

TABLE 4.2
Undesirable Compounds in Food Plants, their Chemical Nature and the Presence or Absence of Transgenic Lines (trans) with Reduced Levels of Each Compound

Compound	Crop	Type	Effect	Trans?
α-Amylase inhibitor	Legumes and cereals	Protein antinutritive	Inhibits starch digestion	No
Trypsin inhibitor	Legumes	Protein antinutritive	Inhibits protein digestion	No
Lectins	Legumes	Protein toxin	Diarrhea, etc	Yes
Lipoxygenase	Soybeans[a]	Protein	Produces off flavors	Yes
Glycoalkaloids	Potato	Alkaloid toxin	Gastroentiritis	No
Tannins	Widespread	Polyphenol antinutritive	Interferes with protein digestion	No
Glucosinolates[b]	Brassicas	Carbohydrate[c] toxin	Goiter, bitter taste	Yes
Cyanogenic glycosides	Cassava	Carbohydrate toxin	Goiter	No
Phytic acid[b]		Carbohydrate antinutritive	Inhibits absorption of phosphorus	Yes
Oligosaccharides	Widespread	Carbohydrate	Flatulence	No

[a] Widespread, but primarily a problem in soybeans.

[b] Primarily of concern in animal feed.

[c] Isothiocyanate, the active form, is not a carbohydrate; glucosinolates also act as antinutrients, inhibiting iodine uptake.

flowering, etc.). However, they are sometimes crucial to the survival of plants, because of their role in deterring insect or vertebrate herbivores or in inhibiting the growth of invading pathogens. Some secondary metabolites, because of their anti-herbivore properties, have adverse effects on humans. We can divide these compounds into two main groups: (1) **toxins**, which have direct negative effects after consumption, and (2) **antinutrients**, which interfere with digestion, absorption, or utilization of nutrients derived from food, without being directly toxic. Numerous other compounds (Table 4.2) are undesirable components of food, in many cases because of their sensory properties (e.g., off flavors). It is theoretically possible to decrease levels of any of these compounds through transgenic technology; some (e.g., proteins) are relatively easy to decrease, by using antisense technology or eliminating the gene coding for the troublesome protein, but others, that are produced through complex pathways (e.g., glucosinolates), are more difficult to tackle using transgenic approaches. Some of these compounds also have positive effects on human health (e.g., glucosinolates); thus it may be possible that transgenics with *increased* levels of the compound are desirable. We will examine three examples of transgenics with modified levels of secondary compounds (glucosinolates, phytic acid, and lipoxygenase).

Glucosinolates are present in many plants, but are particularly important in brassicaceous plants such as cabbage, broccoli, mustard, rapeseed, and canola. They consist of a carbohydrate (often glucose) bound to a sulfur-containing side chain

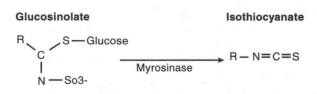

Glucosinolate   **Isothiocyanate**

FIGURE 4.15 Conversion of glucosinolate to isothiocyanate, catalyzed by myrosinase.

(Figure 4.15). Glucosinolates are typically stored in the vacuole, and when cell structure is disrupted, through mastication or processing, the enzyme myrosinase, present in the cytoplasm, comes in contact with the glucosinolate. Myrosinase releases glucose from the compound and isothiocyanate forms from the side chain.

It may seem surprising that people have proposed to use transgenic technology to decrease the level of glucosinolates in plants when you consider that isothiocyanate consumption has been linked epidemiologically to decreased incidence of cardiovascular disease. The reason for the interest in reduced levels of glucosinolates is that some components of animal feed, such as rapeseed, canola, and mustard meal, contain high levels of glucosinolates. Meal is the residue left after the extraction of oil from seeds. The high level of glucosinolates in mustard meal results in a strong bitter taste and unpalatability to animals. The level of glucosinolates is also high enough to cause health problems in animals. This is unfortunate, because mustard meal is 40% protein and has a balanced amino acid composition.

Construction of a transgenic low-glucosinolate line is a daunting prospect. Glucosinolates can be produced from any of four amino acids (tryptophan, methionine, tyrosine, or phenylalanine). A complex set of enzymes is required for modification of these amino acids, covalent addition of glucose, and addition of the sulfur-containing group. Several approaches can reduce glucosinolate levels; one of the most promising strategies is to introduce genes that will lead to diversion of amino acids away from the glucosinolate pathway. It is essential that this diversion lead to the production of innocuous compounds (i.e., safe for the plant, and safe for animals that eat the plant). This strategy has been used successfully to create transgenic canola with glucosinolate levels in the seeds that were 3% of levels in nontransformed seeds. In this case, tryptophan was diverted into the production of tryptamine.

Another strategy is to use antisense technology to inhibit the expression of a key enzyme in the pathway; for example, targeting the enzymes involved in addition of glucose to the modified amino acids could result in decreased glucosinolate production.

Phytate is another example of a compound in plants that interferes with feed efficiency. Many plants store phosphate in the form of phytate (myo-inositol-hexaphosphate). Unfortunately, humans and other mammals are unable to digest phytate, resulting in phosphorus deficiency in animals consuming feed made from soybeans or other plant products, unless the animals are fed expensive, rich sources of available phosphorus. Undigested phytate is released into the environment via manure, and thus constitutes an important source of phosphorus contamination to rivers and other water bodies.

Phosphorus can be released from phytate through the application of **phytase**, an enzyme found in plants and many microbes. In the long run, a cheaper alternative is to integrate a phytase gene into a plant that allows expression in the seeds. To be effective, the phytase should not be active in the seed, because this would probably adversely affect phosphorus stability in the seeds. However, if the phytase became active in an animal's stomach, it could then release phosphorus that could be absorbed by the animal.

Biotechnologists have successfully produced transgenic tobacco and soybeans that contain phytase genes. In the case of transgenic tobacco, the phytase gene was cloned from *Aspergillus niger*, a fungus that is the source of many enzymes used in food processing. It was transferred to tobacco using *A. tumefaciens*, and the resulting tobacco seeds, when tested as a feed supplement, were as affective as direct enzyme application in liberating phosphorus from the feed. Technically, the only glitch encountered was that the amount of glycosylation of phytase differed in tobacco as compared to *A. niger*; however, it is unclear if this significantly affected activity of the enzyme.

One final note about phytate: the presence of phytate in food is also relevant to human nutrition. About 75% of the phosphorus in most plant seeds is in the form of phytate. When humans ingest seeds, this phosphorus is unavailable; furthermore, phytate in solution is a strong chelator of calcium, iron, zinc, and other small cations. Phytate is thought to lead to mineral deficiency in populations that rely on seeds of cereals and legumes as the main food staple. On the other hand, phytate consumption in humans may have beneficial effects on health because of its function as an antioxidant. Any attempt to modify the phytate content of food crops must acknowledge these contradictory effects of phytate on human health.

Our final example focuses on soybeans with reduced levels of **lipoxygenase**. This example is different from previous examples in that plants were modified using conventional breeding techniques. Lipoxygenase causes oxidation of PUFAs during processing of soybeans, leading to development of off flavors. These flavors, characterized as "grassy" or "beany," are thought to be an important factor behind the rejection of soy-based foods by a large segment of the population, particularly in North America.

Three genes code for three isozymes of lipoxygenase, and plant breeders have found alleles of each gene that produce nonfunctional enzymes. The cultivar Century was backcrossed to varieties with these alleles, leading to plants that were homozygous in terms of nonfunctional alleles. Soybeans from these plants were processed into soy flour and soy milk, and a sensory panel compared these products to similar products made using Century soybeans that had functional lipoxygenase genes. The panel rated the soy milk and flour for the following flavors: cereal, beany, rancid, oily, chalky, bitter, and astringent. Several of these flavors (especially beany) were perceived to be less intense in soy products made from soy lacking lipoxygenase. This study illustrates the potential of plant modification to alter sensory properties of foods, and perhaps change patterns of food utilization.

D. PROTEINS IMPORTANT TO BREAD MAKING

Wheat is a crucial plant for humans; after rice, it is the most consumed plant on the planet. As a component of flour it is used to make a wide range of breads, biscuits, and pasta. Given the importance of wheat, it is surprising that transgenic cultivars of wheat have not yet been released. The reason for this is primarily technical; wheat is difficult to transform. The first successful transgenic wheat line was reported in 1993, whereas most other important crops had been successfully transformed about 10 years earlier. Two reasons for the delay in producing transgenic wheat are: (1) *A. tumefaciens* infects wheat cells only under certain conditions (e.g., immature embryos), and (2) it is difficult to obtain regenerated wheat plants from transformed cells, especially if they have been transformed by *A. tumefaciens*. However, these problems have largely been resolved, and transformation of wheat is now relatively straightforward. It is expected that transgenic wheat cultivars will be released in 2005.

What will be the characteristics of these transgenics? The first transgenic cultivars will likely be herbicide resistant; however, it is expected that other transgenics will have improved quality for baking and pasta. **Gluten** has received most of the attention in this context. Gluten refers to a set of proteins that are further grouped as **glutenins** and **gliadins**. These proteins allow wheat doughs to be pounded, kneaded, and stretched, and are also responsible for the formation and stability of air pockets during rising and baking. The chemical nature of glutenins and gliadins, which is dictated by the base sequence of their genes, strongly affects the bread-making, biscuit-making, and pasta-making qualities of the dough. Consequently, a great deal of effort has been devoted to understanding the biochemistry and genetics of glutenins and gliadins. This has led to strategies to modify the base sequence of wheat proteins, with a particular emphasis on high-molecular-wheat glutenins, which are the most important contributors to the extensibility of wheat dough. This is an example of **protein engineering**; this field evolved from our increasing understanding of the relationship between amino acid sequence and protein function. This increased understanding has produced a number of strategies for the improvement of glutenins and many other important proteins.

For example, certain glutenins (e.g., ω-gliadin) are associated with poor bread-making qualities because of their lack of elasticity. Based on comparison with other gliadins, insertion of cysteine residues into ω-gliadin should result in a marked increase in elasticity. How does one go about making such an insertion? Before the advent of molecular biology, the only technique to induce such changes was **random mutagenesis**. This involves the applying radiation or mutagenic chemicals to seeds and hoping that one of the mutations that result will produce a desirable change in a protein.

Random mutagenesis often produces harmful mutations, so methods using recombinant DNA technology are usually preferred. This can be performed via a number of methods. In one frequently used technique, we prepare single-stranded DNA of the gene that is to be modified. We then synthesize an oligonucleotide that has the desired sequence change. The change can be base substitutions, deletions, or additions. The synthetic oligonucleotide is then annealed to the single-stranded DNA of the original gene (it will anneal if the oligonucleotide is still complementary to the original gene through most of its sequence). Once the oligonucleotide has annealed to the original gene, we add the appropriate enzymes (e.g., DNA polymerase) that will result in formation of double-stranded DNA from this hybrid. This

DNA is then transferred into *E. coli* or another suitable host and the desired mutant is then isolated through screening of bacterial clones (see Chapter 3, Section IV.B). A probe based on the oligonucleotide used above could be used to screen the clones. As an alternative to this process, PCR-based methods of site-directed mutagenesis (see Chapter 3, Section VI.D) could be used to introduce extra cysteine-coding sequences.

Protein engineering is forecasted to become increasingly important in future applications of transgenic crops. It requires sophisticated understanding of protein structure and its relation to function. Assessment of the safety of such recombinant proteins also demands a thorough understanding of structure–function interactions.

E. ENGINEERING BETTER STARCH

After protein, starch is the most important contribution of plants to food. It is the main source of energy for humans, and it can also be used as a source of carbohydrates that are useful in food processing. However, extensive chemical and enzymatic processing is required to yield these useful products, because of the chemical and physical characteristics of starch. For example, to obtain glucose from starch derived from corn seeds, you must solubilize the starch by raising the temperature near boiling; then add large amounts of various amylases. Some of these enzymes have a **debranching** function, and others break each linear chain into smaller fragments. (See Chapter 7, Section V.B.1, for a more complete description of starch structure and processing.)

One of the goals of plant biotechnologists is to develop transgenic plants with altered starch structure that would simplify starch processing or improve the characteristics of starch products. We will consider one example: modified plants with increased amounts of **amylopectin**, the branched form of starch. Starch that is high in amylopectin and low in **amylose** (unbranched starch) has superior qualities for certain applications; for example, it is less prone to **retrogadation**, the formation of starch granules from heated starch suspensions. These granules produce an unpleasant gritty texture in foods containing heated, gelatinized starch. Amylopectin can also be used to produce transparent gels, which improves the visual appeal of some pie fillings.

High-amylopectin plants have been obtained using both conventional breeding and transgenic technology. One transgenic approach involved the construction of an antisense gene for **granule-bound starch synthase** (GBSS), the enzyme responsible for amylose synthesis. This strategy has successfully produced amylose-free potato lines. High-amylopectin starch can also be obtained from the waxy wheat line, which was developed through conventional plant breeding; but waxy wheat tends to be difficult to grow, partly because of the complexity of wheat genetics (it is a hexaploid). Transgenic potatoes may be a viable alternative.

F. ALTERNATIVE SWEETENERS

Humans are not the only primates that enjoy sweet fruits. Chimpanzees and monkeys in tropical rainforests are attracted to sweet fruits and are efficient at dispersing seeds from these fruits. This may be linked to the discovery that sweet-tasting proteins that mimic the taste of sucrose have evolved in a number of tropical rainforest plants. The most well characterized of these sweet proteins is **thaumatin**, a product of *Thaumatococcus danielli*, a shrub found in West Africa.

Although not yet approved as a sweetener in the U.S., thaumatin can legally be used in food in Europe. It is an intensely sweet compound that, like other alternative sweeteners, can be used in small amounts. Thus it does not contribute greatly to the calorie count and does not lead to dental caries. This, combined with its "natural" status (most alternative sweeteners are the products of clever organic chemists), makes thaumatin an attractive option for today's health- and calorie-conscious society. It could also be useful for diabetics.

Unfortunately, *T. danielli* is difficult to find and harvest, and when grown in greenhouses or outside its native region it does not produce fruit. This has led to an interest in development of recombinant microbes that contain the gene that codes for thaumatin. A number of bacteria (e.g., *Bacillus subtilis*) and fungi (e.g., *Aspergillus niger* var. *amawori*) have been successfully transformed with the thaumatin gene, but so far, yields of thaumatin have been disappointingly low. The alternative approach of transferring the thaumatin gene to plants such as potato appears more promising and is actively being investigated.

G. Increased Levels of Vitamins and Phytochemicals

Many biotechnologists, frustrated by the low level of consumer acceptance of foods containing transgenic crops, anxiously await the development and release of transgenic crops that will be more attractive to consumers. The so-called "third generation" of transgenic crops is expected to exhibit traits such as increased levels of **phytochemicals**. These compounds are naturally found in plants and have therapeutic use, protect against disease, or enhance human performance. In other words, phytochemicals are the active components of functional foods that are derived from plants. Through epidemiological studies, the consumption of many phytochemicals has been linked to decreased risk of cardiovascular disease (CVD) and specific types of cancer. This protective effect increases with more frequent consumption of the functional food, so increasing the levels of the compounds responsible for the health benefit is an attractive strategy for improving the general health of a population.

It is also desirable to increase the levels of vitamins in certain foods. In particular, many people have difficulty getting enough vitamin E and folates. Furthermore, increasing consumption of these vitamins beyond the recommended daily allowance (RDA) is expected to decrease the incidence of CVD. This would be difficult to achieve with a conventional diet, but it might be possible if specific plant foods were engineered to contain a higher level of these vitamins. This has not yet been achieved for folate, but it has been demonstrated to be technically possible with vitamin E, through the production of transgenic plants of *Arabidopsis thaliana*.

A. thaliana is extensively used in research on the molecular biology and genetics of plants because of its small genome and rapid generation time. It is instructive to examine the strategy that was used to augment vitamin E levels in *A. thaliana* (Figure 4.16), partly because of its relevance to food functionality and partly because

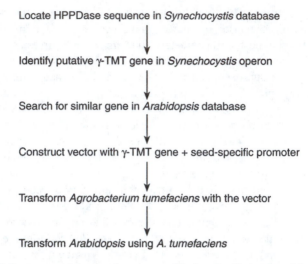

Locate HPPDase sequence in *Synechocystis* database

Identify putative γ-TMT gene in *Synechocystis* operon

Search for similar gene in *Arabidopsis* database

Construct vector with γ-TMT gene + seed-specific promoter

Transform *Agrobacterium tumefaciens* with the vector

Transform *Arabidopsis* using *A. tumefaciens*

FIGURE 4.16 Process used to develop transgenic plants with increased levels of α-toco-pherol. (HPPDase = *p*-hydroxyphenylpyruvate dioxygenase; γ-TMT = γ-tocopherol methyl transferase.)

it is an excellent example of the power of **genomics** (the study of the genetic structure of organisms). But first, we need to review basic aspects of vitamin E. It is a complex of four lipid-soluble compounds (**α-, β-, γ-,** and **δ-tocopherol**) that are present in high concentrations in oilseeds. Hence, humans derive most of their vitamin E from vegetable oils. α-Tocopherol is the only version of vitamin E that is retained by the body; for this and other reasons, it is the form most relevant to human health.

Unfortunately, most oilseeds contain much more γ-tocopherol than α-tocopherol. Changing this requires a sophisticated understanding of the pathways involved in synthesis of tocopherols in plants. It turns out that α-tocopherol is the final step in the tocopherol pathway and that γ-tocopherol is its immediate precursor. In most oilseeds, γ-tocopherol methyl transferase (γ-TMT), the enzyme that converts γ-tocopherol to α-tocopherol, is present in limited amounts. So, increasing the amount of α-tocopherol should be a simple matter of increasing the expression of γ-TMT.

Unfortunately, at the time that this research was done, we knew little about plant γ-TMT or its gene, partly because γ-TMT is membrane bound and thus difficult to purify. To get around this problem, the researchers (Shintani and DellaPenna) started with the DNA sequence coding for one of the well-characterized enzymes of the tocopherol pathway: *p*-hydroxyphenylpyruvate dioxygenase (HPPDase). They then used computer software to search for this gene in the genome of the cyanobacterium *Synechocystis*. This organism was chosen because there is a database containing its complete DNA sequence and because this bacterium would be likely to have all of the genes of the tocopherol pathway grouped in an operon (eubacteria often place genes controlling the enzymes of a pathway in a single chromosomal location; this allows coordinated regulation of gene expression).

Shintani and DellaPenna successfully located the tocopherol operon in *Synechocystis*. They then located a putative γ-TMT gene within this operon. This was possible because certain features of the plant γ-TMT could be predicted, based on the structure of other plant methyltransferases. Thus, certain amino acid sequences (and corresponding DNA base sequences) were expected to be present in γ-TMT.

Once the *Synechocystis* γ-TMT was located, its DNA sequence was used to locate γ-TMT in a computer database of the genome of *Arabidopsis*. This strategy was successful; the γ-TMT gene was located and the sequence was used to construct a plasmid containing the full γ-TMT gene in front of a seed-specific promoter. This plasmid was subcloned into *A. tumefaciens*, which was used to transform *Arabidopsis*. The resulting plants had higher levels of γ-TMT, resulting, in some lines, in >95% of the tocopherol being in the form of α-tocopherol. It is expected that this approach would also be successful in oilseed crops such as soybean and canola, leading to vitamin E–enhanced transgenic oilseeds.

Oilseeds are also promising candidates for development of transgenics with increased levels of health-promoting fatty acids. Vegetable oils are the main source of PUFAs in most populations. Given the strong interest in the health benefits of certain PUFA, and health risks associated with other PUFA, it is not surprising that several research groups are interested in using transgenic technology to modify the fatty acid profile of oilseeds. For example, it is becoming clear that it would be desirable to have vegetable oils containing large amounts of linolenic acid and smaller amounts of linoleic acid. This could conceivably be done by changing the degree of expression of the enzymes responsible for making these fatty acids.

Such an approach has been used to create high-laurate canola. However, this was done not to increase the healthfulness of the oil, but because of the many industrial uses of laurate (e.g., in detergent manufacture). It is expected, though, that in the near future vegetable oils with increased health functionality will be available.

H. Reducing or Eliminating Allergens

For reasons that are unclear, the incidence of pollen-associated allergies in many countries appears to be increasing. Although less well documented, it is also likely that food allergies are becoming more frequent. However, there are many misconceptions among the media and the general population about the nature and extent of food allergies. The formal definition of a food allergy is an adverse reaction to food that is caused by an **immune reaction**. Many reactions commonly assumed to be allergic reactions are actually food **intolerance**, which is often caused by specific enzyme (e.g., lactase) deficiency. True food allergy appears to be relatively uncommon. A recent study of food allergy perception and incidence in the Netherlands found that 12.4% of 1483 questionnaire respondents indicated that they suffered from food allergies. These people were then tested through double-blind, placebo-controlled food challenges. The incidence of true food allergy in this study was only 0.8%. The reasons for the discrepancy between perceived and clinical food allergy are probably related to the popular media and the prevalence of alternative medical practitioners who blame food allergies for a host of medical problems (e.g., complaints of poor general health).

For some people, however, food allergy is a matter of life and death. Peanut allergies, for example, sometimes lead to life-threatening anaphylactic reactions. For these people, as well as for those with milder food allergies, it is desirable to reduce the levels of allergens (compounds that provoke an allergic reaction) in foods. It is difficult to assess the allergenicity of different foods because susceptibility among humans varies with several factors, with age probably the most important. However, common culprits are eggs, fish, celery, carrots, kiwi, cow's milk, legumes, and sunflower seeds. Because it is much easier to transform plants than animals, most discussions of allergen reduction have focused on vegetables.

The best example to date of allergen reduction is in transgenic rice lines that have reduced levels of a 16-kDa (kilodalton, a measure of molecular mass) allergenic protein. This protein was assumed to have some allergenicity because it reacted with antibodies in serum collected from people with rice allergy. The gene for this protein was cloned by the following procedure:

- mRNA was collected from developing rice seeds, and a cDNA library was established using a phage vector (γgt11).
- This library was screened using a nucleic acid probe that was based on the amino acid sequence of one end of the 16-kDa protein.
- A DNA library was then probed to isolate the promoter region of the 16-kDa gene.
- A plasmid vector (similar to pUC) was constructed that contained the 16-kDa promoter in front of the antisense sequence of the 16-kDa gene.
- This plasmid was introduced into rice protoplasts via electroporation; transformed cells were selected through use of an antibiotic marker gene.
- Rice plants were then regenerated from the protoplasts.

Transformed rice plants produced seeds that had about one fifth the amount of allergenic protein as untransformed plants. It is uncertain whether this would be enough of a reduction to prevent reactions among allergenic individuals. It is also possible that other proteins act as allergens, in addition to the 16-kDa protein. The function of the 16-kDa protein is also uncertain; it has sequence similarity with barley trypsin inhibitor and wheat α-amylase inhibitor, and therefore may function as an antiherbivore compound. This would imply that changing the sequence of the 16-kDa protein might *improve* rice digestibility, in addition to reducing allergenicity. Nonetheless, it is essential to learn more about the normal function of this protein, to assess the ramifications of using antisense technology to modify its expression.

Other aspects of transgenic technology may be relevant to food allergies. For example, it has been demonstrated that milk proteins can be expressed in plant tissue; perhaps we could engineer a plant that produces proteins with the nutritional quality of human milk but without the allergenicity associated with cow's milk. Of course, food allergy is important to transgenic technology in another context — regulatory bodies usually carefully consider the potential allergenicity of transgenic proteins before allowing their release and consumption. This important point will be further discussed in Chapter 9.

VI. ALTERNATIVE AGRICULTURAL TECHNOLOGIES

Increasing numbers of growers and corporations are growing and marketing foods that appeal to consumers who are in favor of alternative farming methods, especially those that are less reliant on the use of agrochemicals. Organic agriculture is a striking example of how shifts in consumer attitudes can change the food industry. Twenty years ago organic farming operated at a loosely organized small scale, catering to a small but devoted following. Now, the organic food industry contributes billions of dollars to economies throughout the world. It is no longer solely the domain of small market gardeners, and it is becoming increasingly influenced by large corporations. Although this process may be contrary to the beliefs of many proponents of organic agriculture, it has allowed prices to decrease somewhat, which in turn has expanded the organic food consumer base.

However, there still exists a pronounced price difference between most types of organic food and conventional food. Less extreme modifications of conventional agriculture (use of biofertilizers or integrated pest management [IPM]) may appeal to consumers who are not willing to pay the premium price of organic food but perceive conventional agriculture as overdependent on toxic chemicals. This could be a marketing opportunity for both producers and food companies. Consequently, it is useful to be aware of some of the technologies that benefit both organic farmers and farmers interested in reducing their use of agrochemicals.

Organic farmers do not use synthetic pesticides to control weeds and pests. They also foreswear chemical fertilizers. Instead, weeds are controlled physically through tilling and other methods of disrupting weed growth. Insect pests are usually controlled through a combination of "natural" insecticides (e.g., rotenone), detergents (insecticidal soap), or microbial insecticides (usually spores of *B. thuringiensis*). It is ironic that consumers of organic food have such intense antipathy to food biotechnology, because many organic growers are highly dependent on the ability of microbial biotechnologists to cheaply grow large amounts of *B. thuringiensis* spores. Transgenic crops with Bt genes would also be useful tools for organic growers to control insect pests, but proponents of organic agriculture generally reject any use of recombinant DNA technology.

Chemical fertilizers are replaced by either animal or green manures. Green manures are crops grown on soil to accumulate nutrient-rich organic matter. Legumes are a popular green manure because of their ability to form symbioses with soil bacteria that allow the plants to obtain nitrogen from the atmosphere. Consequently, leguminous green manures can add substantial amounts of nitrogen to a soil. Because nitrogen is the element (after carbon, hydrogen, and oxygen) most required by plants, this is an important consideration.

Biotechnology has much to offer organic farmers — to improve mineral supply to plants and to help control pests. It is beyond the scope of this book to address all aspects of biological pest control (the use of microbes or other biological agents to control pests), but we will briefly examine biofertilizers.

Biofertilizers improve the ability of plants to obtain mineral nutrients from the soil. This technology exploits the natural abilities of microbes such as *Rhizobium*, a bacterium that can form symbioses with the roots of legumes. *Rhizobium* is able to **fix nitrogen** from the atmosphere into ammonium, which can then be used by the plant as a source of nitrogen. Most soils naturally contain *Rhizobium*, but naturally occurring strains are often inefficient fixers of nitrogen. Manipulation of *Rhizobium* populations can correct this problem. The use of efficient *Rhizobium* strains in the cultivation of peas, beans, and other legumes allows decreased reliance on chemical fertilizers. Application of nitrogenous fertilizers is often a large economic burden for farmers, and it can lead to problems such as increased rates of erosion and contamination of underground reservoirs with nitrate. This is a problem for users of underground reservoirs because of hazards associated with the consumption of nitrate in drinking water.

Biofertilizers can also increase the supply of phosphorus to plants. **Mycorrhizal fungi** are extremely common symbionts of plant roots, and they improve plant growth and health through a number of mechanisms. Increased rate of phosphorus supply is usually the most clear-cut benefit to plants from mycorrhizas. Specific groups of soil bacteria can also improve the rate of phosphorus intake by plants. These are the **phosphate-solubilizing** bacteria; their growth and activity lead to increased levels of phosphate in the soil solution. Both of these groups of microorganisms are normally present in agricultural soils, but, as with the case of *Rhizobium*, the most efficient species are not always present, and some soils have low numbers of mycorrhizal fungi and phosphate-solubilizing bacteria.

IPM is similar in some ways to organic agriculture because of its decreased reliance on chemical pesticides and increased reliance on biological control of pests (e.g., using a specific fungal pathogen to control a specific weed). IPM also involves the use of assessment methods that keep track of insects and other pests throughout the growing season. This allows the judicious use of chemical pesticides — they are applied only when a particular pest has reached levels that justify the expense. Consumers throughout the world rate pesticide residues as a major concern, despite reassurances from government regulators and the chemical industry that current pesticide use is safe. Consequently, it represents another potential marketing niche that may be exploited more in the future, if consumer concerns over food safety continue to increase. However, this might pose legal and regulatory problems, because most countries have regulations governing the use of "organic" labels but have no mechanism for labeling food that is produced through reduced use of **agrochemicals** (chemical fertilizers and pesticides). This is important because many farmers find the requirements for organic certification daunting (e.g., a number of agrochemical-free years of production are typically required before a farmer can claim organic status). Alternatives to organic labels could be useful to such farmers.

Unfortunately, IPM and biofertilizers have received little media attention, and the public remains largely unaware of biotechnologies that have the potential to decrease agriculture's reliance on agrochemicals. Similarly, despite many attempts by food biotechnologists to justify the use of transgenic technology based on its potential to decrease pesticide and fertilizer use, most consumers appear unconvinced. The next few years are crucial to plant biotechnologists; they need to demonstrate that plant biotechnology is safe to consumers and can benefit consumers directly, via improved food functionality, for example, and indirectly, through decreased use of agrochemicals that are detrimental to the environment.

RECOMMENDED READING

1. Wilkinson, J., Biotech plants: from lab bench to supermarket shelf, *Food Technol.*, 51, 37, 1997.
2. Dunwell, J. M., Transgenic crops: the next generation, or an example of 2020 vision, *Ann. Bot.*, 84, 269, 1999.
3. Raven, P. H., Evert, R. F., and Eichhorn, S. E., *Biology of Plants,* 6th ed., Freeman, New York, 1999.
4. Wenzel, G., Application of unconventional techniques in classical plant production, in *Plant Biotechnology,* Fowler, M. W. and Warren, G. S., Eds., Pergamon Press, Oxford, 1992, chap. 13.
5. Knorr, D., Caster, C., Dörneburg, H., Dorn, R., Gräf, S., Havkin-Frenkel, D., Podstolski, A., and Werrman, U., Biosynthesis and yield improvement of food ingredients from plant cell and tissue cultures, *Food Technol.*, 77, 57, 1993.
6. Hall, R. D., An introduction to plant cell culture: pointers to success, in *Plant Cell Culture Protocols*, Hall, R. D., Ed., Agritech, Shrub Oak, NY, 1999, chap. 1.
7. Zupan, J., Muth, T. R., Draper, O., and Zambryski, P., The transfer of DNA from *Agrobacterium tumefaciens* into plants: a feast of fundamental insights, *Plant J.*, 23, 11, 2000.
8. McKinnon, G. E. and Henry, R. J., Control of gene expression for the genetic engineering of cereal quality, *J. Cereal Sci.*, 22, 203, 1995.
9. Koziel, M. G., Beland, G. L., Bowman, C., Carozzi, N. B., Crenshaw, R., Crossland, L., Dawson, J., Desai, N., Hill, M., Kadwell, S., Launis, K., Lewis, K., Maddox, D., McPherson, K., Meghji, M. R., Merlin, E., Rhodes, R., Warren, G. W., Wright, M., and Evola, S. V., Field performance of elite transgenic maize plants expressing an insecticidal protein derived from *Bacillus thuringiensis*, *Bio/Technology*, 11, 194, 1993.
10. Armstron, C. L., Parker, G. B., Pershing, J. C., Brown, S. M., Sanders, P. R., Duncan, D. R., Stone, T., Dean, D. A., DeBoer, D. L., Hart, J., Howe, A. R., Morrish, F. M., Pajeau, M. E., Petersen, W. L., Reich, B. J., Rodriguez, R., Santino, C. G., Sato, S. J., Schuler, W., Sims, S. R., Stehling, S., Tarochione, L. J., and Fromm, M. E., Field evaluation of European corn borer control in progeny of 173 transgenic corn events expressing an insecticidal protein from *Bacillus thuringiensis*, *Crop Sci.,* 35, 550, 1995.
11. Hansen Jesse, L. C. and Obrycki, J. J., Field deposition of Bt transgenic corn pollen: lethal effects on the monarch butterfly, *Oecologia*, 125, 241, 2000.
12. Losey, J. E., Rayor, L. S., and Carter, M. E., Transgenic pollen harms monarch larvae, *Nature*, 399, 214, 1999.
13. Wraight, C. L., Zangeri, A. R., Carroll, M. J., and Berenbaum, M. R., Absence of toxicity of *Bacillus thuringiensis* pollen to black swallowtails under field conditions, *Proc. Natl. Acad. Sci. U.S.A.*, 94, 770, 2000.
14. Gould, F., Sustainability of transgenic insecticidal cultivars: integrating pest genetics and ecology, *Annu. Rev. Entomol.*, 43, 701, 1998.
15. Hilder, V. A. and Boulter, D., Genetic engineering of crop plants for insect resistance — a critical review, *Crop Prot.*, 18, 177, 1999.

16. Munkvold, G. P., Hellmich, R. L., and Rice, L. G., Comparison of fumonisin concentrations in kernels of transgenic Bt maize hybrids and nontransgenic hybrids, *Plant Dis.*, 83, 130, 1999.

17. Gressel, J., Molecular biology of weed control, *Transgenic Res.*, 9, 355, 2000.

18. Schuch, W., Improving tomato quality through biotechnology, *Food Technol.*, November, 78–83, 1994.

19. Turner, M. and Schuch, W., Post-transcriptional gene-silencing and RNA interference: genetic immunity, mechanisms and applications, *J. Chem. Technol. Biotechnol.*, 75, 869, 2000.

20. Madhavi, D. L. and Salunkhe, D. K., Tomato, in *Handbook of Vegetable Science and Technology,* Salunkhe, D. K. and Kadam, S. S., Eds., Marcel Dekker, New York, 1998, chap. 7.

21. Redenbaugh, K., Hiatt, W., Martineau, B., and Emlay, E., Regulatory assessment of the FLAVR SAVR tomato, *Trans. Food Sci. Technol.*, 5, 105, 1994.

22. McLaren, D. and Frigg, M., *Sight and Life Manual on Vitamin A Deficiency Disorders (VADD)*, 2nd ed., Task Force Sight and Life, Basel, 2001.

23. Guerinot, M. L., The green revolution strikes gold, *Science*, 287, 241, 2000.

24. Ye, X., Al-Babili, S., Klöti, A., Zhang, J., Lucca, P., Beyer, P., and Potrykus, I., Engineering the provitamin A (β-carotene) biosynthetic pathway into (carotenoid-free) rice endosperm, *Science*, 287, 303, 2000.

25. Waterhouse, P. M., Wang, M.-G., and Lough, T., Gene silencing as an adaptive defence against viruses, *Nature*, 411, 834, 2001.

26. Stuiver, M. H. and Custers, J. H. H. V., Engineering disease resistance in plants, *Nature*, 411, 86, 2001.

27. Lorito, M. and Scala, F., Microbial genes expressed in transgenic plants to improve disease resistance, *J. Plant Pathol.*, 81, 73, 1999.

28. Vandemark, G. J., Transgenic plants for the improvement of field characteristics limiting crop production, in *Molecular Biotechnology for Plant Food Production*, Paredes-López, O., Ed., Technomic, Lancaster, PA, 1999, Chap. 6.

29. Guzmán-Maldonado, S. H. and Paredes-López, O., Biotechnology for the improvement of nutritional quality of food crop plants, in *Molecular Biotechnology for Plant Food Production*, Paredes-López, O., Ed., Technomic, Lancaster, PA, 1999, chap. 14.

30. Duranti, M. and Gius, C., Legume seeds: protein content and nutritional value, *Field Crops Res.*, 53, 31, 1997.

31. Pen, J., Verwoer, T. C., van Paridon, P. A., Beudeker, R. F., van den Elzen, P. J. M., Geerse, K., van der Klis, J. D., Versteegh, H. A. J., van Ooyen, A. J. J., and Hoekema, A., Phytase-containing transgenic seeds as a novel feed additive for improved phosphorus utilization, *Bio/Technology*, 11, 811, 1993.

32. Raboy, V., Low-phytic-acid grains, *Food Nutr. Bull.,* 21, 423, 2000.

33. Davies, C. S., Nielsen, S. S., and Nielsen, N. C., Flavor improvement of soybean preparations by genetic removal of lipoxygenase-2, *J. Am. Oil Chem. Soc.*, 64, 1428, 1987.

34. Utsumi, S., Plant food protein engineering, *Adv. Food Nutr. Res.*, 36, 89, 1992.

35. Schulman, A., Chemistry, biosynthesis, and engineering of starches and other carbohydrates, in *Molecular Biotechnology for Plant Food Production*, Paredes-López, O., Ed., Technomic, Lancaster, PA, 1999, chap. 12.

36. Schuurink, R. C. and Louwerse, J. D., The genetic transformation of wheat and barley, in *Cereal Biotechnology*, Morris, P. C. and Bryce, J. H., Eds., Woodhead Publishing, Cambridge, 2000, chap. 2.

37. Del Vecchio, A. J., High laurate canola: how Calgene's program began, where it's headed, *Inform*, 7, 230, 1996.
38. Henry, R. J., Using biotechnology to add value to cereals, in *Cereal Biotechnology*, Morris, P. C. and Bryce, J. H., Eds., Woodhead Publishing, Cambridge, 2000, chap. 5.
39. Faus, I., Recent developments in the characterization and biotechnological production of sweet-tasting proteins, *Appl. Microbiol. Biotechnol.*, 53, 145, 2000.
40. Shintani, D. and DellaPenna, D., Elevating the vitamin E content of plants through metabolic engineering, *Science*, 282, 2098, 1998.
41. Davey, M. W., van Montagu, M., Inzé, D., Sanmarin, M., Kanellis, A., Smirnoff, N., Benzie, I. J. J., Strain, J. J., Favell, D., and Fletcher, J., Plant L-ascorbic acid: chemistry, function, metabolism, bioavailability and effects of processing, *J. Sci. Food Agric.*, 80, 825, 2000.
42. Matsuda, T., Nakase, M., Adachi, T., Nakamura, R., Tada, Y., Shimada, H., Takahashi, M., and Fujimara, T., Allergenic proteins in rice: strategies for reduction and evaluation, in *Food Allergies and Intolerances*, Eisenbrand, G., Aulepp, H., Dayan, D. D., Elias, P. S., Grunow, W., Ring, J. and Schlatter, J., Eds., VCH Publishers, Weinheim, Germany, 1996, chap. 12.
43. Wüthrich, B., Epidemiology of allergies and intolerances caused by foods and food additives: the problem of data validity, in *Food Allergies and Intolerances*, Eisenbrand, G., Aulepp, H., Dayan, D. D., Elias, P. S., Grunow, W., Ring, J., and Schlatter, J., Eds., VCH Publishers, Weinheim, Germany, 1996, chap. 1.
44. Coleman, E., *The New Organic Grower*, Nimbus, Halifax, Nova Scotia, Canada, 1995.

5 Animal Biotechnology

I. OVERVIEW

On the whole, humans are quite fond of eating meat, fish, and dairy products. In the industrialized world, consumption of some animal products (e.g., red meat) has declined, but this has frequently been balanced by increased consumption of others (e.g., poultry). In the developing world, meat consumption is often viewed as one of the prime benefits of increased personal wealth. Most animal products are excellent, balanced sources of protein and supply vitamins and minerals such as iron. Thus they are considered to be an essential part of the diet by most nutritionists. Animal products can be made into a diverse array of foods, encompassing fresh, cured, and precooked products; this process is a potent and vibrant part of the global economy. Why does meat eating persist, despite the well-known health risks associated with diets with high levels of animal products (especially those that have a high fat content) and the environmental problems associated with animal production (e.g., contamination of water sources by manure from feedlot operations)? The answer is probably partly hard-wired into our genes — humans like to eat meat. The common salivary response to the odor of barbecued meat is perhaps an illustration of this phenomenom.

A diverse range of animals are used as food. Mammalian food animals are divided into two categories: (1) **ruminants** (cattle, goats, sheep) — these animals have a **rumen**, a specialized stomach compartment that incubates ingested plant material for a protracted period of time before allowing it to pass to the intestine; and (2) **monogastrics** (pigs), animals that have only a single stomach compartment. Ruminants rely on bacteria and other microorganisms in their rumen for digestion of cellulose-rich plant biomass. The rumen is composed of a complex, interdependent community of microbes that converts indigestible (to the animal) plant feed into **volatile fatty acids** (e.g., acetate), which are then absorbed by the ruminant and used as a source of energy and carbon.

In addition to these mammalian food animals, several species of birds (chickens, turkeys, ducks, and geese) and their eggs are important animal products. Fish are unique among food animals in that wild fish are more important than domestic fish (**aquaculture**); however, aquaculture is rapidly increasing in popularity, partly because of depletion of global fish stocks. Overfishing, environmental degradation (e.g., increased flow of industrial and domestic sewage to the ocean), and climatic shifts are responsible for the precipitous decrease in certain fish stocks.

Biotechnology plays a role in the production of meat, eggs, dairy products, and fish. **Microbial biotechnology** is very useful, particularly in the production of fermented dairy products; this technology will be discussed in Chapter 7. **Diagnostic**

biotechnology that detects and helps diagnose pathogenic bacteria is increasingly important in the slaughterhouse environment because of the increasing incidence of cattle carrying dangerous bacteria such as *Escherichia coli* O157:H7. Diagnostic applications will be discussed in Chapter 6. Here we will concentrate on the potential of transgenic technology to change food animals. Transgenic feed crops are also relevant to animal production (e.g., transgenic soybeans with increased phytase; see Chapter 4, Section V.C).

Transgenic animals are not currently used by the food industry, but this will likely change soon. The first commercially grown transgenic animal will probably be salmon with boosted levels of growth hormone. These fish will be restricted to the aquaculture industry, and their success hinges partly on regulatory approval and partly on consumer acceptance. Given the outcry against transgenic plants, transgenic fish will likely be met with substantial public hostility, especially from antibiotechnology activist groups.

The lack of transgenic animals in the food industry is at first glance surprising, given the enormous success of transgenic animals in elucidating many aspects of the molecular biology of animals. Transgenic mice in particular have been useful research tools because they have allowed scientists to increase, decrease, or eliminate ("knockout") specific genes and then examine the effects of the gene modification on the animal's physiology or development. However, when a biotechnologist begins research leading to the production of a transgenic animal destined for agricultural use, the research program becomes very different. The end product of the research must be an animal that is improved in some way, in comparison to conventional animals. This improvement must not be accompanied by adverse effects on the health of the animal or on the safety of foods derived from the animal. Another daunting factor is cost; generally it is more expensive to produce transgenic animals than transgenic plants, and this cost rapidly increases with the size of the animal. Indeed, biotechnology companies have shied away from funding research into transgenic farm animals; most researchers in this field are funded by government organizations. Finally, the biotechnologist is virtually guaranteed to face considerable public opposition and regulatory hurdles after spending years developing the transgenic animal. So, perhaps it is not surprising that few laboratories are currently involved in research and development of transgenic animals for the food industry. Despite this pessimistic outlook, though, transgenic biotechnology has promising applications (Table 5.1). The first three traits in Table 5.1 will be discussed later in this chapter. Control of sex ratios in poultry is desirable because producers of broiler chickens use only male chicks and discard female chicks. In contrast, egg producers use only female chicks. This large-scale wastage could be avoided if producers could control the sex ratio. Unfortunately, poultry scientists have not yet deciphered the mechanisms involved in control of sex ratios, making it impossible to develop transgenic chickens with useful modifications of this phenotype.

Many animal scientists hope that transgenic technology will be useful in introducing disease resistance to ruminants, monogastrics, and poultry. This may be an effective approach to control viruses (e.g., the virus that causes foot-and-mouth disease). Many animal viruses are difficult to control using conventional methods (e.g., vaccination), but could be controlled by changing the animal's genome so that

TABLE 5.1
Potential Uses (and Problems) of Transgenic Biotechnology in the Production of Animal-Based Foods

Animal	Transgene	Objective	Problems
Ruminants	Growth hormone	Leaner meat, increased feed efficiency	Many side effects
Ruminants	Modified milk proteins	Improved processing, improved nutrition	Technically difficult
Fish	Growth hormone	Faster growth	Environmental issues
Poultry	?	Control over sex-ratios	Unknown genes are involved
Various animals	Pathogen receptors	Disease resistance	Technically difficult; may carry numerous side effects

it can no longer be infected. For example, viruses get into host cells through binding to specific receptors. If these receptors (usually proteins embedded in the host cell's membrane) could be changed or removed, the virus would be unable to infect the cell. Unfortunately, it is unlikely that this could be done without adversely affecting the animal's metabolism or development — cell surface proteins always have a function of some sort.

This chapter concentrates on the two applications of transgenic technology that are most likely to have an impact on food production in the foreseeable future: fish with enhanced production of growth hormones and ruminants with modified milk proteins. The last two sections address alternative approaches to the development of transgenic animals: the use of embryonic stem cells and the cloning of adult cells.

II. TRANSGENIC FISH

The most famous transgenic animal is the "supermouse," a mouse that had increased expression of growth hormone; these mice, developed in the early 1980s, were twice as large as their nontransgenic littermates. The success of this research likely fostered expectations that farm animals would soon be developed that had boosted levels of growth hormone. Such animals might have leaner meat, better feed efficiency (ability to convert feed biomass into animal biomass), and faster growth. Indeed, transgenic pigs have these improvements, but they are also susceptible to a number of abnormalities, particularly in skeletal development. These problems are likely related to the increased expression of growth hormone, and not to other factors (e.g., the process used to create the transgenic animals), because similar abnormalities can be induced through direct application of growth hormone to animals such as pigs. Because of these serious health problems, transgenic farm animals with increased levels of growth hormone have not yet been released. Interestingly, this has not stopped some meat processors in North America from attempting to label their food as "not genetically modified," a *non sequitur* because currently there are no commercially available transgenic food animals.

Biotechnologists have had much greater success in the development of transgenic fish with boosted levels of growth hormone, although success has been largely limited to salmonid fish (salmon and trout are salmonids important to the food industry). The rationale for these transgenics is that **production efficiency** is the major variable affecting the salmon aquaculture industry. Efficiency is determined largely by feed efficiency but also by growth rates. It has long been known that applying fish growth hormone to salmon results in faster growth and increased feed efficiency, so producing transgenic fish with higher levels of endogenous (from within) growth hormone was a logical step. Salmon are also difficult to rear in aquaculture because of the two distinct phases of their life cycle — initially, they require cold freshwater (which leads to slow growth rates) or heated freshwater (which is expensive), and after several years of this **juvenile rearing**, they need to be transferred to saltwater. Increased levels of growth hormone shorten the juvenile rearing period, thus cutting costs of production.

The overall strategy of creating a transgenic animal is similar to that used to develop a transgenic plant (Chapter 4, Section III.B). The researcher clones the gene of interest and finds out as much as is logistically possible about the gene. For example, knowledge of the DNA sequence that codes for the desired protein and an understanding of promoters, terminators, and regulatory regions of the gene are useful. Promoters and other genetic elements are then chosen and linked to the desired gene. This DNA is then linearized and any other genes (e.g., antibiotic selectable marker genes) are removed. With fish, the linear DNA is then injected into the nuclei of cells in just-fertilized egg cells (with fish, injection into the cytoplasm of the fertilized egg cell is also likely to be successful). The injected DNA will occasionally become permanently integrated into one of the cell's chromosomes. With most animals, this is a rare event and is difficult to control, because the integration mechanism is unknown. Consequently, many attempts at micro-injection of egg cells end in failure. Micro-injection is relatively straightforward with salmonid eggs, partly because large numbers of eggs can easily be obtained, either manually or with the mechanical assistance of a micromanipulator — a device that gives the operator fine-tuned control over the injection process. Note that it is not necessary to integrate desired genes into **vectors** (e.g., plasmids) before micro-injection into nuclei. For this reason, the injected DNA, along with its promoter, is called a **construct**.

Once the embryo develops into an adult, the inserted gene is likely to be active only in certain cells, dictated by the chromosomal location of insertion (different regions of the genome are actively expressed in different cell types). However, the trait should be inheritable, and pure-breeding transgenic lines can usually be obtained after several generations.

A number of lessons were learned during early attempts to generate transgenic salmonids with increased levels of growth hormone. First, biotechnologists used mammalian growth hormone genes and mammalian promoters. In most cases, this did not lead to increased production of growth hormone. The reason for this is unclear, but it may be related to incorrect mRNA processing due to the lack of introns. Molecular biologists have observed that transgenic plants and animals sometimes produce larger amounts of heterologous protein when introns are present in the transgene.

When growth hormone genes from fish are inserted into fish with strong promoters, efficient expression (and, presumably, correct mRNA processing) results. It has also become clear that it is preferable to use promoters derived from fish rather than mammalian promoters. Promoters that demonstrably drive growth hormone expression in transgenic salmon include the ocean pout antifreeze protein promoter and the sockeye salmon metallothionein-B promoter. These promoters are active in the liver; this constitutes a major change from native growth hormone production, which occurs only in the pituitary gland.

Transgenic fish with increased growth hormone grow faster and accumulate biomass more efficiently than conventional fish. However, the technology is not perfect. Depending on the level of transgene expression, some fish may develop **acromegaly**, a condition involving excessive growth of the cartilage in the head region. This can lead to decreased fish viability in extreme cases. When growth hormone is expressed at moderate levels, acromegaly is not seen, but this is accompanied by a more restrained increase in growth rates. Before being released for commercial production, this problem must be solved.

Regulators must also consider environmental risks before allowing release of transgenic salmon. Many environmentalists and ecologists fear that transgenic fish will escape from their aquaculture cages, with unpredictable consequences. A common fear expressed by antibiotechnology activists is that transgenic fish will outcompete their wild counterparts, thus reducing fish diversity. However, most fish ecologists have the opposite concern, predicting that transgenic fish will be relatively unfit compared to their wild counterparts. In this context, "fitness" refers to the overall ability of an individual to reproduce successfully. Even if transgenic fish have reduced fitness, they could still exert harmful effects. If they interbreed with wild salmon, they might introduce deleterious genes into wild salmon populations. This **"trojan gene"** effect could occur if transgenic fish are more successful at mating than wild fish but have decreased viability. Computer simulations have indicated that this could lead to a decline in wild populations.

Transgenic fish with boosted levels of growth hormone may have better fitness than wild salmon, leading to displacement of populations of wild salmon. This is a sensitive issue, because many salmon populations, particularly those of Atlantic salmon, are endangered; any extra stress imposed by escaped transgenic salmon would be undesirable. The major problem with risk assessment of transgenic fish is that the interactions among specific genotypes, fitness, and the environment are poorly understood. Risk assessment is also complicated by the large number of phenotypes that change (Table 5.2) upon the addition of a single gene construct (gene + promoter). Some of these phenotypes would seem to give transgenic fish a competitive advantage over normal fish; for example, increased feeding motivation seems to be a positive trait. However, increased feeding motivation, combined with decreased swimming ability, might make transgenic fish more susceptible to predation.

Those for and against fish biotechnology agree on one point: transgenic fish must be physically and biologically contained, to minimize escapes and the consequences of escapes. Physical containment is achieved through the use of cages and nets, and biological containment can be done either by sterilizing fish chemically

TABLE 5.2
Phenotypes of Salmonids that Change upon
Introduction of a Growth Hormone Gene

Phenotype	Nature of Change
Growth rate	Increased
Smoltification	Earlier
Appetite and feeding motivation	Increased
Metabolism	Increased rate
Swimming ability	Decreased
Cranial morphology	Abnormal
Muscle structure	Increased hyperplasia
Life cycle	Shortened

or by clever breeding manipulation that produces sterile triploid fish (see item 6 in R. O'Flynn et al., 1997, for an explanation of this procedure).

It is important to note that salmon or other fish that have been selectively bred for rapid growth rates using traditional breeding techniques carry similar risks upon escape into the environment. But it is virtually impossible to "follow" them after such an escape; in contrast, transgenic fish carry unique genetic sequences that can be identified through PCR (polymerase chain reaction) and other techniques.

The final point that transgenic fish developers must address is consumer acceptance. The only potential benefit to the consumer of transgenic fish with enhanced growth rates is price. This may not be enough to counter consumer resistance, particularly in the face of vigorous action by antibiotechnology forces, a virtual certainty in the case of transgenic fish.

III. MODIFIED MILK PROTEINS

In the early 1980s, it became clear that it was technically possible to transfer specific genes to mammals using recombinant DNA technology. This led to numerous attempts to introduce new or modified genes into farm animals. With some species (e.g., poultry), transgenic technology is now well established in the laboratory. However, it has proven difficult to produce transgenic animals in cattle, the most important food animal. There are several reasons for this: (1) cattle have long gestation periods, (2) they usually produce only one calf per gestation, and (3) zygotes and embryos of cattle are more difficult to manipulate than those of other farm animals. These factors make it expensive to produce transgenic cattle — large herd sizes are necessary in order to produce enough viable transgenic individuals.

Because of these technical problems, most research into transgenic cattle has focused on the production of human proteins that can be used to treat human disease. This **molecular farming ("pharming")** has the potential to be lucrative for pharmaceutical companies. In some cases, one or two transgenic cows could conceivably produce enough therapeutic protein to serve a $100 million market.

TABLE 5.3
Milk Traits that Could Be Modified
Using Transgenic Technology

Trait	Effect
Human lysozyme	Increased antimicrobial properties
Human lactoferrin	Increased antimicrobial properties
Increased κ-casein	Increased heat stability of milk
Increased β-casein	Improved cheese making
Increased α-casein	Improved nutritional qualities
Addition of plasmin inhibitor	Fewer sensory defects of UHT milk
Addition of β-galactosidase	Decreased levels of lactose in milk
Removal of β-lactoglobulin	Decreased allergenicity
Addition of desaturase	Improved fatty acid profile

Note: Bolded traits have been successfully tested in transgenic animals. Nonbolded traits illustrate potential, but untested applications of transgenic technology. See the text and item 8 in the Recommended Readings List, Mercier and Villotte (1997), for further explanation.

Sheep that have the human gene for α_1-antitrypsin (α_1AT) are examples of this type of transgenic application. About 100,000 people in the U.S. suffer from α_1AT deficiency, a hereditary disease. In this disease, abnormal α_1AT genes produce proteins that do not function properly, leading to inflammation and damage to lung cells, which can lead to emphysema. One possible therapy is to give patients normal α_1AT; however, sufficient amounts of α_1AT cannot be obtained from natural sources. Recombinant α_1AT can be obtained from yeast or bacteria, but in both cases it does not have the proper posttranslational glycosylation. It could be produced in mammalian cell lines, but it has proven difficult to achieve sufficient protein yield using cell culture. For these reasons, biotechnologists have attempted to create transgenic animals that produce human α_1AT. This was achieved with mice in 1990, and in 1991, scientists created transgenic sheep that produced α_1AT in their milk. To date, though, the recombinant proteins produced by these sheep have not been released for human therapy.

Transgenic technology also has the potential to modify **food-related** traits of the milk of cattle and other ruminants. There are a number of objectives to this research (Table 5.3); some are aimed at improving nutritional properties (e.g., increased levels of cysteine-rich α-casein) or decreasing allergenicity (e.g., eliminating or decreasing β-lactoglobulin, which is an allergen, but seems unimportant to the function of milk), whereas others focus on improving processing traits (e.g., increasing the level of β-casein, resulting in milk that forms firmer curds during cheese making).

It is much easier to produce transgenic cattle today than it was ten years ago. This is mainly because of improvements in methods of zygote acquisition, maturation of zygotes, and *in vitro* fertilization of zygotes. As an example, we will examine

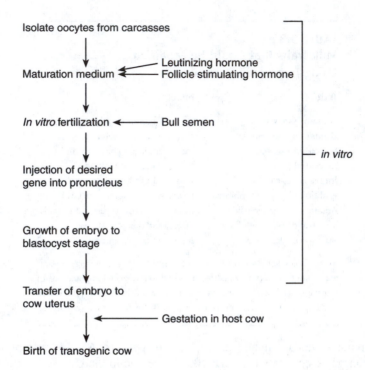

FIGURE 5.1 Process used to introduce a desired gene into cattle through pro-injection of pronuclei.

transgenic cows with an introduced human gene coding for α-lactalbumin, the major whey protein in human milk (Figure 5.1). The objective in this case was to produce milk more similar to human milk than conventional cow's milk.

The first step was to obtain large numbers of **oocytes** (immature, unfertilized egg cells) from freshly slaughtered carcasses. The oocytes were then incubated in a **maturation medium**, which contains bovine **leutinizing hormone** and **follicle-stimulating hormone**. These hormones triggered maturation of the oocytes, which is a necessary prelude to fertilization. The oocytes were then fertilized *in vitro*, using semen collected from a bull.

Once the oocyte has been fertilized, it is ready to begin embryonic development. This is a crucial stage in the transgenesis process. Before the fertilized zygote begins to divide, the foreign DNA construct must be injected into one of the **pronuclei**. This term is given to the nuclei originating from sperm and egg cells that unite to form a diploid nucleus. This is an opportune time to inject foreign DNA, because if DNA is injected at a later stage, after cell division has begun, only part of the embryo will contain the transgene. However, if the fertilized zygote is transformed, all of the cells of the embryo and the adult will contain the transgene.

In the α-lactalbumin example, zygotes were incubated after DNA injection in a medium that mimics the environment of the oviduct, where initial growth of the zygote normally occurs. After 7 to 8 days, when the embryos had reached the **blastocyst** stage (a hollow ball of cells), each embryo was transferred to the uterus of a cow. After the completion of gestation (about 39 weeks), calves were born and were tested for the presence of the transgene. Nine individuals (five male, four female) were transgenic, and after six months, one of the females was induced to begin lactation. Human α-lactalbumin was present in her milk at a concentration of 2.4 mg/mL.

This project illustrates why it is so expensive to produce transgenic cattle. The researchers fertilized 20,918 oocytes; they then injected DNA into 11,507 of the resulting zygotes. This led to 1,011 embryos that were of good enough quality and at the correct stage for implantation in cow uteri. Embryos were implanted in *478 cows* — a large herd indeed. This produced 155 successful pregnancies and 90 calves. Of these 90 calves, 9 were transgenic. It is expected that this low success rate is unlikely to improve quickly in the coming years; however, past experience in this field demonstrates that the scope of new achievements is difficult to predict (the surprising announcement of cloning of adult sheep cells in 1997 is a good example).

IV. THE SEARCH FOR EMBRYONIC STEM CELLS

It is unfortunate that mice are not suitable for milk or meat production. Embryonic stem (ES) cell lines that can be used to produce germline chimeras have been isolated only from mice, despite numerous attempts to isolate them from agriculturally important animals. What does ES mean and why is it so important? ES cells are isolated from early-stage mouse embryos. With the proper treatment, they can be developmentally **arrested**; this means that they do not differentiate further and do not complete their normal program of embryogenesis. However, they remain capable of cell division (proliferation), allowing researchers to obtain large numbers of ES cells. The other key feature of ES cells is that after a period of growth in culture, they can be transplanted into another early-stage (**blastocyst**) embryo. They will then become part of that embryo and remain capable of differentiating into a number of different cell types and tissues (in other words, they are **pluripotent**).

These characteristics are useful because they allow sophisticated manipulation of the ES cell's genome *in vitro*. Specific genes can be deleted, added, or modified, and genes can be inserted in specific locations in the genome (**targeted insertion**). In contrast, pronuclear injection (described in the previous section) allows only the addition of genes, and there is no control over the site of insertion of the new genetic material.

When ES cells with altered or additional genes are transplanted into a blastocyst, this gives rise to a **chimeric** embryo. Some of the cells of the embryo will be unaltered, and some of them will have originated from cell division and differentiation of the ES cells. This embryo will eventually develop into an adult chimera, which has a similar mixture of cells and tissues originating from unaltered or ES cells. This is referred to as a **somatic chimera**. A **germline chimera** is a somatic chimera that produces gametes (sperm or oocytes) that are derived from ES cells. This is a crucial characteristic because it allows the use of animal breeding procedures to produce pure breeding lines with the altered characteristic.

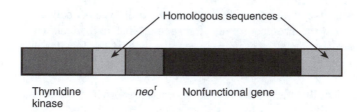

FIGURE 5.2 DNA construct that can be used to make knockout mice that lack a specific gene. (neo^r = neomycin resistance)

As previously noted, germline chimeras have so far been obtained only from mice. However, somatic chimeras have been obtained from mice, rabbits, and pigs. Many biotechnologists are confident that germline chimeras will eventually be obtained from agricultural animals such as pigs and cattle; if this happens, transgenic food animals will be much easier to produce, and the range of possible genetic modification will be vastly increased.

As an example of the use of ES cells to make a transgenic animal, we will examine the procedure used to make a **knockout** strain, in which a specific gene has been removed from the animal's genome. The biotechnologist produces a gene construct (Figure 5.2) that has a modified, nonfunctional version of the gene. On both sides of this, there are regions of sequence **homology** (identical to the sequences on either side of the native gene). A **neomycin-resistance** gene is also placed between the homologous sequences. Another gene, coding for **thymidine kinase**, is placed outside of the area between the homologous sequences.

A population of ES cells is then electroporated to induce uptake of this construct. As an aside, the process of DNA uptake by mammalian cells is termed **transfection** rather than **transformation**, because transformation in the context of mammalian cell culture has traditionally referred to a cellular transition to a cancerous state.

A small proportion of cells take up the foreign DNA and insert it into random points in the genome (**random insertion**) (Figure 5.3). A very small proportion of cells, through the process of **homologous recombination**, replace the targeted gene with the modified gene present in the construct. It is important to note that *only the sequence between the homologous regions* in the construct is inserted. Thus, the nonfunctional gene in the construct precisely replaces the native gene. Note also that this means that the thymidine kinase gene *will not be inserted into the chromosome*.

The flanking homologous sequences are essential for homologous recombination. This procedure works because animals, and most other eukaryotes, have DNA repair enzymes that recognize DNA segments that are homologous and splice them together.

The cells are then grown in a medium containing neomycin. All cells that lack the integrated construct will then die, because they lack the neomycin-resistant gene. Surviving cells are then transferred to a medium containing the antiviral agent gancyclovir. Cells that have randomly inserted constructs will have the thymidine kinase gene. This enzyme phosphorylates gancyclovir; the resulting compound is a

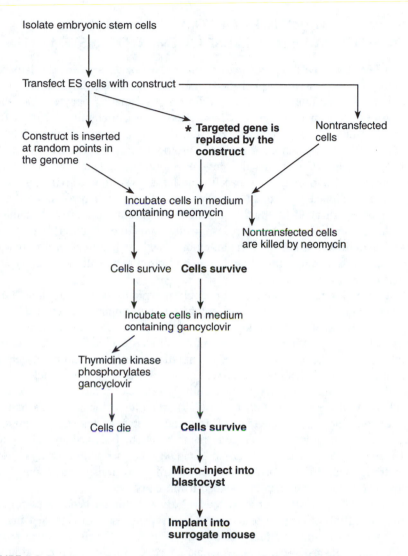

FIGURE 5.3 Process used to make knockout mice through homologous recombination. The pathway leading to knockout mice is bolded and the homologous recombination is indicated by the asterisk (*).

nucleotide analog that will be incorporated into replicating DNA by DNA polymerase. However, it cannot function in transcription, and the cell will die. This will eliminate all cells with randomly inserted constructs. Cells with targeted insertion (i.e., those that result from homologous recombination), though, will lack the thymidine kinase gene, because homologous recombination results in integration of only the DNA between the homologous sequences. The thymidine kinase gene on the construct lies outside of the homologous sequences, so it will not be integrated.

V. TRANSFER OF SOMATIC NUCLEI:
AN ALTERNATIVE TO THE USE OF EMBRYONIC STEM CELLS

It is a rare event when a sheep is able to dominate newspaper headlines, as happened in February 1997, when *Nature* published the first account of cloning of adult somatic cells, in "Dolly." Wilmut and colleagues did this by isolating oocytes from sheep and then removing their nuclei (this is referred to as **enucleation**). Nuclei from mammary gland cells of an adult sheep were then transferred into the enucleated cells (this was done by a process similar to protoplast fusion). The cells were then cultured in a medium that promoted embryo development, and the resulting embryos were transplanted into recipient sheep. Viable lambs were born, giving rise to the first mammals cloned from adult somatic cells. This was a revolutionary event because it proved that cell differentiation in mammals is not irreversible. A mammary cell nucleus, when placed in the cytoplasmic environment of an unfertilized oocyte, was able to **dedifferentiate** and form an embryo with its hundreds of different cell types; it also was able to form a functioning embryo that had all these cell types in the right place at the right time.

Adult cell cloning is a technique that could potentially be used to clone human beings; consequently, it is highly controversial, and much debate has ensued on how to prevent such use of the technology. However, adult cell cloning has many applications that are less controversial. For example, similar techniques can theoretically be exploited to use primary cultures of mammalian cells as the source of genomic DNA for a developing embryo. In this case, the oocytes would be enucleated and replaced with a nucleus from a primary culture. Primary cultures can be obtained from many adult organs, and they can usually be cultured for numerous passages (moving from one flask to another). This means that gene targeting is possible, because the procedures of homologous recombination, followed by selection of transfected cells that are used with ES cells, can also be used with primary cultures. Nuclei from successfully transfected cells can then be transferred to enucleated oocytes, and embryos can then be induced from the oocytes.

In June 2000, another paper appeared in *Nature* that described successful gene targeting using this strategy. The human α-antitrypsin gene was inserted into the sheep genome at the end of a procollagen gene (Figure 5.4). This particular procollagen gene was selected because it is well characterized in sheep and is constitutively expressed in fibroblast cells (these cells were desirable for a number of reasons). The α-antitrypsin transgene was then expressed as a fusion protein linked to the procollagen protein. Fusion proteins result from translation of mRNA transcribed by adjacent genes controlled by the same promoter. Consequently, they are a composite of two proteins. Transfected cells were introduced into enucleated oocytes, and the resulting embryos were transferred to recipient ewes. Three viable lambs were born; after 1 year, one lamb was induced to lactate, and the milk contained human α-antitrypsin at a concentration of 650 μg/mL.

This technology is not perfect; both cloned and transgenic sheep suffer from several abnormalities, including a tendency to obesity. A large proportion of embryos derived in this way die during gestation, often because of abnormal kidney or brain

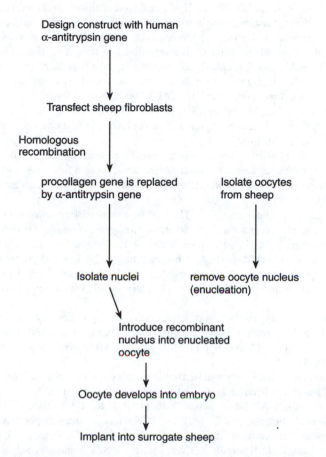

FIGURE 5.4 Introduction of the human α-antitrypsin gene into sheep using homologous recombination and enucleated oocytes.

development. These problems must be solved before the technology can be widely applied to the development of transgenic animals with altered properties for food and agriculture. Nevertheless, the process of adult cell cloning has given biotechnologists an important option in projects aimed at improvement of food animals.

RECOMMENDED READING

1. Sang, H., Transgenic chickens — methods and potential applications, *Trends Biotechnol.*, 12, 415, 1994.
2. Sheldon, B. L., Research and development in 2000: directions and priorities for the world's poultry science community, *Poultry Sci.*, 79, 147, 2000.
3. Devlin, R. H., Transgenic salmonids, in *Transgenic Animals: Generation and Use*, Houdebine, L. M., Ed., Harwood Academic Publishers, Amsterdam, 1997, chap. 19.

4. Dunham, R. A. and Devlin, R. H., Comparison of traditional breeding and transgenesis in farmed fish with implications for growth enhancement and fitness, in *Transgenic Animals in Agriculture*, Murray, J. D., Anderson, G. B., Oberbauer, A. M., and McGloughlin, M. M., Eds., CAB International, Wallingford, U.K., 1999, chap. 6.

5. Muir, W. M. and Howard, R. D., Possible ecological risks of transgenic organism release when transgenes affect mating success: sexual selection and the Trojan gene hypothesis, *Proc. Natl. Acad. Sci. U.S.A.*, 96, 13853, 1999.

6. O'Flynn, F. M., McGeachy, S. A., Friars, G. W., Benfey, T. J., and Bailey, J. K., Comparisons of cultured triploid and diploid Atlantic salmon (*Salmo salar* L.), *ICES J. Mar. Sci.*, 54, 1160, 1997.

7. Devlin, R. H., Risk assessment of genetically-distinct salmonids: difficulties in ecological risk assessment of transgenic and domesticated fish, in *Aquaculture and the Protection of Wild Salmon*, Gallaugher, P., Ed., Simon Fraser University, Vancouver, 2000.

8. Mercier, J.-C. and Vilotte, J.-L., The modification of milk protein composition through transgenesis: progress and problems, in *Transgenic Animals: Generation and Use*, Houdebine, L. M., Ed., Harwood Academic Publishers, Amsterdam, 1997, chap. 70.

9. Murray, J. D. and Maga, E. A., Changing the composition and properties of milk, in *Transgenic Animals in Agriculture*, Murray, J. D., Anderson, G. B., Oberbauer, A. M., and McGloughlin, M. M., Eds., CAB International, Wallingford, U.K., 1999, chap. 14.

10. Eyestone, W. H., Production of transgenic cattle expressing a recombinant protein in milk, in *Transgenic Animals in Agriculture*, Murray, J. D., Anderson, G. B., Oberbauer, A. M., and McGloughlin, M. M., Eds., CAB International, Wallingford, U.K., 1999, chap. 13.

11. Anderson, G. B., Embryonic stem cells in agricultural species, in *Transgenic Animals in Agriculture*, Murray, J. D., Anderson, G. B., Oberbauer, A. M., and McGloughlin, M. M., Eds., CAB International, Wallingford, U.K., 1999, chap. 4.

12. Wilmut, I., Schnieke, A. E., McWhir, J., Kind, A. J., and Campbell, K. H. S., Viable offspring derived from fetal and adult mammalian cells, *Nature*, 385, 810, 1997.

13. McCreath, K. J., Howcroft, J., Campbell, K. H. S., Colman, A., Schnieke, A. E., and Kind, A. J., Production of gene-targeted sheep by nuclear transfer from cultured somatic cells, *Nature*, 405, 1066, 2000.

6 Diagnostic Systems

I. WHY ARE DIAGNOSTIC SYSTEMS NEEDED?

A. OVERVIEW AND GLOBAL PERSPECTIVE

It is impossible to completely eliminate pathogens from the food supply; some pathogens (e.g., *Bacillus cereus*) are common in soil and on vegetation, and food handlers often carry others (e.g., *Staphylococcus aureus*), even if they follow standard hygiene practices. Lapses in worker hygiene or sanitation in food processing plants results in a dramatic expansion of the range of potential pathogens, because of the broad distribution of pathogenic bacteria and viruses. This ubiquity of pathogens makes it essential that the food industry have access to efficient diagnostic tools that allow detection and identification of pathogens.

Diagnostic tools are also essential for clinicians, to help in diagnosis of food-borne illness, which is usually classified as **gastroenteritis** (inflammation of the stomach or, small or large intestines). Many pathogens cause similar symptoms, including severe abdominal pain, diarrhea (watery stools), and vomiting. Identifying the culprit is important because effective treatments vary among the pathogens; for example, antibiotic treatment of *Escherichia coli* O157:H7 is not usually effective, whereas antibiotic treatment of *Clostridium difficile* infections is effective and necessary.

Accurate identification of the cause of food-borne illness is also useful to public health officials attempting to identify the source of an outbreak, and for epidemiologists interested in long-term trends, such as the increasing frequency of *E. coli* O157:H7 infections. Public health officials identify sources by testing the stools of affected people for the presence of pathogens. Once a pathogen has been identified, investigators carefully interview victims and attempt to uncover commonalities. For example, all the victims may have eaten at the same restaurant the previous day or may have attended a recent family reunion. In some cases it is essential to identify all the people affected by an outbreak, so that they can be closely monitored for the eruption of severe symptoms. For example, children under the age of 5 are at risk for developing potentially life-threatening kidney failure after infection by *E. coli* O157:H7; it is essential to identify this pathogen as soon as possible.

Diagnostic speed is also essential in food processing plants. If pathogen contamination can be detected before a product lot leaves the plant, it may be possible to avert an outbreak. However, if enough time elapses before contamination is detected, it may be too late to prevent an outbreak, depending on the nature of the product and its distribution system. Current methods of microbial identification often require 2 to 3 d, which increases the risk that contamination of food by pathogens will be undetected until it is too late.

TABLE 6.1
Emerging Pathogens and Examples of Outbreaks

Pathogen	Outbreak Location	Year	Food Implicated	Number Affected
E. coli O157:H7	Osaka, Japan	1996	Radish Sprouts	>8000
Cyclospora	U.S. & Canada	1996	Raspberries	1465
Cryptosporidium	U.S.	1996	Apple cider	160
Listeria monocytogenes	Nova Scotia, Canada	1981	Sauerkraut	41
Calicivirus	Cruise ship	1993	Fresh-cut fruit	217

Food-borne pathogens and difficulties in detection and identification are a continuing worry for the food industry and public health authorities, partly because the incidence of emerging pathogens (Table 6.1) is increasing. *E. coli* O157:H7, for example, was first implicated as a food-borne pathogen in 1982. Since then, it has caused a number of large outbreaks throughout the world. Because the principal reservoir of this bacterium is in cattle, outbreaks are usually linked to undercooked beef or to contamination of food or water by cattle manure. This pathogen causes a wide range of severity of symptoms, but it is most serious in the elderly, in whom it often leads to severe dysentery, and in the very young, in whom it may cause kidney failure. *E. coli* O157:H7 poses a challenge to developers of diagnostic tests because the **infective dose** (number of bacteria required to cause illness) is extremely small (as low as 50 organisms). Fortunately, it can be differentiated from nonpathogenic strains by its inability to ferment sorbitol. To make matters more confusing, a number of other strains of *E. coli* cause similar illnesses as *E. coli* O157:H7 but are able to ferment sorbitol. Efficient identification of these strains is difficult with current technologies.

Cyclospora is another example of an emerging pathogen. This protozoan is similar to *Cryptosporidium*, which has a much longer history as a cause of food- and water-borne illness. Both are protozoans belonging to the same group (apicomplexans) as *Plasmodium*, the cause of malaria. In 1996, *Cyclospora*-contaminated raspberries imported from Guatemala led to 1465 cases of cyclosporiasis, which is characterized by diarrhea that lasts from 1 to 6 weeks. It is unclear how the raspberries became contaminated, but the most likely explanation is that water used to mix insecticide and fungicide formulations was contaminated with oocytes (dormant egg-like structures produced by apicomplexans) of *Cyclospora*. This organism is difficult to detect in clinical (stool) or environmental (soil or water) samples — unlike most bacteria, it cannot be grown on agar media. It can be detected only by microscopic examination of stained samples. The development of more sensitive diagnostic tests would increase our power to monitor the distribution of this parasite and understand how it contaminates food.

Another reason for concern about food-borne pathogens is that the incidence of food-borne illness is increasing worldwide. This is partly because of increased reporting of gastroenteritis and acknowledgment that many cases formerly ascribed to "stomach flus" are in fact caused by food-borne viruses such as the Norwalk virus (a calicivirus). But the true incidence also seems to be increasing, particularly in industrialized countries. One reason is that consumers demand access to fresh produce year round, which leads to large-scale importation of fruits and vegetables (*cyclospora* in raspberries is an example of this).

Food-borne pathogens are also a major concern in the developing world. Diarrhea caused by pathogenic microbes and parasites is estimated to lead to more than 2 million deaths yearly in the developing world. Solving this immense problem requires development and distribution of inexpensive water purification and sewage treatment systems. In the developed world, water purification is achieved through centralized plants that filter particles, absorb chemicals, and chlorinate the water before sending it through a system of pipes to homes and businesses. Unfortunately, this type of water purification is prohibitively expensive for many communities in the developing world; some researchers recommend that interim, inexpensive systems (e.g., chlorination by homeowners) be implemented to treat water until community-level systems can be built. Also, the World Health Organization (WHO) is currently setting up improved systems for reporting of food- and water-borne diseases and outbreaks, as well as databases focusing on the distribution of the important food-borne pathogens throughout the developing world. The development of cheap, accurate diagnostic systems would greatly aid this important goal.

B. DIAGNOSTICS AND HAZARD ANALYSIS CRITICAL CONTROL POINTS

Food companies have traditionally relied on **end product** inspection and testing to ensure that the food that leaves the factory does not contain an excessive load of nonpathogenic and pathogenic microbes. Theoretically, 100% of products can be visually inspected, but human frailties (e.g., distractability, boredom) decrease the efficiency of inspection. Furthermore, many microbial defects cannot be detected by visual inspection. Hence, destructive sampling of end products is often required. In principle, this is a simple process — a defined proportion (subset) of products is removed at the end of the production line and the level of microbial contamination is assessed. Typically, microbes are put into suspension by grinding each sample in a Stomacher™ or similar apparatus. A dilution series of this suspension is then plated onto an agar medium, and after a suitable period (24 h at 37°C), bacterial or fungal colonies are counted. In some cases, enrichment and selective media are used to detect specific pathogens (e.g., *Salmonella*) (Figure 6.1). This technology can be quite powerful; in some cases, cultural methods can distinguish closely related strains. For example, *E. coli* O157:H7 can be distinguished from nonpathogenic strains through the use of selective and differential media (differential media produce a detectable change in appearance/color of either the agar medium or the colonies when the target bacterium is present).

Microbiologists have also devised clever techniques to increase the speed and accuracy of culture-based diagnostics. For example, de Boer and colleagues in the Netherlands developed a medium for the detection of *Salmonella* that is selective and differential and makes use of *Salmonella*'s motility to help identification. *Salmonella* are visible as a migrating pink zone within a green medium. This medium allows identification of *Salmonella* 1 d earlier than usually possible.

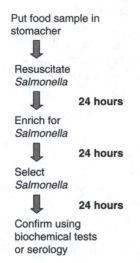

FIGURE 6.1 Detection of *Salmonella* in food using cultural methods.

Cultural methods such as this are a reliable indicator of microbiological quality in food, but there are many disadvantages of end product microbial analysis. Financial concerns sometimes make it difficult to test an adequate number of samples, because cultural methods are labor and materials intensive. If too few samples are assessed, the risk of missing contaminated product increases. This risk also increases if contamination occurs sporadically, rather than on a regular basis. Also, cultural methods are slow, particularly if the aim is to identify specific pathogens. For example, conventional detection of *Salmonella* (see Figure 6.1) requires that a food sample be incubated in three successive media for a total time of 72 h. Additional tests that confirm the presence of *Salmonella* are then required. This confirmation can be **biochemical**; the colonies that are presumed to be *Salmonella* are typically inoculated into a series of media containing a range of different carbon sources (e.g., glucose, sucrose, mannitol, proline, etc.). The bacteria respiring or fermenting (depending on the diagnostic system) these carbon sources produce a "pattern of utilization" that can be used to identify the bacteria. Unfortunately, the growth period required for these tests further adds to the delay in assessment of contamination. Antibody-based tests, in contrast, can confirm the presence of *Salmonella* immediately. However, if negative, they do not provide any other information that could be used for identification, unlike biochemical tests.

Resuscitation and selective media for cultural methods of identification are necessary because it is very difficult to detect small numbers of specific microbial species, especially if there is a large background of nonpathogenic bacteria and fungi. A food that has just been cooked at a high temperature is unlikely to have a large background, but many other foods (e.g., yogurt and cheese) typically harbor large populations of useful or harmless bacteria and fungi.

Because of the time, labor, and financial constraints of end product testing, and because of its intrinsic inefficiency (in most cases, it is impossible to take enough samples to achieve a satisfactory level of protection), most government regulators

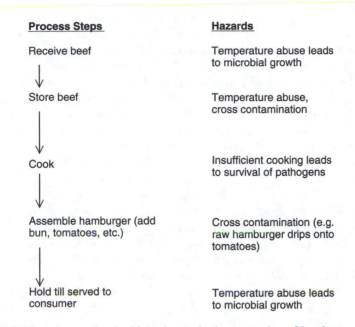

FIGURE 6.2 Hazards associated with each step in the processing of hamburgers at a fast-food restaurant. For each step, a strategy would be devised for control of hazards and for monitoring. The aim of monitoring is to ensure that control is achieved on a day-to-day basis.

and standardization organizations (e.g., International Organization for Standardization [ISO]) agree that **hazard analysis critical control points (HACCP)** systems can increase food safety in production lines. HACCP processes aim to **prevent contamination** of food by pathogens, rather than simply detect contamination after it has happened.

HACCP analysis begins with a thorough understanding of each step in the process. Next, the hazards associated with each step are identified (Figure 6.2). Although devised for industrial food processing plants, HACCP principles can also be applied to domestic, restaurant, catering, or even agricultural processes. Common hazards include contaminated raw materials (e.g., *Salmonella* and *Campylobacter* in raw chicken), cross contamination (e.g., contamination of vegetables by contact with raw meat), or temperature abuse that allows microbial growth. For example, if foods are not refrigerated during holding times, bacteria such as *S. aureus* or *B. cereus* could grow and produce toxins. The HACCP plan then outlines steps that can be taken to control these hazards. In each process, certain steps are crucial, in terms of food safety. They are designated as **critical control points (CCPs)**. CCPs often consist of heat treatments that eliminate or reduce the number of microbes. Inadequate performance of a CCP (e.g., heating at too low a temperature) could lead to potential contamination of the food and potential transmission of pathogens to the consumer.

Once the hazards and CCPs in a process have been identified, they must be **monitored**. This can often be done by checking the operating parameters (e.g., temperature) of processing machinery. It is often useful to monitor control points using microbial assays, aimed at detection of specific pathogens, or assessment of overall numbers of contaminating microbes (microbial load). However, time and labor constraints associated with traditional cultural methods may make it difficult to effectively monitor the efficacy of a HACCP system. For example, HACCP plans for a poultry processor could include periodic tests for the presence of *Campylobacter* in chicken parts. This is statistically feasible in most regions because of the high frequency of *Campylobacter* contamination in chickens entering a processing plant. However, primarily because of the time involved in testing for *Campylobacter*, monitoring is usually based on total plate counts, which estimate the total numbers of contaminating microbes. Unfortunately, total plate counts are unlikely to be an accurate indicator of *Campylobacter* contamination. However, if rapid diagnostic tests were available, they could be used to effectively monitor levels of *Campylobacter* and other pathogens. This is one example of the need for rapid diagnostic tests in the food industry. Ideally, detection of pathogens should take less than 24 h, which allows food processors to take corrective action if contamination occurs. Some biotechnology companies are also attempting to develop **on-line** methods of microbial monitoring that could be used to **continuously** monitor microbial parameters; these will be discussed later in this chapter, in the context of **biosensors**.

HACCP programs should always undergo a **verification** procedure. Diagnostic methods are useful here because they can tell HACCP developers if the program is successful in preventing food contamination by pathogens. Regulatory authorities sometimes require HACCP verification as part of the certification process.

C. Nonpathogen Diagnostics

Most commercial diagnostic tests are aimed at the detection of pathogens, but a number of tests target nonpathogenic microbes, especially spoilage agents (Table 6.2). Food processors often incur heavy losses through microbial spoilage. Many of the topics discussed above in the context of pathogens (outbreak investigation, HACCP) are also applicable to food spoilage. For example, if a large lot of prunes is spoiled by growth of a xerophilic (dessication-loving) fungus, the company involved needs to quickly identify the fungus responsible and the factors that allowed it to contaminate and grow in the product. Then, a HACCP plan can be developed to prevent further incidents.

Mycotoxin contamination is also a great worry to the food industry. Some of the fungi that commonly contaminate food crops (e.g., *Aspergillus flavus* and *Fusarium* spp.) produce potent mycotoxins (aflatoxins and tricothecenes, respectively) that are a significant public health hazard. Consequently, governments throughout the world have established allowable limits for several mycotoxins in food. For example, the U.S. Food and Drug Administration (FDA) allows 20 μg/kg (ppb) contamination of aflatoxins in food (this is primarily aimed at nuts and nut products). Allowable levels for animal feed are substantially higher (e.g., 300 ppb in animal feed for cattle bound for slaughter). Due to a lack of universal agreement on safe levels of aflatoxins, allowable limits vary among countries. In Canada, for example, a limit is set for nuts destined for human consumption (15 ppb) and for all animal feeds (20 ppb). In the European Union (EU), draft regulations call for extremely low limits (6 ppb)

TABLE 6.2
Diagnostic Applications Relevant to Food Production and Processing

Goal of Diagnostic	Example	Current Technology	Upcoming Developments
Assess microbial contamination	Bacteria on carcasses	Plate counts	Flow cytometry, impedimetry, biosensors
Identify pathogen	E. coli O157:H7	Selective media, immuno-assay, biochemical, DNA tests	Bioluminescence, biosensors
Identify spoilage agent	Yeasts	Biochemical identification, immuno-assay	DNA tests
Investigation outbreak	Tracing pathogen strains to source	Biochemical tests, immuno-assay	DNA tests (RAPD,[a] RFLP,[b] pulsed-field electrophoresis, 16S rRNA[c] sequence analysis)
Monitor microbial growth	Yeast fermentation	Plate counts	Biosensors, flow injection, flow cytometry
Monitor hygiene	Work surfaces in processing plant	Plate counts/ATP using bioluminescence	Increased specificity (combining hygiene tests with pathogen identification)
Process monitoring	Glucose consumption in bioreactor	Analytical chemistry, biosensors	More biosensors
Detect toxin residues	Aflatoxin	Immuno-assay	Biosensors
Detect pesticide residues		Immuno-assay	
Detect residues linked to allergy/intolerance	Gluten	Analytical chemistry	Antibody based
Detect adulteration	Pork in all-beef products	Detection of unwanted protein	Detection of unwanted DNA
Detect transgenic crops in food	Herbicide-resistant soybeans	DNA test or immuno-assay	Improved sensitivity and reliability of DNA-based tests

[a] RAPD, randomly amplified polymorphic DNA.
[b] RFLP, restriction fragment length polymorphism.
[c] rRNA, ribosomal RNA.

of aflaxotins in nuts for human consumption. Evidently, we need efficient diagnostic tests to monitor levels of aflatoxins and other mycotoxins, especially considering the scope of international trade in commodities (e.g., wheat, peanuts) that are susceptible to mycotoxin contamination. Currently, most mycotoxin detection is done through antibody-based tests or analytical chemical techniques (e.g., high-performance liquid chromatography [HPLC]).

Food adulteration is an ongoing problem in the food industry, particularly in processed meats. For example, unscrupulous companies incorporate pork or other meats into products labeled "all beef." Diagnostic tests that detect such adulteration are commercially available, and are usually based on the ability of antibodies to differentiate between muscle proteins from different species. Sometimes, processors inadvertently allow contamination of food by unwanted ingredients. For example, a food that is labeled "gluten-free" may become contaminated by gluten-containing wheat introduced through one of the food ingredients. Because the major market for gluten-free products is people who have gluten intolerance, the presence of contaminating wheat is clearly "intolerable." Diagnostic tests have the potential to reduce the risk of this happening.

In 1999, polymerase chain reaction (PCR)-based diagnostic tests were used to detect **Starlink** corn in food. This cultivar of transgenic corn had been approved for feed use by the FDA, but not for food. The ensuing controversy, which reverberates to this date, demonstrated the difficulties associated with segregation of transgenic crops, as well as the urgent need for effective and cheap diagnostic tests for transgenic crops. Reliable tests to detect transgenic crops will increase in importance in the future, as the EU and other countries decide on allowable limits of transgenic crops in food. Current proposals call for a 1% limit for transgenic residues in food. If the level is greater than 1%, the food label will be required to indicate that the food contains transgenic (genetically modified) organisms.

Finally, diagnostic techniques are used in breweries, cheese factories, and dairy plants to identify useful microbes. These processes use specific strains of yeasts and lactic acid bacteria, and the presence of the wrong strain can adversely affect product quality. This is a challenging problem for developers of diagnostic methods, because many strains of lactic acid bacteria are physiologically and genetically very similar, making it difficult to differentiate among them using cultural methods. Yeast strains used in the brewing and wine-making industries are similarly difficult to differentiate. DNA-based methods are increasingly used in this context.

II. DIAGNOSTIC BIOTECHNOLOGY

A. SCOPE

Biotechnology can dramatically reduce the time and labor required to detect and identify microbes. The following applications of biotechnology to diagnostics are discussed in this chapter.

- **Gene probes** to detect pathogenic microbes
- **PCR** to detect contaminating pathogens or the presence of transgenic crops
- **DNA chips** and **micro-arrays**
- **Antibodies** in diagnostic systems

- **Bioluminescence** to monitor **hygiene** and contamination by specific microbes
- **Biosensors** in the food industry

Antibody-based assays that facilitate detection are available for most major pathogens and are particularly important for identification of food-borne viruses. Nucleic-acid-based kits for the detection of *Salmonella, Campylobacter, E. coli, S. aureus*, and *Listeria monocytogenes* are also commercially available. Nucleic-acid-based techniques are also becoming increasingly important in the typing (strain identification) of bacteria.

B. NUCLEIC ACID PROBES

DNA and RNA probes can be highly specific; they can be designed so that they hybridize only to the target species or strain. All that is required is identification of a DNA sequence that is unique to the target microbe. The use of nucleic acid probes is well established in molecular biology (see Chapter 2, Section V.H.3). Biotechnologists have long recognized the potential of probes in diagnostic systems, and this potential is now starting to be realized. However, we needed to transform traditional hybridization procedures such as southern blotting, which are labor intensive and "fussy," into more user-friendly and safer techniques. For example, when DNA probes were first used in molecular biology, they were labeled with radioactive isotopes (e.g., ^{32}P). This works well in a research laboratory, where the problems associated with the use of radioisotopes (occupational safety, disposal of wastes) can be safely and adequately addressed. However, few quality control laboratories in food companies have the necessary equipment for radioisotope work. The requisite training of personnel in the use of radioactive compounds is an additional problem for food companies. For these reasons, **nonradioactive** detection systems are used in commercial versions of probe-based assays.

For a diagnostic test to achieve widespread use in the food industry, it must also be easy to use. Probe-based tests have achieved this goal through immobilizing probes to inorganic supports (**dipsticks**) that allow the user to easily manipulate the probe (e.g., wash off unhybridized DNA) without damaging or losing it. This is referred to as **solid-phase** hybridization; other hybridization systems (e.g., solution-based) are also possible.

The principle behind the use of DNA probes is quite simple (see Figure 2.15). Short single strands of DNA that are *complementary* to genes present in a pathogenic microbe are synthesized. The food sample must then be treated so that any microbial cells are lysed, releasing their DNA. Microbial DNA is then treated to convert it from double strands to single strands, and the probe is added. **Hybridization** (annealing of complementary strands) then occurs between the single-stranded DNA probe and single-stranded DNA released from pathogenic microbes present in the food. Probe DNA that has not hybridized must then be removed, usually by washing the sample, and the presence of hybridized DNA probes can then be detected.

To understand how the above process works, we can examine the Gene Trak™ system for detection of *Salmonella* in food samples (Figure 6.3). This system uses

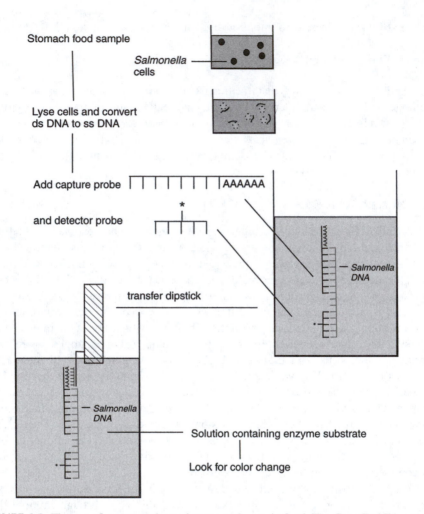

FIGURE 6.3 The use of a gene probe to detect a pathogen in food. The Gene Trak™ system for the detection of *Salmonella* is used as an example. The capture and detection probes anneal to different regions of the ribosomal DNA gene of *Salmonella*. The dipstick is used to remove the capture probe–*Salmonella* DNA–detection probe complex. The complex is then placed in the appropriate solution that will reveal the presence of the detection probe. In earlier versions of this system, detection was based on fluorescein in the probe binding to antifluorescein antibodies, which in turn bind to enzyme-linked antibodies. The enzyme then catalyzes formation of a colored end product.

a **capture probe** and a **detection probe**. These probes hybridize to different regions of the genes coding for ribosomal RNA (rRNA) in *Salmonella*. rRNA genes were selected because DNA probes work best if the target DNA is highly copied (i.e., many copies of the gene are present in the cell). Around 5000 copies of rRNA genes are present in *Salmonella*. The large amount of rRNA in a bacterial cell is also important because most of the binding between probe and target is through rRNA binding to the probe. It is also essential that the DNA probes be specific to *Salmonella*;

they must not bind to related bacterial species such as *E. coli*. Specific rRNA sequences were found by comparing the rRNA sequences of a large range of bacteria and selecting sequences that are specific to *Salmonella*.

So, how does Gene Trak work? Enrichment of *Salmonella* is required, in order to produce detectable amounts of *Salmonella* DNA. However, time savings are achieved after enrichment, because time-consuming growth in selective media is not required. Thus, Gene Trak can detect *Salmonella* after 48 h (sometimes after 24 h), at least 1 d sooner than through conventional methods. After enrichment, bacteria in the food sample are lysed using NaOH, and the capture and reporter probes are added.

Note that the capture probe contains a **poly A sequence** — a sequence of adenine residues. Note also that the reporter probe is covalently bound to fluorescein. Both probes hybridize to complementary regions of DNA released from *Salmonella* cells. The next step requires the removal of **unbound capture and reporter probes**. This is done by adding a dipstick that contains strands of **poly T** (a sequence of thymine residues). The poly T strands hybridize to the poly A tails on the capture probe. The crucial point is that if the capture probe has hybridized to a strand of *Salmonella* DNA, then that strand of *Salmonella* DNA has probably also hybridized to the reporter probe. Therefore, when the dipstick is removed from the solution, it will carry with it: (1) unhybridized capture probes and (2) strands of *Salmonella* DNA that have hybridized to both capture and reporter probes.

The presence of unhybridized capture probe on the dipstick is irrelevant, because detection of the captured *Salmonella* strand is based on the presence of fluorescein in the reporter probe. Unhybridized capture probes will not contain fluorescein and will not be detected. The final step involves detection of fluorescein. Several alternatives are possible; the Gene Trak method uses an antibody-based system. Antibodies that bind specifically to fluorescein are added. The antibodies are covalently linked to an enzyme (horseradish peroxidase) that can convert a synthetic substrate (chromogen) into a colored end product. Thus, *Salmonella* is detected through a change in color of the solution.

The principle advantage of this system is the reduced time required for positive identification of *Salmonella*. Another advantage is the assay's specificity; several bacteria found in food (*Citrobacter, Enterobacter, Escherichia, Klebsiella,* and *Proteus*) are very closely related to *Salmonella* and are often able to grow in media that are "selective" for *Salmonella*. Furthermore, some isolates of *Salmonella* are atypical and do not produce the usual colony characteristics of *Salmonella* on differential media. The main problem with the Gene Trak system is that enrichment of *Salmonella* is still required, making "instant" identification of *Salmonella* impossible. However, enrichment is in a sense useful, because it makes it unlikely that dead *Salmonella* will be detected, unless they are present in high numbers. This is important, because dead *Salmonella* do not pose a health hazard, and the detection of dead *Salmonella* constitutes a false positive that may be wasteful of a company's resources. Because it is still possible that dead bacteria could lead to a false postitive, positive probe results are usually confirmed through traditional cultural methods.

Similar probe-based tests are available from at least one other company (Accuprobe®) for a range of clinically and food-relevant pathogens.

C. EXPLOITING THE POLYMERASE CHAIN REACTION

PCR has the potential to allow extremely rapid (several hours) identification of pathogenic microbes. As explained in Chapter 3, PCR can quickly amplify specific sequences of DNA and can theoretically amplify a single copy of a DNA sequence of a sample. Thus, PCR has the potential to realize a long-sought goal of food diagnostics: the detection of a single pathogenic microbe in a 25-g food sample within a few hours. Unfortunately, the use of PCR in food samples is technically challenging; food usually contains a large background of plant or animal DNA that might interfere with the PCR, as well as chemicals that inhibit the PCR, sometimes by directly inhibiting DNA polymerase. These problems have been addressed for certain foods, through dilution of inhibitors or concentration of bacterial cells. A number of strategies are possible for concentration, including centrifugation of liquid samples or liquefaction of solid foods, followed by an **affinity** separation technique. One affinity technique involves covalently binding pathogen-specific antibodies to magnetic beads. The magnetic beads are then added to a suspension of a food sample, and, after incubation, the beads (and any beads bound to pathogens) are separated magnetically. The presence of low numbers of pathogens can then be revealed through PCR. Magnetic beads have also been used to increase the efficiency of *E. coli* O157:H7 detection using cultural methods.

PCR-based diagnostic kits for detection of food-borne pathogens (e.g., *L. monocytogenes*) are now commercially available. One company (Biotechon) has developed a system that allows detection of amplified products during amplification. This is done through the use of fluorescent probes that are added to the PCR mixture. One probe is a "donor" compound and another probe is an "acceptor." The donor compound absorbs light energy of a specific wavelength and transfers it via resonance energy to the acceptor compound. This results in emission of light at a higher wavelength than the original light, which is detected electronically. This fluorescence happens only when the donor and acceptor compound are in close proximity, as when both probes are hybridized to amplified DNA specific to the target bacterium.

It is too early to tell if the food industry will embrace PCR-based diagnostics. Considerable expenses are involved in the acquisition of thermal cyclers and the training of personnel in their use. Also, these techniques need to be extensively **validated**, through comparison of their ability to detect low levels of contamination in food to those of conventional culture-based methods. One recently published study (Bellin et al., 2001) demonstrated successful use of LightCycler™ to detect a food-borne pathogen (strains of *E. coli*, including O157:H7, that produce Shiga toxins). However, this study detected the bacterium from pure cultures, not from food.

The ability of PCR to detect **dead** organisms is a problem. Tiny amounts of DNA released from pathogens killed by heat treatments, for example, would be amplified by PCR. This is less of a problem if enrichment precedes PCR, but that extends the procedure to at least 1 d. In some situations the presence of **live** *or* **dead** pathogens is an important indicator of food safety. For example, if PCR detects *Clostridium botulinum* in a food sample, it is a cause for concern whether the bacterium is alive or dead, because dead bacteria may have produced botulinum toxin before dying. The toxin could then persist in the food, causing a potential for serious illness.

One final note on PCR: several alternative amplification systems are actively under development. For example, nucleic acid sequence based amplification (NASBA®) uses three viral enzymes to amplify either RNA or DNA targets. The main advantage of this technique is that it is **isothermal** (occurs at a constant temperature), avoiding the need for expensive thermal cyclers. So far, NASBA diagnostics have mainly been aimed at identification of viruses, but applications for the identification of *Campylobacter* and *L. monocytogenes* are also under development.

D. DNA Chips and Micro-Arrays

Few techniques have caused as much excitement among microbiologists as DNA chips and micro-arrays. Both are intrinsically miniaturized extensions of conventional tests of nucleic acid hybridization. Imagine a membrane that is used for a **dot blot** (Chapter 2, Section V.H.3). A sample of DNA is placed onto the membrane; labeled probe is then added; and hybridization (if present) is detected. Now consider a slightly different scenario: a number of oligonucleotides (each representing a different gene) are immobilized onto separate points of a membrane. A bacterial culture is then exposed to a chemical that results in labeled messenger RNA (mRNA) transcripts. The bacteria are lysed and then placed on each oligonucleotide on the membrane. After washing, hybridization between labeled mRNA and immobilized oligonucleotides can be detected. This **macro-array** of oligonucleotide probes allows simultaneous detection of expression of a number of genes.

Now imagine a similar array of immobilized oligonucleotides on a 1×1 cm square on a glass microscope slide. Further miniaturization can be achieved with tiny wells etched into a circuit board. Oligonucleotides are then immobilized onto these wells, and the pattern of probe immobilization (i.e., which probes are loaded into which wells) can be controlled electronically. These are often referred to as **DNA chips** or **micro-arrays** (arrays on glass slides are also commonly referred to as micro-arrays). The great advantage of these systems is that they require only small amounts of resources. DNA chips can also be developed into **laboratories in a chip**, wherein an experimental routine, perhaps involving heating of reagents and mixing of several different chemicals, or even electrophoresis, can occur at a micro scale. If successfully applied to diagnostic testing, micro-arrays embedded into silicon chips could allow efficient testing of large numbers of samples and could even be used to amplify sample DNA using PCR. Theoretically, with one test the investigator could detect the presence of DNA of many different pathogens in a food sample.

Micro-arrays are currently commercially available for assessment of **global gene expression** (the total pattern of gene expression by a cell). Up to 8000 genes can be simultaneously tested for hybridization, allowing investigators to dissect changes in gene expression after different experimental treatments. This type of research has many potential applications in food microbiology (e.g., determining pattern of gene expression in a pathogenic bacterium triggered by exposure to acidic preservatives). Micro arrays also have great potential as diagnostic systems, but this is currently in the research and development phase.

E. ANTIBODY-BASED DIAGNOSTIC SYSTEMS

1. Applications of Antibodies to Diagnostics

Antibodies are proteins produced by the mammalian immune system. Their biological function relies on their ability to **bind specifically** to proteins and other compounds. Antibodies are produced by **B-cells** when foreign compounds invade the body of a mammal. For example, the presence of *S. aureus* in the blood will trigger the production of antibodies that bind specifically to proteins on the surface of *S. aureus* cells. Live microbes are not essential for this process; microbial products (e.g., exotoxins) or components (e.g., cell wall fragments) also provoke the production of specific antibodies. There are five classes of antibodies — IgA, IgD, IgG, IgM, and IgE. IgG is the most frequently used in diagnostic tests; it is a y-shaped protein made up of a **constant** region (the stalk of the y), which is the same in all IgG molecules, and a **variable** region (the top of the y), which is a mixture of regions that are similar among different molecules and regions that are specific to each molecule. Thus the variable regions give antibodies their specificity of recognition. The molecule (usually a protein) that is recognized by an antibody is an **antigen**. Antibodies exhibit variable levels of **specificity** (ability to bind to one antigen and not to others) and **affinity** (a measure of the strength of antigen–antibody binding). In a diagnostic test, it is important that the antibodies that are used have high specificity (thus avoiding **cross reactions** with similar antigens) and high affinity, which allows the use of low concentrations of the antibody. Cross reactions can lead to expensive false-positive reactions; for example, if an antibody is used in a diagnostic test to detect *Vibrio vulnificus*, it must not bind to other, nonpathogenic species within the *Vibrio* genus.

Antigens have numerous **epitopes** representing different regions of antibody specificity. When a pathogen invades the human body, the immune system responds by producing antibodies to proteins and other **immunogenic** (capable of provoking an immune response) compounds of the pathogen. A number of antibodies are also produced against each immunogenic molecule, each binding to different regions of the molecule.

Antibodies, as part of the normal immune response (Figure 6.4), are essential in the protection of the human body from invasion by pathogens. Antibodies increase the ability of immune cells to engulf and destroy bacteria, stimulate the production of chemicals that kill fungi, and **neutralize** viruses by coating viral proteins that are required for entry into host cells. Furthermore, antibodies neutralize some toxins; this is why people who have deep skin wounds are given an injection of antibodies that specifically bind to tetanus toxin. This binding prevents the toxic action of any toxins produced by *Clostridium tetani*, a microbe that is able to invade and colonize deep wounds.

The specific binding properties of antibodies are invaluable to biotechnologists, particularly as a key part of diagnostic systems. For example, antibodies against flagellar proteins of *Salmonella* are used in the food industry and in hospitals to definitively identify *Salmonella*. The process is similar to that described in

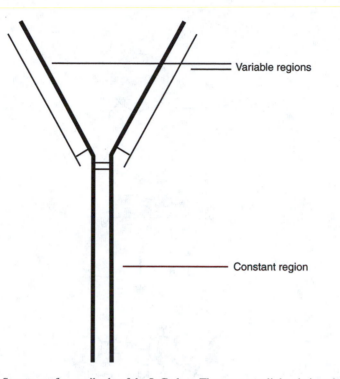

Variable regions

Constant region

FIGURE 6.4 Structure of an antibody of the IgG class. There are two light-chain polypeptides and two heavy-chain polypeptides (bolded). The protein is stabilized by disulfide links between polypeptides. The variable regions are responsible for antigen binding and give the antibody its specificity of binding. Each variable region can bind to one antigen molecule.

Figure 6.1, except that the presence of *Salmonella* is confirmed by an antibody-based test instead of a biochemical test. **Latex agglutination** tests are popular (Figure 6.5). Antibodies that bind specifically to flagellar proteins of *Salmonella* are first adsorbed onto latex beads. A drop of solution containing these beads is then added to a drop from a culture that has been tentatively identified as *Salmonella*. If *Salmonella* is present, antibodies on the latex beads will bind to *Salmonella* cells. This will create a network that results in the agglutination of *Salmonella*. This agglutination is readily visible as a suspension of particles in the drop.

Antibody-based systems have been developed for a number of important food-borne pathogens such as *E. coli* O157:H7, the strain of *E. coli* that causes hemorrhagic symptoms following infection. Many of these systems do not rely on agglutination for detection; other strategies, such as enzyme-linked immunosorbent assays (**ELISA**s), are often more sensitive (i.e., they can detect smaller numbers of pathogens or smaller amounts of toxins) and are quantitative (i.e., they allow measurement of the numbers of pathogens or the concentration of toxin).

A number of different strategies are possible with ELISA systems. In **sandwich** ELISAs (Figure 6.6), the antibody is coated on the bottom of wells in plastic multi-well plates (96-well plates are often used). This is the **capture** antibody. For example,

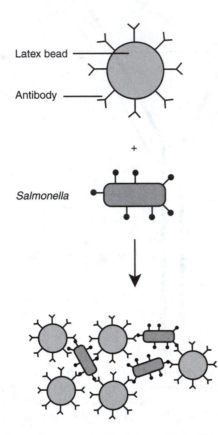

Latex bead

Antibody

+

Salmonella

FIGURE 6.5 Latex agglutination assay. Antibodies specific to *Salmonella* are bound to latex beads. These antibodies link *Salmonella* cells to a network of latex beads, causing agglutination (clumping).

antibodies that bind specifically to **aflatoxin** (a potent **mycotoxin**) could be coated onto a 96-well plate. Samples of cereal grains (after grinding and other pretreatments) could then be added to the wells of the plate. The capture antibody would essentially capture any aflatoxin present in the sample. The plates would then be thoroughly rinsed so that the only substance remaining from the grain sample would be aflatoxin caught by the capture antibody. A second antibody would then be added (the **detection** or **reporter** antibody) that binds to a different epitope of the antigen than the capture antibody. This antibody would be covalently bound to a compound that allows detection. For example, it could be linked to an enzyme such as horseradish peroxidase or other enzyme that can convert a substrate into a visible end product. Not surprisingly, then, the next step in the ELISA is to wash off any detection antibody that has not bound to aflatoxin. A chromogenic substrate is then added, to reveal detection antibody–aflatoxin–capture antibody complexes. The amount of colored end product is directly proportional to the amount of aflatoxin present in the original grain sample.

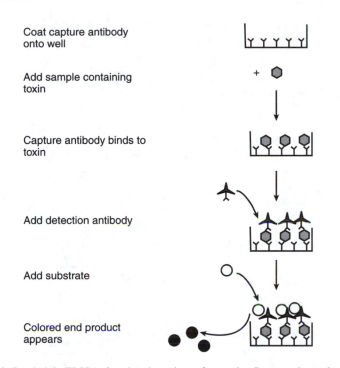

Coat capture antibody
onto well

Add sample containing
toxin

Capture antibody binds to
toxin

Add detection antibody

Add substrate

Colored end product
appears

FIGURE 6.6 Sandwich ELISA for the detection of a toxin. Increased numbers of toxin molecules will lead to increased formation of colored end product.

Another popular strategy is **competitive** ELISA (Figure 6.7). As an example, consider a competitive ELISA aimed at detecting *Salmonella enteritidis*. Instead of immobilizing a capture antibody to the wells of an ELISA plate, the **target antigen** is attached to the plate. The food sample, after enrichment, is then added to the wells, and antibodies against the target antigen are added at the same time. The immobilized antigen and any *S. enteritidis* cells in the food sample will compete for binding sites on the antibodies. If *S. enteritidis* is present, fewer antibodies will bind to the immobilized antigen. After the wells are washed, a second antibody is added; this antibody binds to any antibody that is in a particular **class**. If the first antibody was mouse IgG, the second antibody would bind to the constant region of any mouse IgG molecule. The second antibody is also covalently bound to an enzyme that is able to catalyze a reaction leading to a colored end product. This allows visualization of the amount of immobilized antigen that bound to the first antibody. Thus, low levels of color occur when *S. enteritidis* is present in the original food sample.

Competitive and sandwich ELISA systems are available for the detection of a range of food-borne pathogens; they typically are able to detect 10^3 to 10^5 cells per milliliter. Thus, if a 10-g food sample containing less than 10^4 cells of a pathogen is suspended into 100 mL of buffer, and then tested by ELISA (or an agglutination test), the pathogen will not be detected. Clearly, this is not sensitive enough for most

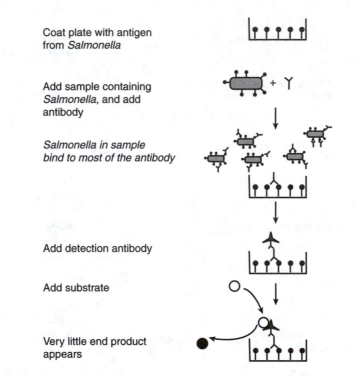

Coat plate with antigen from *Salmonella*

Add sample containing *Salmonella*, and add antibody

Salmonella in sample bind to most of the antibody

Add detection antibody

Add substrate

Very little end product appears

FIGURE 6.7 Competitive ELISA for the detection of *Salmonella*. As the density of *Salmonella* increases, less antibody is available to bind to the antigen coating the wells. This leads to less binding of detection antibody to antibodies in the wells and fewer colored end products.

food applications. Consequently, enrichment is usually required before the use of ELISA or other immunological assays.

2. Manipulating the Immune Response

As we have seen, antibodies are part of the mammalian immune response to the presence of foreign antigens. How can we manipulate this response in order to obtain antibodies that can be used in a diagnostic test? Basically, there are three approaches to the production of antibodies: **polyclonal, monoclonal,** and **recombinant antibodies**. Before 1975, the only practical method was to inject microbes or purified components of microbes into animals such as rabbits or mice. After a prolonged period (at least 3 weeks), antibodies could be purified from blood of the animals. This strategy has several disadvantages. Many animals are required for large-scale antibody production, which is expensive. Even with purified compounds (e.g., a specific protein antigen found on the surface of a microbe), this approach yields an antiserum composed of a mixture of antibodies that bind to different epitopes of the antigen. Hence, these are called **polyclonal antibodies**. Some of these antibodies may have high affinity and specificity, but others may have less ideal binding properties. Antibodies with a low degree of specificity in a diagnostic assay would give a large proportion of false positives, whereas antibodies with low affinity would

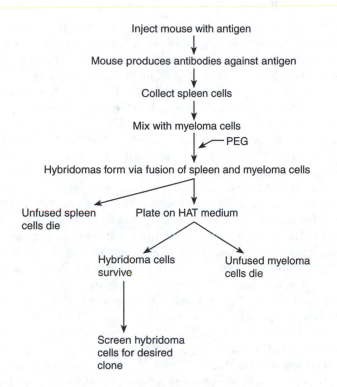

FIGURE 6.8 Procedure used to isolate hybridoma cells producing monoclonal antibodies. (PEG = Polyethylene glycol; HAT = hypoxanthine aminopterin thymidine).

be less sensitive (i.e., unable to detect low numbers of pathogens). Finally, antisera are very susceptible to variation from animal to animal. This causes inconsistency in the ability of the antibodies in the antiserum to detect their target. Despite these problems, polyclonal antibodies are sometimes used to develop diagnostic assays.

Fortunately, **monoclonal antibodies** can also be obtained against most antigens. Unlike polyclonal antibodies, which are derived from several lineages (clones) of B-cells, monoclonal antibodies, as their name suggests, arise from only one clone. This means that the antibodies produced by each cell of the clone are identical and bind to the same epitope of the antigen. Thus, there is little batch-to-batch variation in the binding efficiency of monoclonal antibodies. Another advantage is that, once established, monoclonal antibodies are produced by cells grown in culture. Antibodies can be harvested from spent (i.e., used) media of cell cultures and then purified by a number of methods (e.g., column chromatography). Therefore, the production of large amounts of monoclonal antibodies does not require the use of large numbers of animals nor the long incubation times needed to produce polyclonal antibodies.

However, animals must be used to isolate clones of B-cells that produce the desired antibody (Figure 6.8). Usually, a mouse is injected with a compound or mixture of compounds derived from the target organism. If the target is a pathogen, one might use purified flagellar protein or another protein that is present on the surface of the organism. If the aim is to develop an assay for toxin detection, the toxin itself can be injected, as long as the toxin does not adversely affect the mice. If mice are vulnerable to the toxin, the toxin can be inactivated by chemical treatment (e.g., denaturation with formaldehyde). With luck, this will result in loss of the biological activity of the toxin but maintenance of its three-dimensional structure. The binding of antibody to antigen is based on structural interactions, and conservation of this structure is essential.

Several weeks later, the mouse is killed and its spleen is removed. Spleen cells are then mixed with **myeloma** cells — tumor cells that are able to grow indefinitely in cell cultures. Polyethylene glycol (PEG) is then added to the mixture. Recall from Chapter 4, Section II.D.2, that PEG induces the fusion of plant protoplasts. It has a similar effect on animal cells, and cell fusion occurs in the mixture of myeloma and spleen cells. The aim at this stage is to fuse **B-cells** from the mouse spleen with myeloma cells. The resulting cells (**hybridomas**) will be able to produce antibodies (derived from the B-cells) and grow indefinitely in culture (derived from the myeloma cell). Unfused B-cells cannot be used directly to produce monoclonal antibodies, because they are not **immortal**; they die after several weeks in culture.

Undesirable fusion products also occur. Myeloma cells fuse with myeloma cells and B-cells fuse with B-cells. Desired hybridoma cells are isolated using an elegant selection scheme. The myeloma cells used are mutants lacking a crucial enzyme that normally allows cells to import hypoxanthine, a compound that can be converted into nucleotides required for DNA synthesis. However, this enzyme is present in B-cells. After fusion, the cells are grown in a medium that forces them to depend on hypoxanthine in the medium as a source of nucleotides. The only cells that survive and grow will be B-cells and hybridomas. B-cells that do not fuse with myeloma cells do not survive for very long, because of their limited ability to divide.

The surviving hybridomas are diluted and seeded, so that individual cells are placed in a separate well (e.g., one cell per well of a 96-well plate). Each hybridoma cell then divides, giving a suspension of identical progeny cells. Thus, a series of clones is produced, each of which is able to divide indefinitely in culture and produce only one kind of antibody. This is a crucial point: each B-cell produces only one type of antibody. Therefore, if a clone is transferred into a larger culture vessel, the cells will continue to divide and produce **monoclonal** antibody indefinitely. These antibodies have a single specificity. This means that all antibodies produced by a particular clone recognize and bind to the same epitope.

At this point, it is necessary to **screen the clones** to identify the clones that produce useful antibodies. One screening method involves ELISA; the antigen is immobilized onto a multi-well plate (this is often done through simple absorption), and spent medium from each clone is added to each well. Each sample of spent medium contains antibodies. After rinsing, useful clones can be identified through the addition of an enzyme-linked secondary antibody (e.g., if mice were used for the primary immunization, we could use an anti-mouse antibody produced by a goat). The valuable clones are transferred to larger culture vessels, and large amounts of antibodies can then be collected from the culture medium. The immortal properties of the myeloma cell allow these antibody-producing cells to be produced indefinitely.

One problem associated with the production of monoclonal antibodies is that mammalian cells are much more difficult to grow than most bacteria and fungi. Nevertheless, many biotechnology companies have successfully marketed diagnostic systems based on monoclonal antibodies.

The third option for obtaining antibodies is through gene cloning methods. **Recombinant antibodies** can be obtained by isolating mRNA from immunized or nonimmunized mouse cells. If the mouse is not immunized, B-cells with a broad range of specificities are obtained, whereas an immunized mouse will provide a more restricted range of antibody specificities. Once the mRNA is isolated, cDNA is made using reverse transcriptase, and PCR is used to amplify antibody genes. The next step is usually to insert these genes into a phage vector, creating a library of antibody genes. If the vector is constructed so that recombinant proteins are produced from the antibody genes, then the library can be screened for the desired antibodies. Once the phage containing the desired antibody is isolated, it can be used to produce large amounts of monospecific antibodies, much like monoclonal antibodies.

There are several advantages to the recombinant approach. It is fast and extremely flexible. Antibody genes can be directly manipulated to find, for example, the smallest amino acid sequence that confers specific recognition of the antigen. Also, because the antibody genes are already cloned, sequencing is straightforward; this may allow design of new sequences that have improved characteristics (e.g., improved specificity or greater heat stability).

3. Future Applications of Immuno-Assays

There is currently great interest in the development of immuno-assays, as seen in Table 6.3, which was compiled from a search of the Current Contents™ database for the period January 2000 to July 2001, using the keywords "food" and "immuno-assay"). To make this survey representative of current diagnostic development goals, we excluded studies that simply used currently available immuno-assays to detect residues in food and studies that were veterinary in scope (e.g., detection of brucellosis in cattle). Of the 76 studies, 33 were aimed at the detection of specific pesticides in food, soil, or water samples. Eight studies attempted to develop immuno-assays for the detection of mycotoxins and six for the detection of antibiotic residues. Surprisingly, only nine studies targeted detection of bacterial pathogens or toxins. Five studies were aimed at assessment of food quality (not related to pathogens). To summarize, the current thrust of diagnostic development appears to be related to the detection of undesirable residues in food; another trend is toward the development of diagnostic techniques that can be incorporated into an electronic detection system. These are **biosensors** and will be further discussed in the next section.

Detection of pesticides, antibiotics, and mycotoxins using antibodies is difficult because these compounds are usually low-molecular-weight nonprotein compounds. These sorts of compounds tend to have low **immunogenicity** (they do not elicit antibody production). However, if they are conjugated to other, larger compounds, their immunogenicity often increases dramatically. The small compound is said to act as a **hapten** in this context.

TABLE 6.3
Diagostic Immuno-Assays under Development

Detection Goal	Number of Studies	Types of Immuno-Assay
Pesticide	33	Agg,[a] cELISA,[b] biosensor,
Mycotoxins	8	ELISA, fluorescence, biosensor, cELISA, immuno-affinity
Antibiotic residues	6	Biosensor, cELISA
Salmonella spp.	6	cELISA, ELISA, automated assay, biosensor, agg
Allergen	5	ELISA, luminescence, biosensor, dot blots
Meat identification	3	cELISA, ELISA
Staphylococcal enterotoxin	2	Biosensor, fluorescence
Vitamins	2	Biosensor
Plant toxins	2	cELISA, fluorescence
Chitin oligos?	1	cELISA
Heat treatment of milk	1	cELISA
Lipid oxidation in meat	1	cELISA
Listeria spp.	1	ELISA
Spoilage microbes	1	ELISA
Gluten	1	ELISA
Transgenic crops	1	ELISA
Algal toxins	1	Biosensor
Norwalk virus	1	ELISA

[a] agg, agglutination.
[b] cELISA, competitive ELISA.
Note: Current Contents™ was searched using the keywords "food" and "immuno-assay" for the period January 2000 to July 2001.

Why bother to develop immuno-assays for these compounds, given the problems developing antibodies specific for them? Conventional methods of pesticide detection rely on time-consuming methods such as HPLC or gas chromatography (GC), which require expensive instrumentation. This limits the number of laboratories that are equipped to detect pesticide residues and the number of samples that can be tested. Consequently, it is difficult to resolve the current high level of consumer uncertainty about the safety of pesticide residues in food. Also, government regulators throughout the world are concerned with the sporadic occurrence of high levels of pesticide residues, particularly on fruits or vegetables that are consumed without processing. The development of easy, rapid diagnostic tests for pesticides would allow more frequent testing of food commodities and more efficient tracing of pesticide residues as they enter the soil and water environment.

F. LUMINESCENCE AND DIAGNOSTICS

1. Hygiene Assessment

Many food processors would like to be able to monitor the level of microbial contamination of surfaces and equipment. For example, the adequacy of equipment cleaning practices can be assessed and monitored by measuring the microbial load of the equipment after cleaning. This can be done by traditional cultural methods, but this approach suffers from the same problems described earlier in this chapter. Namely, it is laborious, time consuming, and expensive.

Recently, several biotechnology companies have marketed devices that allow the measurement of **adenosine triphosphate (ATP)**, an indicator of microbial load. This approach was pioneered by microbial ecologists, who have long grappled with the problems associated with measurement of microbial biomass and activity in natural systems such as soils and sediments. The ATP approach to measuring microbial biomass is possible because all organisms use ATP as the **energy currency** of the cell. To grow, microbes must take in energy sources (e.g., carbohydrates such as glucose) from their environment and convert this energy into ATP. The resultant ATP is then used to fuel a multitude of processes, such as cell wall growth, protein synthesis, and membrane formation.

Because all cells use ATP, the presence of ATP on an inert surface or equipment component indicates the presence of microbes. The detection of ATP tells the user nothing about the identity of the organism (it could be *Salmonella*, yeasts, food particles, human saliva, etc.); it is considered to be only an indicator of hygiene. Significant levels of ATP on a surface, for example, may indicate that the cleaning process is inadequate and must be revised. This is particularly useful in dairy operations, where hygiene is critical and visual monitoring of hygiene is insufficient.

The detection of ATP typically involves taking a swab of the area of concern and the lysis of any swabbed microbial cells. If ATP is present, an enzyme called luciferase (isolated from fireflies), causes the emission of light, through the following reaction:

$$\text{luciferin} + \text{ATP} + \text{Mg}^{2+} \xrightarrow{\text{luciferase}} \text{oxyluciferin} + \text{AMP} + \text{CO}_2 + \textbf{light}$$

The amount of light produced is directly proportional to the amount of ATP. Because ATP detection relies on the sensing of low levels of light, specialized instruments (luminometers) are required to monitor hygiene using ATP.

2. Novel Applications of Luminescence

Luminescence has great potential in other areas of food safety assurance. Most of these applications involve engineered use of bacterial luciferase genes, which have been successfully transferred to *Bacillus* spp., *Listeria* spp., *Staphylococcus* spp., *Aeromonas* spp., and lactic acid bacteria (e.g., *Lactococcus lactis*). Although the light-emitting reaction catalyzed by bacterial luciferase is different from the eukaryotic reaction (one key difference is that ATP is not used directly), it is similar in that the intensity of light emission is related to the cellular viability and energy status. This has led to novel applications of recombinant strains of bacteria relevant to food. Bacteria with recombinant luciferase genes can be used to assess the efficacy of cleaning regimes in industrial food processing operations. Luminescent bacteria are applied to preparation surfaces, cleaning is completed, and bacterial light emission (if present) is detected. This allows an assessment of efficacity against a specific microbe in **real time** (virtually immediate); in contrast, a conventional plate count assessment would require at least 24 h.

Another example of luminescence technology is the use of recombinant phage with the bacterial luciferase gene. This is an exciting technology because each bacterial species has a set of specific phage that are unable to infect other bacteria. Some phage have a broader host range, which can be useful. For instance, a phage that infects bacteria within the Enterobacteriaceae family would be a useful monitor of a range of food-borne pathogens, including *E. coli* and *Salmonella*. Such a phage could be incubated briefly (1 h) with a food sample, and if enteric bacteria were present, they would be infected by the phage, resulting in expression of phage genes, including luciferase. This would result in detectable light emission.

G. BIOSENSORS

1. Applications of Biosensors

The term "biosensors" has been used to describe a number of distinct diagnostic systems; this has made it difficult to define the term precisely. However, we can make the following generalization: biosensors have a **biological sensor** that is connected to a **transducer**. A transducer is a device capable of converting signals from the biological sensor into signals (usually electrical) that can be easily recorded and stored. For example, a number of biosensors use the specificity of antibody–antigen binding to detect pathogens in food samples. When the pathogen is present, it binds to the antibody. The key event follows: binding of antigen to antibody produces an electrical signal that can be detected and recorded. Biosensors, then, are an example of what science fiction authors describe as "cybernetics" — the fusion of organic matter to electronic circuitry. Although less dramatic than such fictional constructs as the "Borg" of Star Trek fame, biosensors offer many examples of inspired cooperation among microbiologists, biochemists, physicists, and electronic engineers.

Biosensors have many applications in clinical settings (e.g., diagnosis of food-borne pathogens from stool and other samples) and in maintenance of food safety (e.g., assessment of microbial loads or detection of specific pathogens in food). For example, it is theoretically possible to design biosensors that are sensitive (e.g., able to detect one pathogen in 25 g of food), selective (able to discriminate specific pathogens from a large background of nonpathogenic microbes), fast (real time), automated, portable, and inexpensive. To date, this potential has not been realized, but research in biosensor development is highly active and steady improvements in design are predicted.

Biosensors also have many applications that are not related to detection of pathogens. For example, one of the problems with large-scale cultivation of microbes (see Chapters 7 and 8) is that it is difficult to monitor certain aspects of microbial growth. The rate of formation of a desired product often can be monitored only through the analysis of samples taken from the bioreactor (vessel used to grow large microbial cultures). It would be highly preferable to monitor product formation **continuously**, perhaps by having a product-specific electrode in the interior of the bioreactor. Biosensors are available for certain products (e.g., lactic acid) and a number of other bioreactor parameters (e.g., glucose consumption, cell growth, and viability).

Continuous monitoring is also useful for many food safety or spoilage applications, particularly in the processing of liquids (milk, beer, etc.), where it is desirable to monitor microbial numbers **in line** (in piping systems used to transfer liquids from one vessel to another or to packaging processes). For example, postpasteurization contamination of beverages and foods is a significant cause of spoilage and has been implicated in outbreaks of food-borne illness. In 1987, postpasteurization contamination by *S. enteritidis* of milk used to make ice cream led to one of the largest recorded outbreaks of food-borne illness in the U.S. In-line detection could prevent this sort of accident, as well as the more common problem of spoilage arising from post pasteurization contamination (e.g., reduction of shelf life of milk by postpasteurization contamination by psychotrophic microbes).

Contamination of processing equipment is often difficult to eradicate. A continuous monitoring system that detects microbial growth in the lines is much better than monitoring based on examination of discrete samples or equipment swabs. If microbial growth could be immediately detected in transfer lines, the process could be stopped and the contamination eliminated, thus avoiding the production of large amounts of contaminated product.

Conventional diagnostic systems based on enrichment and selective culture are not adaptable for continuous monitoring. Instead, they require the collection of discrete sampling units (**batch samples**). Each sample is then cultured in the appropriate media. Continuous monitoring using a culture-based system would require an infinite (or at least very large) number of sampling units, whereas biosensors can continuously monitor without the collection of discrete samples.

Biosensors can also be used for batch sampling. One important application of biosensors is to speed up pathogen identification using culture-based methods. For example, spoilage of fresh meat is an economically important problem and is often linked to the presence of high levels of a variety of spoilage microbes. Conventional monitoring is done through plate counting (total bacterial count) after incubation on nonselective media that allow a wide range of bacteria to grow. However, this requires at least 24 h — not ideal for preventing meat spoilage. Biosensors have been designed that assess levels of microbial contamination after short (1 h or less) incubation of meat samples.

It is also often desirable to continuously monitor physical processes in the food industry. Physical parameters such as temperature can easily be monitored continuously, allowing immediate adjustment if the temperature strays outside a defined range, and also providing a record of temperature changes. A continuous record can be useful if product quality declines; deviations in the temperature of the process could be a causal factor. Biosensors can be used to monitor some physical–chemical processes (e.g., CO_2 production).

2. Types of Biosensors

Biosensors can be classified according to the type of sensor, the transduction strategy, and the directness of the assay. **Affinity-based** biosensors rely on specific recognition between the sensor and the target. Antibodies are most commonly used, but nucleic acid hybridization, similar to that used in gene probe assays, and receptor–ligand interactions (e.g., insulin binding to insulin receptor molecules) are also used to create affinity-based biosensors. Antibody-based biosensors have been developed for most of the major food-borne pathogens (e.g., *Salmonella*, *E. coli* O157:H7, and *L. monocytogenes*).

How is antibody–antigen binding detected and converted into an electrical signal? One approach is to immobilize antibodies to the surface of **piezoelectric crystals**, which are very sensitive to changes in mass. When antigens bind to the antibody, they increase the mass of the antibody–crystal complex. Piezoelectric crystals are unusual in that the application of external forces (e.g., gravity pulling on an attached antibody) leads to oscillation and a detectable electrical potential. When an antigen binds to the antibody, it increases the mass of the complex, which changes the frequency of oscillation of the crystal. This change of frequency can be detected electronically.

Enzyme-linked antibodies can also be used in biosensors. For example, a biosensor that detects *S. aureus* uses antibodies immobilized to an electrode (Figure 6.9). These antibodies "catch" the bacteria; antibodies that bind to another epitope of *S. aureus* are then added. These antibodies are covalently linked to horseradish peroxidase (HRP). The electrode is then moved to a solution containing amino salicylic acid (AMSA). Another enzyme (glucose oxidase) is also immobilized onto the electrode; the sole purpose of glucose oxidase is to generate hydrogen peroxide from glucose and oxygen. HRP catalyzes the reaction between hydrogen peroxide and AMSA to form 5-ASA-quinoneimine (ASAQ), which is then reduced by an electron supplied by the electrode. This current of electrons to ASAQ is detectable electronically.

One of the chief problems with affinity biosensors has been regenerating electrodes between samples. All bound antigens must be removed from the electrode to restore its sensitivity. It is difficult to achieve this without the use of harsh chemicals (e.g., 8 M urea) that tend to decrease the longevity of the electrode.

This is one of the reasons for increased interest in affinity biosensors based on nucleic acid hybridization. Unlike antibody–antigen binding, hybridization occurs over a relatively long molecular distance and is stabilized by frequent hydrogen bonds. In contrast, antibody–antigen binding typically occurs within short amino acid sequences and involves a mixture of hydrogen bonds, ionic attractions, and nonpolar interactions. Besides stability, nucleic acid hybridization is also attractive because it can easily be undone through adjustment of the ionic environment of the solution. The design of affinity biosensors using nucleic acid hybridization usually

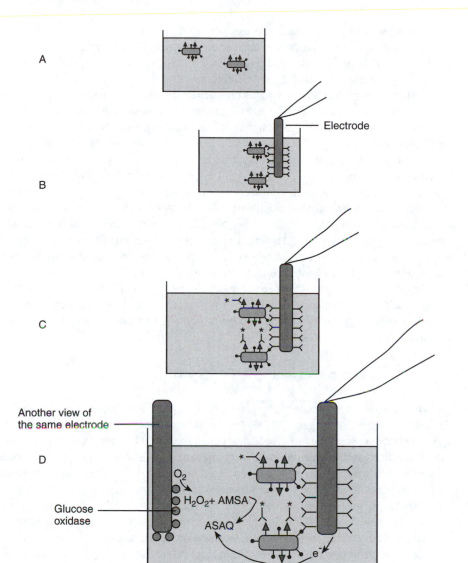

FIGURE 6.9 A biosensor for the detection of *S. aureus*. Cells of *S. aureus* (A) bind to antibodies on the electrode (B). A second antibody, which binds to a different epitope on the cell, is added (C). This antibody is covalently linked to horseradish peroxidase (HRP), designated as "*". The electrode is then removed and placed in a solution of amino salicylic acid (AMSA). (D). The electrode is also coated with glucose oxidase. This enzyme generates hydrogen peroxide (H_2O_2). HRP catalyzes the reaction between AMSA and H_2O_2 to form 5-ASA-quinoneimine (ASAQ). An electron from the electrode then reduces ASAQ. Current flow from the electrode increases as the numbers of *S. aureus* increase.

involves immobilization of oligonucleotide probes onto an electrode. Hybridization can be detected by similar interfaces as with antibodies (e.g., piezoelectric crystals that react to the change of mass induced by hybridization with target nucleic acids).

Optical sensors have also been successfully used. These sensors rely on changes in the optical qualities of substances coating an electrode. For example, hybridization might cause a change in the refractive index of a coating substance, which is then detected by the use of sensitive light detectors.

Another group of biosensors relies on microbial **metabolism** to monitor microbial growth or levels of microbial contamination. The most common strategy is to use **oxidoreductases** in cells as a signal of microbial growth. Usually, oxidoreductases change the structure of a **mediator** chemical that in turn triggers a response from the transducer. In one system, bacteria are immobilized onto an electrode that is coated with p-benzoquinone, a mediator. Dehydrogenase, a type of oxidoreductase found in most microbial cells, transfers an electron to the mediator. The reduced mediator then passes on this electron to the electrode, resulting in a detectable current. Although effective, this and other biosensors based on microbial metabolism function well only when large densities of microbes are present. This may limit their use in food sampling to circumstances in which low levels of contamination are acceptable but high levels are not (e.g., fresh meat).

Biotechnologists have also been successful in the design of biosensors that detect changes in **metabolites** such as glucose or lactate. Changes in the concentration of glucose can be continuously monitored by biosensors that use the enzyme **glucose oxidase**. This enzyme catalyzes the following reaction:

$$\text{glucose} + \text{oxygen} \xrightarrow{\text{glucose oxidase}} H_2O_2 + \text{gluconate}$$

The key part of a glucose biosensor that uses glucose oxidase is a system for detecting changes in oxygen or H_2O_2 levels. Several strategies are possible; many use fluorescent compounds to detect changes in oxygen. One design uses tris (1,10 phenanthroline) ruthenium chloride, a compound that has different fluorescence characteristics when exposed to oxygen. This biosensor has a fiber optics system that allows excitation of the ruthenium and detection of the resulting fluorescence.

Changes in glucose are important in the production of glucose syrups from starch. The conversion of starch to glucose is a complex process driven by a number of microbe-derived enzymes, and the concentration of glucose in starch undergoing processing is an important parameter. Microbial production of food-related metabolites is another process that can be monitored by the use of glucose biosensors, because many of the feeds used in bioreactors are rich in glucose. Glucose is also an important food ingredient in many processed foods (e.g., candies and other confections). The ability to monitor glucose without setting up extensive analytical facilities is attractive as a means to efficiently maintain quality control.

Despite the utility of glucose biosensors in food biotechnology, the driving force behind their development has been the enormous market of diabetics who need to frequently monitor their blood glucose. Recent clinical studies have demonstrated that tight control over blood glucose, achieved through frequent testing, results in a lower incidence of health complications. This has increased the need for easy, quick monitoring systems, and has also driven research into implantable systems that ultimately will lead to an **artificial pancreas**. This would consist of a subcutaneous biosensor (probably using glucose oxidase) that continuously monitors blood glucose, connected to an electronic system that controls operation of an insulin pump. Such a device would create a "closed loop," so that insulin-dependent (type I) diabetics would not need to monitor blood glucose through skin pricks nor inject insulin. Achieving this goal will require significant improvements in biosensor design, which will probably be adaptable for use in the food industry.

Improvements in our ability to miniaturize electronic components and to develop "laboratories in a chip" are also expected to lead to more frequent application of biosensors in the food industry. Applications are currently not common, mostly because of low sensitivity, interference by compounds in food matrices, and difficulties regenerating electrodes. The potential benefits, though, are significant enough to justify continued research and development of these elegant diagnostic systems.

RECOMMENDED READING

1. de Boer, E. and Beumer, R. R., Methodology for detection and typing of foodborne microorganisms, *Int. J. Food Microbiol.*, 50, 119, 1999.
2. Ray, B., *Fundamental Food Microbiology*, CRC Press, Boca Raton, FL, 1996.
3. Karch, H., Bielaszewska, M., Bitzan, M., and Schmidt, H., Epidemiology and diagnosis of shiga toxin-producing *Escherichia coli* infections, *Diagn. Microbiol. Infect. Dis.*, 34, 229, 1999.
4. Curry, A. and Smith, H.V., Emerging pathogens: *Isospora*, *Cyclospora* and microsporidia, *Parasitology*, 117, S143, 1998.
5. Reiff, F. M., Roses, M., Venczel, L., Quick, R., and Witt, V. M., Low-cost safe water for the world: a practical interim solution, *J. Public Health Policy*, 17, 389, 1996.
6. Herwaldt, D. L., Ackers, M.-L., and the *Cyclospora* Working Group, An outbreak in 1996 of cyclosporiasis associated with imported raspberries, *New Engl. J. Med.*, 336, 1548, 1997.
7. Mortimore, S. and Wallace, C., *HACCP: A Practical Approach*, Chapman & Hall, London, 1994.
8. Savage, R. A., Hazard analysis critical control point — a review, *Food Rev. Int.*, 4, 575, 1995.
9. Mozola, M. A., Detection of microorganisms in foods using DNA probes targeted to ribosomal RNA sequences, *Food Biotechnol.*, 14, 173, 2000.
10. Lantz, P.-G., Hahn-Hägerdal, B., and Rådström, P., Sample preparation methods in PCR-based detection of food pathogens, *Trends Food Sci. Technol.*, 5, 384, 1994.
11. Bellin, T., Pulz, M., Matussek, A., Hempen, H. G., and Gunzer, F., Rapid detection of enterohemorrhagic Escherichia coli by real-time PCR with fluorescent hybridization probes, *J. Clin. Microbiol.*, 39, 370, 2001.
12. Talary, M. S., Burt, J. P. H., and Pethig, R., Future trends in diagnosis using laboratory-on-a-chip technologies, *Parasitology*, 117, S191, 1998.
13. Umek, R. M., Lin, S. W., Vielmetter, J., Terbrueggen, R. H., Irvine, B., Yu, C. J., and Kayyem, J. F., Yowanto, H., Blackburn, G. F., Farkas, D. H., and Chen, Y. P., Electronic detection of nucleic acids: a versatile platform for molecular diagnostics, *J. Mol. Diagn.*, 3, 74, 2001.
14. Lee, H. A. and Morgan, M. R. A., Food immuno-assays: applications of polyclonal, monoclonal and recombinant antibodies, *Trends Food Sci. Technol.*, 4, 129, 1993.

15. Paraf, A., A role for monoclonal antibodies in the analysis of food proteins, *Trends Food Sci.*, 3, 263, 1992.

16. Spinks, C. A., Broad-specificity immuno-assay of low molecular weight food contaminants: new paths to Utopia! *Trends Food Sci. Technol.*, 11, 210, 2000.

17. Walker, A. J., Jassim, S. A. A., Holah, J. T., Denyer, S. P., and Stewart, G. S. A. B., Bioluminescent *Listeria monocytogenes* provide a rapid assay for measuring biocide efficacy, *FEMS Microbiol. Lett.*, 91, 251, 1992.

18. Kodikara, C. P., Crew, H. H., and Stewart, G. S. A. B., Near on-line detection of enteric bacteria using *lux* recombinant bacteriophage, *FEMS Microbiol. Lett.*, 83, 261, 1991.

19. Deshpande, S. S., Principles and applications of luminescence spectroscopy, *Crit. Rev. Food Sci. Nutr.*, 41, 155, 2001.

20. Ivnitski, D., Abdel-Hamid, I., Atanasov, P., and Wilkins, E., Biosensors for detection of pathogenic bacteria, *Biosensors Bioelectron.*, 14, 599, 1999.

21. Rogers, K. R., Principles of affinity-based biosensors, *Mol. Biotechnol.*, 14, 109, 2000.

22. Jaremko, J. and Rorstad, O., Advances toward the implantable artificial pancreas for treatment of diabetes, *Diabetes Care*, 21, 444, 1998.

7 Cell Culture and Food

I. OVERVIEW

Cell culture has facilitated basic research into animal, plant, and microbial cell metabolism. The ability to grow uniform cultures of cells is particularly useful for the study of cellular metabolism and behavior outside of the complex multicellular environment of a plant or animal. This is relevant to food; for example, the study of human intestinal epithelial cells has elucidated the mechanisms of nutrient digestion and uptake. Such studies would be difficult or impossible in intact animals.

With regard to unicellular organisms, cell culture has been equally important, and the study of isolated species or strains of bacteria and fungi has provided great insight into basic microbial structure and function. This is also relevant to food. For example, discovering the nature of toxins produced by *Escherichia coli* O157:H7 has helped us understand the characteristics that make this strain so much more dangerous than other strains of *E. coli* and has given us an important diagnostic tool — that is, instead of trying to detect the bacterium, in some cases it may be easier to detect the toxin.

A wide range of cell culture applications are directly relevant to food production and processing. Many of these, such as the use of yeasts to ferment foods into ethanol, have been employed by humans throughout recorded history and are often called **traditional biotechnology**. Others, such as the use of bacteria to produce flavor enhancers such as glutamate, have existed only in the 20th century. In this chapter, we examine both traditional and modern microbial food biotechnologies. The brewing and dairy industries, the main bastions of traditional biotechnology, have widely embraced many aspects of what is usually considered modern biotechnology. For example, both industries use advanced control and diagnostic systems in bioreactor design and both have made use of recombinant DNA technology to modify cells. Recombinant yeast and bacteria are technically ready to be used by the dairy and brewing industries, but are currently awaiting public acceptance before widespread introduction. Consequently, it is important to include these traditional industries in a discussion of modern biotechnology. Furthermore, they are among the most successful biotechnologies. Economically, the brewing industry and associated distilling and wine-making industries constitute a greater force than all other biotechnologies. For example, in Canada, biotechnology companies (excluding brewing and dairy companies) had a total revenue of $1.67 billion in 1993. In 1996, Canadians spent $8.8 billion on beer alone. In the U.S., 2.5 million people are employed in the brewing industry, and in Germany, 108 million hectoliters of beer were consumed in 1996. These figures demonstrate that traditional yeast fermentation is an important economic force globally. Other biotechnologies, especially those applied to human therapeutics, are likely to increase in value, closing the economic gap between them and traditional biotechnologies. For example, in 1997, global revenues from recombinant therapeutic proteins were $17 billion. Nonetheless, traditional biotechnologies will remain a potent economic force in the future.

Yeast cells ferment carbohydrates to ethanol, forming the basis of the brewing industry; lactic acid bacteria (LAB) ferment carbohydrates to lactic acid, a necessary part of the production of cheese, yogurt, and various other dairy products. Both ethanol and lactic acid are end products of energy-generating fermentation pathways. However, other aspects of microbial metabolism have been exploited in the food industry. Microbial **enzymes** are used in a number of food processes, including starch processing, where amylases convert starch into glucose and other useful compounds, and cheese making, where lipases are often used to improve cheese flavor. Microbial **polysaccharides** (e.g., xanthan gum) are widely used in the food industry as thickeners and texture modifiers. Amino acids (e.g., **glutamate**) derived from microbial cells are also important. Finally, microbial **biomass** is sometimes used directly as food. Yeast biomass is an economically important product, used for bread rising and also directly as a food (e.g., Vegemite™). Meat substitutes can also be made from microbial biomass — Quorn™, made from biomass of *Fusarium graminearum*, is an example, as is tempeh, which is composed of soybeans and biomass of *Rhizopus oligosporus*.

The primary objective in this chapter is to explain how cell culture can be used to produce these valuable products. Another aim is to develop an understanding of the strategies that can be used, and that will be used in the future, to improve the efficiency of production of cellular products. Although animal cell products (e.g., antibodies) have application to the food industry, and plant cell products (e.g., flavor compounds such as vanillin) may be important in the future, we will focus on microbial cell products, because these have so far been the most useful to the food industry (Table 7.1).

II. BREWING

A. INTRODUCTION

Most cultures have at least one traditional alcoholic beverage that is produced from grains, vegetables, or fruits. Usually *Saccharomyces cereviseae* is the driving force behind the production of ethanol. Although it is immensely popular throughout the world, ethanol is also very controversial; consumption of alcoholic beverages is forbidden in certain religions, and there is no doubt that abuse of ethanol-containing beverages can lead to addiction and tragic consequences. On the other hand, millions of people enjoy alcoholic beverages without any negative effects, and indeed, evidence is accumulating that the regular consumption of small amounts of alcoholic beverages (e.g., red wine) can be beneficial to human health.

Beer is the most widely consumed alcoholic beverage — approximately 1.2 billion hectoliters (1 hL = 100 L) of beer are consumed annually worldwide. Conventional beer brewing makes elegant use of the biological properties of three organisms: *S. cereviseae,* barley (*Hordeum sativum*), and hops (*Humulus lupulus*).

TABLE 7.1
Microbial Cells Used to Make Products that Are Valuable to the Food Industry

Product	Use	Producing Organism	Type of Organism
Beer	Beverage	*Saccharomyces cereviseae*	Fungus (yeast)
Wine	Beverage	*S. cereviseae*	Fungus (yeast)
Vodka	Beverage	*S. cereviseae*	Fungus (yeast)
Baking yeast	Bread making	*S. cereviseae*	Fungus (yeast)
Yogurt	Food	*Lactobacillus bulgaricus*	Bacterium
Buttermilk	Beverage	*Lactobacillus acidophilus*	Bacterium
Cheese	Food	*Streptococcus thermophilus*	Bacterium
Glutamate	Flavor enhancement	*Corynebacterium glutamicum*	Bacterium
Lysine	Feed/food supplement	*Brevibacterium* spp.	Bacterium
Glucoamylase	Starch processing	*Aspergillus niger*	Fungus
Glucose isomerase	Fructose production	*Streptomyces* spp.	Bacterium (actinomycete)
β-amylase	Starch processing	*Bacillus subtilis*	Bacterium
Invertase	Starch processing	*S. cereviseae*	Fungus (yeast)
Xanthan gum	Thickener	*Xanthomonas campestris*	Bacterium
Pullulan	Thickener	*Aureobasidium pullulans*	Fungus

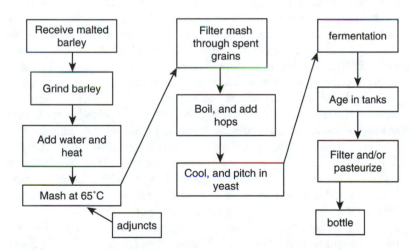

FIGURE 7.1 Overview of the brewing process.

The overall strategy and sequence of events in the beer-brewing process has not changed in the last century; however, the details of the process and the extent of the brewer's control over the process have changed enormously. We will examine the brewing process as well as biotechnological approaches to improvements in the brewing process.

Brewing can be divided into four discrete stages (Figure 7.1): **malting, mashing, fermentation,** and **packaging**. During this process, starch in the barley seed is **converted** into fermentable sugars (malting and mashing), which are then fermented by brewing strains of *S. cereviseae* into ethanol. Fermentation is usually divided into two stages: the **main fermentation**, which results in rapid formation of ethanol, and **secondary fermentation**, wherein the beer **matures** and improves in flavor. Packaging usually involves filtration and carbonation of the beer, followed by injection of the beer into bottles or cans.

Of course, beer is not the only fermented beverage. Wine is also a fermented product, and brandy, whisky, vodka, and rum are all examples of distillates of fermented beverages. The overall process of fermentation is similar in these products. The events prior to fermentation, however, vary widely. In general, beverages that are made from starch-rich substrates require enzymatic pretreatment, because *S. cereviseae* is incapable of breaking down starch. In contrast, grapes do not require pretreatment because they are rich in sucrose and other carbohydrates that can be directly fermented by *S. cereviseae.*

Before proceeding further, it is worthwhile to consider the range of beer types that consumers enjoy. We can define beer as a fermented beverage made using yeast and malted grains, but this does not come near to describing the incredible range of beverages that are referred to as "beer." If we restrict our discussion to North American and European beers, we can identify the following major groups:

- *Lagers* are made by bottom-fermenting strains of *S. cereviseae.* Lager fermentation typically occurs at cool temperatures (4 to 12°C). Lagers are quite different in North America and Europe. North American lagers tend to have little hops, are not bitter, and are made from a mixture of barley, corn, and rice malts. In contrast, European lagers tend to be more bitter, because of higher levels of hops. Lagers usually have a long (up to 3 weeks) period of maturation (secondary fermentation) before bottling.

- *Ales* are made from top-fermenting strains of *S. cereviseae.* These are fermented at a warmer temperature (12 to 18°C) and typically have a short maturation period (e.g., 1 to 3 d). Because of the warmer fermentation temperature, ales often have a more complex range of flavors than lagers (e.g., fruity flavors arising from yeast-derived esters). The many different styles of ales include strong ales (alcohol content >5%), stouts, porters, pale ales, and bitters. These ales vary in color from yellow to black and have a large range of flavor depth and complexity.

- *Dry and light beers* can be made from lagers or ales, although lagers are more commonly used. They typically have a simpler taste than other beers, because of low levels of nonfermentable carbohydrates in the final product. Both styles are becoming increasingly popular, particularly in North America. The basic difference between dry and light beers is that light beer initially contains fewer fermentable substrates. This results in less alcohol production and fewer calories. Dry beers, in contrast, have similar alcohol and calorie contents to conventional beer but have different sensory qualities. In particular, they are perceived to be less filling than conventional beer.

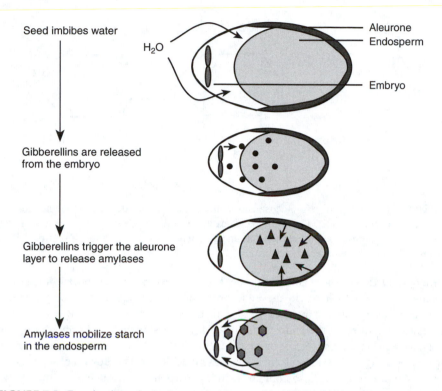

Seed imbibes water

H_2O

Aleurone
Endosperm

Embryo

Gibberellins are released
from the embryo

Gibberellins trigger the aleurone
layer to release amylases

Amylases mobilize starch
in the endosperm

FIGURE 7.2 Germination of a barley seed. After seed imbibition, gibberellins (circles) diffuse from the embryo to the aleurone layer (outer layer of the endosperm). The aleurone layer responds by releasing amylases (triangles) and other degradative enzymes. This initiates starch breakdown, releasing glucose (hexagons) and other carbohydrates for use by the embryo.

This array of products can be made from the same basic set of ingredients because of: (1) inherent flexibility in the traditional brewing process (e.g., the degree of roasting that occurs during kilning has dramatic effects on beer color and flavor); (2) variation in metabolism of yeast strains; and (3) biotechnological refinements to the brewing process (e.g., use of extra enzymes during mashing).

B. MALTING

Brewing begins with malting, and malting begins with the addition of water to barley grains (seeds). This is referred to as **steeping**. The grains soak in the water and are then spread out on the floor of a humidity-controlled room for at least 48 h. The seeds imbibe water, which triggers germination. Several crucial events occur during germination (Figure 7.2), culminating in the release of degradative enzymes, including α- and β-amylases, proteinases, glucanase, and phosphatase. From the seed's perspective, the purpose of this release of degradative enzymes is to mobilize carbohydrates, amino acids, and phosphate from the **endosperm**, the major storage organ of the seed. These nutrients are then used by the growing embryo to fuel the

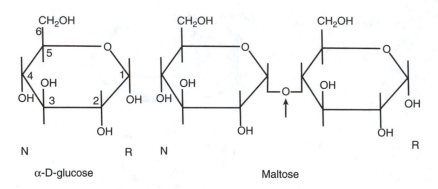

FIGURE 7.3 Structure of α-D-glucose and maltose. Each molecule has a reducing end (R) and a nonreducing end (N). The arrow indicates the position of the α-1,4 linkage in maltose. The numbering convention for carbons in glucose is shown. According to convention, "1" is given to the most oxidized carbon.

initial production of shoot and root tissue. For the brewer, these events are fortuitous, because the carbohydrates in the endosperm will ultimately be one of the major sources of energy and carbon for yeast cells during fermentation. However, carbohydrates are stored in the endosperm in the form of **starch**, and *S. cereviseae*, the yeast used in brewing, is unable break starch into fermentable sugars such as glucose or maltose (Figure 7.3). Starch must be enzymatically broken down before *S. cereviseae* can use it. Barley also contains β-linked carbohydrates such as cellulose and glucans that are less affected by conventional malting or mashing. *S. cereviseae* cannot use these carbohydrates, so they often persist after fermentation and are present in the finished product. In some cases, these carbohydrates are desirable, because they add body to the beer. However, when the desired beer is light in flavor and body, glucans in particular are troublesome. This is one reason why brewers in North America often replace a proportion of the barley malt with the endosperm of rice or corn, which has low levels of β-glucans.

It is clear, then, why the release of amylases from the endosperm is crucial to brewing. Equally important is control over the timing of the enzymatic breakdown of starch. The maltster usually does not let germination proceed past 48 h. The main reason for this is that continued germination results in unacceptable losses of carbohydrates, driven by vigorous growth of the embryo. **Kilning** the seeds prevents this loss. Dry heat is applied, driving moisture from the seeds. This kills the embryos but will not damage the amylases and other enzymes, as long as the seeds do not become excessively hot (>60°C). Once most of the water has been driven from the seeds, the malted barley is very stable — it can be stored for long periods, providing that the relative humidity of the storage environment is low enough to prevent condensation of water on the seeds. This is an important consideration, because malted barley is very susceptible to microbial attack; dry conditions must be maintained.

Another important consideration is the quality of the barley grains. Plant breeders have developed barley cultivars that are particularly well suited to malting. Such cultivars germinate uniformly, so that at the time of kilning, amylases have been released in all of the seeds and endosperm starch has been **partially converted** to simpler carbohydrates. This is important because mashing requires a large number of amylases. Furthermore, partial conversion during malting speeds up the total conversion of starch during mashing. These are not the only reasons for seeking uniform germination. Brewers want all of the seeds to proceed substantially through the germination process because malting results in favorable changes in the flavor of the barley grains. These flavor changes are vital to the quality of many beers.

The quality of the seed is also important. It should be clean and free of soil and insects. High grades of malting barley will have a high germination rate (proportion of seeds that germinate), resulting in the accumulation of large amounts of enzyme during malting. If the seed is too old and has a low germination rate, there is unlikely to be sufficient enzyme present during mashing. Old seed is also more likely to be contaminated with fungi that can cause severe problems during malting. During the initial stages of malting, the conditions are ideal for microbial growth, and high levels of contamination of the barley grains usually result in extensive spoilage of the germinating seeds.

C. MASHING

Unlike malting, which usually requires dedicated malting operations, mashing typically occurs in the brewery. First, malted barley is ground, releasing starch grains from the seeds. Water is then added to the ground malt, and the mixture is slowly heated to 65°C. Some mashing processes have a prolonged incubation at 40°C to allow proteolysis and protein reduction in the mash. This is necessary if large amounts of protein are present in the barley seed that would result in haze of the finished product if the proteins were left intact. The proteases in barley malt are inactivated by the usual mash temperature (65°C), so slightly prolonged (e.g., 30 min) incubation at a lower temperature is advised if haze is a recurring problem.

Eventually, though, the temperature will be increased to 65°C. This is the optimal temperature for β-amylase, which is an exoenzyme that removes maltose (a disaccharide made up of two α-linked glucose molecules) from the nonreducing ends of starch polymers. The action of α-amylase is also important — it is an endoenzyme that attacks starch *within* the polymer, generating a large number of linear starch chains that are efficiently attacked by β-amylases. Mashing is said to result in *total* conversion, but this is not quite true. Especially when barley malt is used, a substantial proportion of the starch is converted into short chains of glucose (**dextrins**) that *S. cereviseae* is unable to use.

The conversion of starch to simpler carbohydrates occurs much more quickly in the mash than in germinating seeds. Within an hour, very little starch remains, and the resulting **wort** is an excellent medium for yeast fermentation. However, the wort at this stage is not completely ready for yeast inoculation. First, the wort is filtered, either through an artificial filter or through a **lauter tun**, which is a filtration bed of barley husks and seed coats. The filtered wort is then heated to boiling, for four reasons:

- To inactivate amylases and other enzymes.
- To precipitate proteins, which will settle out. This will reduce the level of protein haze in the final product.

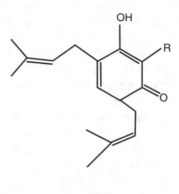

Humulone

FIGURE 7.4 General structure of humulones. These compounds provide much of the bitter flavor of hops. Essential oils, which are important aroma compounds, have a similar structure but are smaller and more volatile. The "R" group is a hydrocarbon chain of varying length.

- To allow efficient extraction of resins and essential oils from added hops.
- To reduce the level of microbial contamination in the wort.

D. HOPS

H. lupulus, the hop plant, is a member of the Cannabinaceae, a family that is renowned for the production of compounds that have strong effects on humans. *Cannabis sativa* (marijuana) contains secondary compounds that affect the human nervous system. Secondary compounds are, by definition, not involved in primary plant metabolism. Therefore, they are not required for basic processes related to the acquisition of energy and carbon. Why do plants bother to produce secondary compounds? Many secondary compounds are toxic to animals and thus are thought to be an important defense against herbivorous animals.

Fortunately for humanity, many secondary compounds are not toxic to humans and are actually highly prized as flavor compounds. Most spices contain flavorful secondary compounds. Hop flowers have many secondary compounds in the form of **resins**; this includes the **humulones** and **essential oils**, which are particularly important in brewing (Figure 7.4). These lipophilic compounds are insoluble in water; they solubilize only through prolonged boiling. They are responsible for much of the bitterness, flavor, and aroma of beer. One of the most crucial jobs of a brewmaster is to add the correct *amount* of the correct **cultivar** of hops that will give the desired flavor to a beer. Many of the beer brands produced in North America contain very little hops, because of consumer demand for beer that is not bitter. However, many of the beers produced by microbreweries for niche markets in North America and many European beers contain substantial amounts of hops. In these beers, the type and amount of hops that are added to the wort are extremely important.

Some of the resins that are released from hops have antimicrobial properties, particularly against Gram-positive bacteria. Originally, this was probably one of the most important reasons for adding hops to wort. Before the 20th century, brewing was a risky business; contamination of the yeast inoculum inevitably occurred, and the antimicrobial properties of hops were doubtless essential to the brewer. The antimicrobial properties of hops are less important to modern brewers.

E. PRIMARY FERMENTATION

Fermentation is the crucial part of the brewing process. The whole point of brewing is to produce a beverage with ethanol that is pleasing to the senses, and fermentation fulfils that goal. Primary fermentation also has a special place in the history of science. The process of glycolysis and fermentation was first elucidated in *S. cereviseae*. This was the first biochemical pathway to be deciphered, and it laid a foundation for the development and unfolding of biochemistry in the 20th century. *S. cereviseae* is still an important model organism, although its role has shifted toward helping unravel the molecular processes that govern cell behavior, especially surrounding cell division.

In terms of the technical aspects of fermentation, a few steps are required before adding the yeast inoculum. After boiling, hops are removed from the wort, and the wort is cooled and transferred to a bioreactor. The appropriate lager or yeast strain of *S. cereviseae* is then added (**pitched**) to the wort. Hundreds of yeast strains have been isolated that have slightly different growth and fermentation characteristics. The choice of strain sometimes has important effects on the final flavor of the beer.

Brewers usually add large amounts of yeast to the wort (e.g., 0.5 L of yeast suspension per hectoliter of wort), so that the yeast does not require a long lag period to grow to heavy populations. As yeast growth begins, so does fermentation. Fermentation is an anaerobic process with the purpose of generating cellular energy through the reduction of organic electron acceptors. This reduction is usually linked to the catabolism of carbohydrates such as glucose (Figure 7.5), although other compounds, such as amino acids, can usually be shunted into fermentation pathways. The fermentation of glucose by yeasts can be summarized by the following equation, which was derived by Gay-Lussac in 1810:

$$\text{Theoretical yield:} \quad \underset{\substack{\text{glucose}\\180\text{ g}}}{C_6H_{12}O_6} \longrightarrow \underset{\substack{\text{ethanol}\\92\text{ g}}}{2C_2H_5OH} + \underset{\substack{\text{carbon dioxide}\\88\text{ g}}}{2CO_2}$$

The wort should be well aerated or oxygenated during the initial stages of fermentation. Although fermentation is an anaerobic process, *S. cereviseae* requires oxygen for the production of ergosterol, the compound in fungal plasma membranes that confers strength and rigidity. If the initial stages of fermentation are aerobic, the yeasts will be able to synthesize enough ergosterol to supply cellular maintenance needs throughout the fermentation period.

If aerobic conditions occur, then why don't the yeasts use respiration to oxidize carbohydrates? Indeed, it would appear to be foolish for a yeast cell to use the inefficient, low-energy-yielding fermentation pathway in preference to respiration. The solution to this conundrum lies in the relationship between yeast metabolism

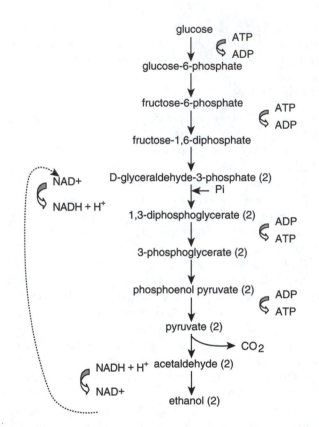

FIGURE 7.5 Fermentation of ethanol. Glycolysis is followed by conversion of pyruvate into ethanol and carbon dioxide. This process allows regeneration of NAD$^+$ and efficient elimination of waste products of glycolysis (ethanol diffuses freely across membranes). Note that a small amount of adenosine triphosphate (ATP) is gained through fermentation. (NAD, nicotinamide adenine dinucleotide; ADP, adenosine diphosphate; Pi, phosphate.)

and the local concentration of carbohydrates. Many strains of *S. cereviseae* are unable to respire in the presence of **large concentrations of glucose and other carbohydrates**. Such conditions are found in wort. Mitochondrial structure is disrupted, and the yeast cells switch to fermentation as the predominant energy-yielding pathway. This effect, known as the **Crabtree effect**, leads to vigorous fermentation by brewing yeasts despite the presence of oxygen.

F. Secondary Fermentation

Once fermentation is complete and all of the fermentable carbohydrates have been converted to ethanol by the yeasts, the beer is usually transferred to another tank, cooled, and stored. Yeasts are still present at this point. The length of time required for maturation varies from several days to several months. A number of chemical changes occur in the beer; some are a consequence of metabolism by residual yeasts that are unable to grow because of nutrient starvation but are still capable of limited metabolism. Other changes are strictly due to chemical reactions occurring in the maturing beer. The maturation period is important, because the changes that occur virtually always result in a better-tasting beer.

The most important change that occurs during secondary fermentation is the reduction in levels of **diacetyl**. This compound has a strong buttery taste that is desirable in butter but undesirable in beer. During and after yeast growth, diacetyl forms from α-acetolactate through a nonenzymatic process (α-acetolactate is an intermediate in several amino acid biosynthetic pathways). Yeast can convert diacetyl to acetoin, an unflavored compound, but this conversion is delayed by the slow rate of formation of diacetyl from α-acetolactate. Hence, secondary fermentation some-times requires several weeks. In many ales, conversion of diacetyl to acetoin is less critical, partly because of the presence of strong flavors that mask the flavor of diacetyl.

At the end of the maturation period, the beer is packaged, either in kegs or in bottles — in some traditional operations, secondary fermentation occurs in the kegs or bottles. Modern breweries carbonate the beer before packaging. Fermentable carbohydrates (usually glucose) are added to the kegs or bottles; fermentation by residual yeasts will then result in CO_2 production and a carbonated final product.

G. Biotechnological Improvements: Catabolite Repression

Catabolite repression can cause prolonged fermentation times. Wort usually has large amounts of glucose, maltose, and maltotriose. Glucose is a better source of carbon because it can be directly funneled into the fermentation pathway. Consequently, many yeast strains preferentially ferment glucose. Other carbohydrates will be used only when glucose is completely used up (thus use of certain catabolites is *repressed* by other catabolites). This results in an undesirable decline in the rate of ethanol production. Brewers are acutely aware of the economic costs of delays in the brewing process. When summed over an entire year, small delays can cause significant losses in brewery output.

First, we will describe catabolite repression in *E. coli*, because it is best under-stood in this organism. In *E. coli*, and the Enterobacteriacea in general, catabolite repression occurs when **glucose** is present in the growth medium. High levels of glucose inhibit **adenylate cyclase**, resulting in low levels of **cyclic adenosine mono-phosphate** (cAMP). This depresses expression of genes necessary for catabolism of other carbohydrates (e.g., maltose). However, when glucose is depleted from the medium, adenylate cyclase is no longer inhibited. Levels of cAMP rise, and cAMP interacts with **catabolite activator protein** (CAP). This interaction allows CAP to bind to promoters upstream from genes and operons that code for enzymes that catabolize carbohydrates. Following this binding of CAP to the promoters, RNA polymerase binds efficiently to the promoters, and transcription can occur.

At first glance, catabolite repression seems to be a cumbersome and unnecessary method of gene regulation. However, it is important to carefully consider the natural environment of *E. coli* — the mammalian gastrointestinal (GI) tract. Nutrient supply in this environment is unpredictable and sporadic. When nutrients appear in the GI tract, intense competition ensues among the resident bacteria. Glucose is an excellent source of energy for organisms because it can be directly shunted into glycolysis; other carbohydrates usually have to be enzymatically modified before they can enter glycolysis or other energy-generating pathways. Consequently, *E. coli* can maintain a higher growth rate when using glucose than when it is using other carbohydrates. In the highly competitive environment of the GI tract subtle changes in growth rate can determine the fate of a bacterium. It makes sense, then, for *E. coli* to attempt to "grab" glucose when it becomes available and to "ignore" other carbohydrates while glucose is present. Catabolite repression is similarly important for *S. cereviseae*, which must compete with many other microbes in its natural environment (plant surfaces and sucrose-rich fruits).

Because of the importance of yeast metabolism to the brewing and baking industries, a number of researchers have studied catabolite repression (also known as **glucose signaling**) in yeast. Overall, the phenomenom is similar in yeast as in the Enterobacteriaceae. A pronounced **diauxic shift** occurs when cells exhaust glucose from the medium. This refers to the process of shutting down genes involved in glucose uptake and utilization and the activation of genes involved in uptake and utilization of other sugars, such as maltose. However, the mechanism of glucose signaling appears to be very different in yeast. cAMP is involved in regulation, but the pattern is different. In *E. coli*, cAMP levels increase when glucose is depleted, and this triggers transcription of genes involved in utilization of other sugars. However, in yeast, cAMP levels increase when glucose is present in the medium. This shuts down transcription of genes that code for proteins involved in uptake and use of maltose and other sugars. The mechanism of this regulation is unclear, but several complex models have been proposed.

Because the molecular mechanisms behind catabolite repression in yeast are still unclear, it is impossible to precisely and directly alter yeast genes in order to modify catabolite repression. Instead, catabolite repression has been eliminated in certain yeast strains using a mutagenesis approach (this approach has been successful with both brewing and baking strains of *S. cereviseae*) (Figure 7.6). Yeast cells are exposed to a mutagen (a compound that causes random changes in the DNA sequence) and then to 2-deoxyglucose, a synthetic analog of glucose. 2-Deoxyglucose cannot be used as a source of energy and carbon by yeast cells. Cells with normal "wild-type" catabolite repression are unable to use alternative carbon sources because of the inhibiting effect of 2-deoxyglucose. However, if the biotechnologist is lucky, one of the many mutants that occur after mutagenesis will have impaired catabolite repression, perhaps due to a mutation in a gene coding for a key regulator of gene expression. Such a mutant may be able to use other carbohydrates, despite the presence of 2-deoxyglucose.

This strategy has been successful in producing mutant yeast strains that can use glucose, maltose, and maltotriose simultaneously, thus avoiding the lag phase that normally occurs as glucose levels decline.

H. HIGH GRAVITY BREWING

One strategy that has been employed to improve fermentation efficiency is to use worts that have high **specific gravity**. Initial specific gravity of a wort is determined

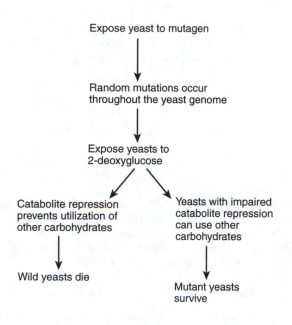

Expose yeast to mutagen

Random mutations occur
throughout the yeast genome

Expose yeasts to
2-deoxyglucose

Catabolite repression
prevents utilization of
other carbohydrates

Yeasts with impaired
catabolite repression
can use other
carbohydrates

Wild yeasts die

Mutant yeasts
survive

FIGURE 7.6 The use of a mutagenesis strategy to isolate yeast strains with impaired catabolite repression.

by the concentration of solutes in the wort. Specific gravity is an important measure because of its direct relationship to the final ethanol concentration of the beer. Worts with higher than normal initial specific gravity result in beer with increased ethanol content. This beer can then be diluted, giving a larger volume of beer per fermentation run.

This high-gravity approach is not always successful. If the osmotic potential of the wort is too high, a **stuck fermentation** may occur, because of inhibition of yeast metabolism. If the concentration of ethanol exceeds the tolerance of the yeast, then fermentation will halt. Consequently, one of the aims of brewing research has been to increase the tolerance of yeast strains to high initial levels of solutes and high levels of ethanol toward the end of primary fermentation. With some strains, simple adjustments in the brewing environment allow successful high-gravity brewing. For example, the combination of increased initial oxygen concentration, increased temperature during fermentation (from 14 to 25°C), and increased density of yeast inoculum leads to successful high-gravity brewing in certain yeast strains. Other studies have demonstrated that establishing high levels of oxygen and **free amino nitrogen** (a good source of nitrogen for yeast cells) is crucial to success in high-gravity and very-high-gravity brewing systems. Another crucial development has been the isolation of yeast strains that tolerate high levels of ethanol (as much as 14%) and that also perform well in the brewing environment. With these technological improvements, high-gravity brewing is increasing in popularity among brewers, because of the substantial cost savings.

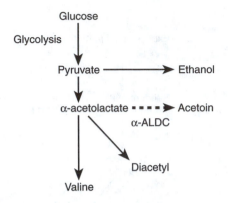

FIGURE 7.7 Recombinant approach to reducing diacetyl production by brewing yeast. The gene for α-acetolactate decarboxylase (α-ALDC) is transferred to a brewing strain of *S. cereviseae*. Diacetyl is normally produced from α-acetolactate. α-ALDC diverts α-acetolactate toward the production of acetoin, dramatically decreasing production of diacetyl. (Modified from Onnela, M.-L., et al., *J. Biotechnol.,* 49, 101, 1996.)

I.   THE β-GLUCAN PROBLEM

β-Glucans are polysaccharides that are present in the cell wall of barley seeds. They are released into the wort during mashing, and, although β-glucanases are also present in the mash, they are usually ineffective, because they lack thermotolerance and are inhibited by the temperatures used for mashing. β-Glucans lead to increased viscosity of the wort, which increases the time required for filtration of the mash, especially when filtration occurs through a lauter tun.

One approach to tackle this problem is to modify barley so that it produces a thermotolerant β-glucanase. This cannot be done through conventional breeding, so it is necessary to take a transgenic approach. Unfortunately, it was not possible until relatively recently (1990) to produce transgenic barley, primarily because of difficulties in tissue culture of this plant. In the early 1990s, though, several research groups successfully transformed barley, and transgenic plants with thermotolerant β-glucanases appeared in the literature. One major obstacle has been that only certain barley varieties are amenable to transformation, and these varieties are not necessarily the best varieties for brewing purposes. Consequently, extensive backcrossing will be required to achieve satisfactory cultivars with thermostable β-glucanases.

J.   GETTING RID OF DIACETYL

As previously discussed, a major objective of secondary fermentation is to decrease levels of diacetyl below 0.5 mg/L (below this threshold, diacetyl flavor is undetectable). Unfortunately, this sometimes leads to extended incubation in large stainless steel tanks, which is expensive to the brewer. One strategy for accelerating the rate of diacetyl removal is to introduce the enzyme α-acetolactate decarboxylase (α-ALDC). This enzyme catalyzes the conversion of α-acetolactate to acetoin, thus hampering diacetyl production (Figure 7.7).

α-ALDC is not found in *S. cereviseae*. However, it is present in a number of bacteria, including *Klebsiella terrigena* and *Enterobacter aerogenes*. Several research groups have successfully introduced genes coding for this enzyme into brewing strains of *S. cereviseae*, and when used to ferment wort, these recombinant strains produce beer with undetectable levels of diacetyl after primary fermentation. Thus secondary fermentation is unnecessary with these yeast strains. A number of other recombinant yeast strains are also available; some have introduced amylase genes, leading to improved utilization of carbohydrates, and some have improved flocculating ability ("super-flocculent yeasts"). These latter strains are useful because they are easier to separate from beer after fermentation because they form thick flocs (clumps) that settle to the bottom of the fermentation tank. The beer can then be removed, leaving the sediment behind. Super-flocculent yeasts also tend to be stickier in general, and thus are more easily immobilized to cellulose beads, wood chips, or other substances. Scientists are very interested in using immobilized yeasts for either primary or secondary fermentation, because theoretically this could shorten substantially the time required for fermentation and make it more energy efficient. Brewers could use immobilized yeasts in a **continuous** fermentation system rather than the almost universal **batch** system (see Chapter 8, Section III.B, for a discussion of continuous bioreactor operation).

III. DAIRY BIOTECHNOLOGY

A. INTRODUCTION

Few members of the public realize that when they are consuming yogurt they are consuming live bacterial biomass, suspended in a mixture of acidified milk mixed with bacterial slime layers. Perhaps it is fortunate the public remains ignorant of this; many people would likely consider the idea of eating bacterial biomass to be somewhat disturbing. Nonetheless, yogurt is an excellent example of the utility of large-scale cell culture in the dairy industry. Yogurt, cheese, and buttermilk cannot be made without the participation of a specific group of Gram-positive eubacteria: the lactic acid bacteria. These bacteria are unusual in that they are relatively acid tolerant — they can survive and grow in acidic environments, some as low as pH 3. Other important characteristics of LAB include the following:

- They lack pathogenicity. They have a long history of safe use in foods and do not cause disease, except in rare and unusual circumstances. Some of the LAB can cause spoilage (e.g., *Pediococcus*), but this results in food deterioration, not danger to human health.
- They can ferment lactose and other carbohydrates into lactic acid. When inoculated into milk, LAB produce lactic acid, which lowers the pH of the milk. If enough lactic acid accumulates, casein in the milk coagulates, resulting in a semisolid product. In combination with chymosin (rennin), the resulting **curd** is quite firm, and when pressed and ripened, the curd develops into **cheese** (see Chapter 3, Section XI.A, for an explanation of

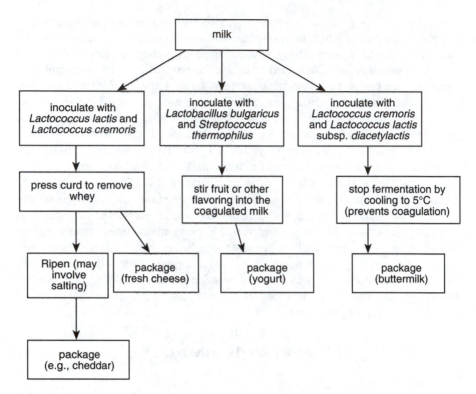

FIGURE 7.8 The use of lactic acid bacteria to transform milk into a variety of value-added fermented foods.

the action of chymosin). Products with a variety of consistencies are produced by different species; for example, sour cream, buttermilk, yogurt, feta cheese, and cheddar cheese are produced with help from LAB, but they have extremely different textures and consistencies. Each of these products is made with a specific suite of LAB (Figure 7.8).

- Certain LAB can improve the flavor of dairy products. For example, the diacetyl that gives buttermilk its characteristic flavor is produced by *Lactococcus cremoris*.
- Partly because of their pH-lowering fermentation products, and partly due to the common production of bacteriocins (compounds that inhibit other bacteria), growth of LAB in milk extends the shelf life of the milk and increases protection against pathogenic bacteria.

As is the case with the brewing industry, the overall strategies used for the production of fermented dairy products have not fundamentally changed from traditional practices. Cheese in particular has a long history of use by humans; it is likely that human consumption of animal milk did not occur to a great extent before cheese making was invented, because of the extremely short shelf life of unrefrigerated milk. Cheese has become immensely popular in many cultures, and cheese making has evolved into a complex art that can produce cheeses with a diverse range of flavors and textures. Cheese shares with wine and few other foods the distinction of being widely popular among average consumers as well as having cult-like aficionados who elevate it to the highest levels of gastronomy.

B. STARTER CULTURES

The major application of large-scale cell culture to dairy biotechnology is in the inoculation of starter cultures of LAB to milk, followed by the transformation of the milk into a variety of products. One of the major trends within the dairy industry is toward the use of defined starter cultures, instead of using undefined mixtures of bacteria. This allows improved consistency; a defined mixture of LAB should give consistent levels of lactic acid production and flavor compounds, leading to decreased variability in the time required for production and in the characteristics of the final product.

The use of defined starter cultures requires extensive knowledge of the biological characteristics of individual lactic acid strains. Our understanding in this area still has large gaps, but we can make some comments about specific LAB:

- *Lactococcus lactis* subsp. *lactis* and *Streptococcus thermophilus* are homofermentative, meaning that they produce only one end product of fermentation. *S. thermophilus* is unusual in that it is **thermophilic**, and thus can withstand higher temperatures (it grows well at 40 to 45°C). Most other bacteria used in starter cultures are **mesophilic** and require temperatures around 30°C. Lactic acid is the only product of fermentation. These bacteria efficiently acidify milk and are extensively used in starter cultures. *L. lactis* is usually included in cheese starter cultures, and *S. thermophilus* is usually used in yogurt starter cultures. *S. thermophilus* is also used in the production of certain hard cheeses (e.g., Emmenthal and Gruyère), because their production requires a heat treatment (~45°C) that nonthermophilic starters cannot withstand.
- *Leuconostoc mesenteroides* subsp. *cremoris* is a heterofermenter, meaning that it produces more than one end product of fermentation. It produces lactic acid and various flavor compounds (e.g., acetic acid). It is normally used in combination with homofermenters in starter cultures, although it can be used alone as a starter culture for certain cheeses (e.g., Brie and Camembert).
- *Lactobacillus delbrueckii* subsp. *bulgaricus* and subsp. *lactis* are homofermenters, producing lactic acid via fermentation. However, they are also important producers of flavor compounds and are popular additions to starter cultures. *L. delbrueckii* subsp. *bulgaricus* is usually part of yogurt starter cultures.
- *Bifidobacterium* spp. are obligate anaerobes found in the GI tracts of many animals. They are heterofermenters and are included in starter cultures because of their production of flavor compounds (e.g., acetic acid) and

because they are considered to be probiotic (health promoting). The probiotic aspects are linked to the ability of bifidobacteria to colonize the
human GI tract.

In cheese production, the main purpose of the starter culture is to acidify the
milk, thus promoting coagulation of the curd. However, in ripened cheese, the starter
organisms also play a role in modifying the texture and flavor of the cheese during
ripening. Ripening can occur in many ways, but it often consists of cutting blocks
of curd, floating them on brine solutions (this increases the salt concentration within
the curd, which discourages the growth of spoilage microbes), and then incubating
in a cool, humid environment for weeks or months. During this time, enzymes
present in the cheese modify the texture and flavor of the cheese. Because the main
source of enzymes is from either starter organisms or microbes that survived milk
pasteurization, the nature of the microbial flora in ripened cheese is of utmost
importance. Recently, there has been a trend toward the use of defined cultures of
bacteria or fungi during the ripening process, rather than relying on starter bacteria
or "contaminating" bacteria to affect the desired changes.

An increasing trend has been to consider the health-promoting aspects of the
bacteria that are used in starter cultures (these are called **probiotic** bacteria). Most
attention has focused on *Bifidobacterium* spp. and *Lactobacillus rhamnosus* GG.
These bacteria are able to colonize the human intestine and are thought to provide
protection against certain intestinal pathogens (e.g., *Clostridium difficile*). Although
they do not contribute greatly to acid or flavor production, they may make the final
product more attractive to health-conscious consumers. More conventional starter
organisms such as *Lactobacillus acidophilus* and *Lactobacillus bulgaricus* also
appear to have probiotic potential, although the benefit to the consumer may lie in
modulation of the immune system.

C. PHAGE

The biggest problem encountered during the fermentation of milk is **starter failure**
due to bacteriophage. Phage can spread rapidly through a fermentation vat and
completely stop the acidification of milk, leaving the milk vulnerable to spoilage by
competing microorganisms. Pathogenic organisms such as *Staphylococcus aureus*
may also grow to high densities, creating a potential public health problem through
the accumulation of toxins.

The source of phage is either the milk or the starter culture. Many phage are
heat resistant and are unaffected by pasteurization; for this reason, fermentation
processes may need to include a more rigorous heat treatment of the milk (90°C for
30 min). In most cases, though, the emphasis is on prevention of phage infection of
the starter cultures. This is usually accomplished by using **phage-inhibitory media**
(see below) to grow starter organisms, if the cheese maker produces its own inocula
of starter cultures for its milk fermentations. A popular alternative is to purchase
freeze-dried cultures that can be directly inoculated into the fermentation vat, thus
avoiding potential phage contamination during in-house starter cultivation.

Fortunately, phage infection does not automatically lead to starter failure. Like many other viruses, there is strain-to-strain variation in **virulence**. **Virulent** phage proceed through the lytic cycle (see Chapter 2, Section IV.A), killing the host bacterium, whereas **temperate** phage have a period of dormancy that does not immediately lead to death of the host cell. Some strains exhibit intermediate levels of virulence and only cause starter failure if fermentation vats are not completely sanitized between production runs. Such phage often replicate to high numbers (e.g., 10^8 per milliliter) without significantly affecting the growth of LAB or, more importantly, the rate of lactic acid production. However, if this level of inoculum is still present when a new batch of milk is added, then starter failure is likely.

Lysogenic strains can also cause starter failure, because mutant, virulent viruses occasionally arise from such strains. Also, most lysogenic strains have a low frequency of spontaneous induction of lytic phage; consequently, most cultures that carry lysogenic phage also have "free" phage that have been released via the death of phage-containing cells that have been induced into the lytic cycle. In many countries (e.g., France) **artisan** cultures, used by small-scale manufacturers of specialty cheeses, are a potent reservoir of lysogenic phage and can lead to dispersal of phage to other cheese makers.

Thus, phage is a prevalent and potentially devastating problem for cheese makers and other manufacturers of fermented milk products. Fortunately, two strategies are available to combat phage:

- **Phage-resistant strains** (also known as bacteriophage-insensitive mutants, or **BIMs**). LAB are not universally susceptible to phage. Genetically derived resistance in some cases is due to well-characterized **restriction–modification** systems (the combination of restriction enzymes that degrade phage DNA and methylases that protect bacterial DNA from the restriction enzymes — see also Chapter 3, Section IV.A). In other cases, the resistance is less well understood, but it appears to be related to bacterial interference with phage absorption or injection. Many of the resistance genes are plasmid encoded and therefore can be transferred to other LAB.
- **The use of phage-inhibitory media to cultivate starter organisms**. It has been known since the 1960s that the addition of calcium chelators to milk inhibits phage multiplication. Calcium is required for phage multiplication, and the addition of chelators results in suppression of a wide range of phage. The only drawback to this system is that the chelators often inhibit the growth of LAB as well.

D. Recombinant Lactic Acid Bacteria

1. Improved Starters

Researchers have shown great interest in the last decade in developing techniques to transform *Lactococcus* spp., *Lactobacillus* spp., and other LAB. This is partly due to their importance to the dairy industry, but also because of their importance to other food and beverage fermentations, such as the use of *Lactobacillus plantarum* in sauerkraut production and the use of *Leuconostoc* spp. to drive the malolactic fermentation in wine. In the latter process malic acid is converted to lactic acid, decreasing the sour taste of certain wines.

Unfortunately, the LAB are more difficult to transform than bacteria such as *E. coli*. However, considerable progress has been made, and successful reports of recombinant LAB are appearing with greater frequency in the literature. As an example, we will discuss the production of stable recombinants of *Lactobacillus casei* that have immunity to phage A2.

Plasmid vectors are available for the transformation of LAB, but plasmids have strong disadvantages for use as cloning vectors for LAB utilized in food fermentations. The main reason is that plasmids are unstable (can be lost by the host bacterium) in the absence of selective pressure that favors plasmid retention. In routine cloning experiments with plasmid vectors, this problem is solved by inserting antibiotic-resistant genes into the vector. Antibiotics in the growth media then exert selective pressure for maintenance of plasmids within host cells (see Chapter 3, Section IV.B). Obviously, it is impractical (and illegal in most countries) in a food fermentation setting to add large amounts of antibiotics. There is also a less obvious problem — antibiotic-resistant genes are considered to be **non-food-grade** DNA. The incorporation of this DNA would mean that the LAB would lose its **GRAS** (generally recognized as safe) status and would not be usable in food fermentations. LAB used in food fermentation have GRAS status, which means that the U.S. Food and Drug Administration (FDA) does not consider these bacteria to be food additives. Their use in food is less restricted than would be the case if they were considered to be additives.

The best way to circumvent the problem of plasmid instability is to **integrate** desired DNA sequences into the bacterial genome. A Spanish research group recently demonstrated that this is feasible, by inserting a gene from phage A2 into the chromosome of *L. casei*. The resulting recombinant strain is immune to infection by phage A2. Two plasmid vectors were used (Figure 7.9). One plasmid (pEM76:cI) contained the phage gene (*cI*) as well as a gene (*int*) that directs integration of associated DNA into a specific region of the *L. casei* genome. DNA sequences designated as *six* were also present in this vector; they flanked a sequence of DNA containing several antiobiotic-resistant genes. The second plasmid (pEM68) contained a gene coding for **β-recombinase**. This gene codes for an enzyme that deletes DNA between *six* sequences.

L. casei was first transformed by electroporation with pEM76:cI (Figure 7.10). The integrase coded by *int* then catalyzed insertion of this plasmid into the chromosome of *L. casei*. The resulting transformants could be selected because of their antibiotic resistance. Note that pEM76:cI does not have an *ori* for *L. casei*; hence, cells would grow in the presence of the antibiotic only if pEM76:cI was integrated into the genome.

The next step was to remove the antibiotic-resistant gene once recombinant bacteria were isolated. This was achieved by transforming these bacteria with pEM68. The β-recombinase then directed the removal of the antibiotic-resistant genes because they were flanked by *six* sequences. This led to bacteria that were

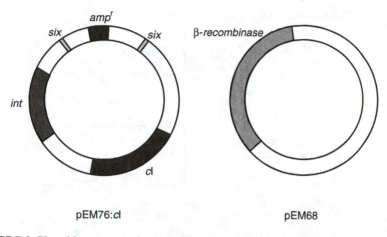

FIGURE 7.9 Plasmid vectors used to transform *Lactobacillus casei* with stable phage resistance. (amp^r, ampicillin resistance.) (Data from Martin, M. C., et al., *Appl. Environ. Microbiol.*, 66, 2599, 2000.)

sensitive to the antibiotics but had the phage A2 gene stably integrated into the bacterial genome. When challenged with phage A2 in milk fermentation assays, the recombinant bacteria were immune. This occurred because virally infected cells are often immune to infection by other viruses. In this case the presence of a single viral gene was sufficient to generate a similar degree of immunity.

2. Recombinant Lactic Acid Bacteria as Vaccines

LAB have great potential to deliver vaccines that protect against human disease. Many have GRAS status, making them amenable to **oral** delivery of vaccines. Oral vaccines are preferable to injected vaccines because they are better suited to distribution and use, particularly in the developing world. They also tend to stimulate the mucosal branch of the immune system. Because most pathogens enter the human body by breaching mucosal defenses (e.g., intestinal or lung mucosal epithelia), mucosal immunity is considered to be more protective than a more generalized immune response.

Food-grade LAB are also attractive as vaccine delivery systems because they are closely related to several pathogens, including *Streptococcus pyogenes* and *Streptococcus pneumoniae*. Thus, they have a similar genetic background to these pathogens. The importance of this similarity is illustrated by a project that involved transfer of *S. pneumoniae* genes required for capsule production to *L. lactis*. These capsules form a polysaccharide covering for *S. pneumoniae* and are key virulence factors (traits related to the ability of a pathogen to invade and cause damage in a host), because they help the bacterium evade engulfment by phagocytic cells of the immune system. The genes were successfully inserted into *L. lactis* via a plasmid vector, and the resulting recombinant bacteria, after injection into the peritoneal cavity of mice, provoked an antibody response against *S. pneumoniae* capsules.

When a similar strategy was attempted with *E. coli*, intact capsules were not assembled and thus were incapable of causing the desired immune response. Evidently,

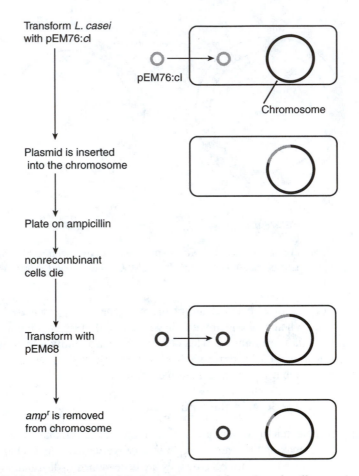

Transform *L. casei*
with pEM76:*cI*

pEM76:cI

Chromosome

Plasmid is inserted
into the chromosome

Plate on ampicillin

nonrecombinant
cells die

Transform with
pEM68

amp[r] is removed
from chromosome

FIGURE 7.10 Method used by Martin et al., to transform *Lactobacillus casei* with stable phage resistance. (Data from Martin, M. C., et al., *Appl. Environ. Microbiol.,* 66, 2599, 2000.)

the genetic and cytoplasmic environment of *L. lactis* was more similar to that of *S. pneumoniae*, allowing correct assembly of the polysaccharide capsule.

We have discussed only two of many potential examples of the utility of recombinant LAB. The broad use of these transformants, though, will depend on further research into technical aspects of transformation and expression of heterologous proteins in LAB and on public acceptance of products derived from recombinant LAB.

IV. AMINO ACIDS:
NUTRITIONAL BOOSTS AND FLAVOR ENHANCERS

A. OVERVIEW

Amino acid harvesting from microbial cultures is an economically important industry. The Japan Amino Acid Association estimates that the worldwide production of amino acids is at least 1.66 million tons per year. Why produce such vast amounts of amino acids when they are part of the normal human diet and can be obtained from an incredible variety of animal and plant sources? The answer lies in six main areas:

1. **Flavor enhancers. Monosodium glutamate** is a flavor enhancer that is added to many processed foods (e.g., dehydrated soups). Glutamate production is estimated to be greater than 1 million tons per year.

2. **Supplements.** Many animal feeds are low in specific amino acids such as **lysine** and **methionine**. The addition of purified amino acids from microbial sources is a relatively inexpensive way of improving the nutritional quality of feed. Worldwide production of lysine and methionine is estimated to be at least 800,000 tons per year. Humans also consume amino acid supplements; from a nutritional perspective, this is unnecessary for people that ingest protein from a variety of sources. However, some alternative health practitioners recommend consumption of amino acids for a variety of ailments. Indeed, certain amino acids have pharmacological effects, because of their structural similarity to human metabolites (e.g., tryptophan and serotonin).

3. **Sweeteners. Aspartame** is a popular ingredient of soft drinks and many processed foods. It is a methyl ester of aspartyl-phenylamanine, which is made using aspartate and phenylalanine.

4. **Parenteral nutrition (e.g., intravenous infusions).** Although this is less important economically, the development of technology for the production of many amino acids has been spurred by the need for nutritive, injectable solutions for people who are unable to ingest food.

5. **Industrial uses.** Certain amino acids have nonfood uses. For example, phenylalanine is used as a component of detergents, and as a chelator during water purification.

6. **Pharmaceuticals.** Amino acids such as L-proline and D-alanine are used as building blocks for the construction of pharmaceuticals.

These applications have led to the establishment of industrial operations that produce billions of dollars worth of amino acids every year. These industries did not develop overnight, though. With the exception of glutamate, it has proven to be extremely difficult to find microorganisms that naturally overproduce amino acids. Consequently, much effort has been expended to modify microbes in order to achieve overproduction. The most successful method so far has been to isolate **mutants** that have defective control over amino acid production. A number of recombinant DNA approaches have also been successful. The latter process is an example of **metabolic engineering** and is only possible if we understand the pathways of amino acid biosynthesis and related pathways (e.g., those supplying required coenzymes and substrates). The relationships between biosynthetic pathways and other pathways are also important, because metabolic engineering toward amino acid overproduction will inevitably disrupt overall metabolic flow. This can be visualized in terms of overall carbon and nitrogen; if substantial amounts of carbon- and nitrogen-containing compounds are diverted toward the production of a specific amino acid, that will reduce the flow of carbon and nitrogen to other pathways. Successful metabolic engineering requires a complete understanding of the ramifications of such disruptions.

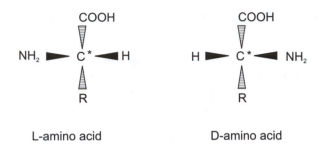

L-amino acid   D-amino acid

FIGURE 7.11 Stereochemistry of amino acids is dictated by the position of the *chiral carbon* (*) and its relationship with the nitrogen in the amino group. Solid lines indicate bonds that project out of the page, and dashed lines project in the opposite direction. D- and L-amino acids are present in many organisms (e.g., carnosine in the muscle tissue of cows and other vertebrates), but only L-amino acids are constituents of proteins.

In many cases, we do not have a sufficient understanding of amino acid metabolism to precisely engineer amino acid overproduction via recombinant DNA technology. This is why mutagenesis approaches are still popular, because they can be applied to "black box" systems that are incompletely understood. This section examines the use of both mutagenesis (proline) and recombinant DNA approaches to achieve amino acid overproduction. We will also discuss examples of the use of natural amino acid overproducers (*Corynebacterium glutamicum* and glutamate) and the use of immobilized enzymes to produce amino acids (aspartate).

One question frequently arises: Why don't we obtain amino acids from meat or other protein-rich foods? In fact, this is a valid approach that is sometimes successful. The major problem is that natural protein sources contain a wide range of amino acids; separation of a specific amino acid from all the others is expensive. Another option is to produce the amino acid synthetically using the many elegant methods developed by organic chemists. The major problem with this strategy is that it is difficult to synthesize specific stereoisomers of an amino acid. Amino acids exist as D and L stereoisomers (Figure 7.11); D-amino acids are present in all cells but not as a constituent of protein. For many of the applications of amino acids, only one of the isomers is useful. The reason for this is clear in the case of amino acids used as feed supplements or infusions — if the ultimate aim is to supplement a human or animal with amino acids that will be used to build protein, then it is essential to use L-amino acids. In other applications, the stereochemistry is equally important. For example, L-glutamate is an effective flavor enhancer, but D-glutamate is completely ineffective. For these reasons, it is usually preferable to use microbes or microbial enzymes to produce most amino acids, so that rigid control over the optical characteristics of the amino acid can be achieved.

B. CHOICE OF MICROBE

It is easy to convince microbes to produce large amounts of an end product of fermentation (e.g., ethanol). Simply supply sufficient fermentable substrates and the correct environment. In contrast, the overproduction of amino acids is a challenging

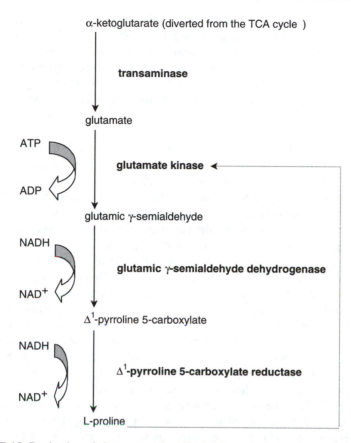

FIGURE 7.12 Production of glutamate and proline by *Serratia marcescens* and the associated energy costs. The cell must invest energy in the form of ATP, NADH, and α-ketoglutarate. Energy must also be expended to synthesize four enzymes. Finally, energy is lost because of the diversion of α-ketoglutarate from the TCA cycle. The dashed arrow indicates feedback inhibition by proline.

proposition. Fermentation end products normally leave cells quickly, either through diffusion across membranes (e.g., ethanol) or through active transport (e.g., lactic acid). Thus it is not difficult to separate product from cells, if desired. However, microbes do not normally secrete amino acids, because they are usually required within the cell. Secretion of amino acids is energetically unwise for microbes, because amino acid synthesis typically requires the coordinated activity of several enzymes (Figure 7.12). This makes amino acid production an expensive process that is necessary, but must be tightly controlled, so that costly, unnecessary overproduction does not occur. The secretion of amino acids *serves no purpose* for a microbe, and therefore it is not surprising that the isolation of a bacterium that naturally overproduces an amino acid is extremely rare. Consequently, biotechnologists have been forced to artificially manipulate amino acid metabolism to achieve overproduction.

The choice of organism is critical to the development of an overproducing microbe. *E. coli* would seem to be an ideal microbe for the development of an overproducing strain because it is such a well-characterized organism. However, *E. coli* is unsuited for most amino acid applications, with some important exceptions (e.g., aspartate). Amino acid production in *E. coli* is controlled through regulation of enzyme **activity, transcription of mRNA, and translation**. First, we need to understand how this is achieved:

1. **Regulation of enzyme activity through feedback inhibition**. Many enzymes have an **allosteric** site, which is spatially separated from the active site. The end product of the pathway specifically binds to the allosteric site, which results in a conformational change that reversibly inactivates the enzyme. The affected enzyme is typically at the *beginning of the pathway*. This allows the pathway to be shut off, without risking the build-up of intermediates of the pathway. In the case of proline synthesis (see Figure 7.12), proline allosterically inhibits glutamate kinase.

2. **Regulation of transcription** (e.g., repression). This is conceptually similar to feedback inhibition; the difference is that enzyme activity is unaffected, but transcription of enzymes is repressed (Figure 7.13). The end product of an amino acid synthetic pathway typically binds reversibly to a **co-repressor** protein. This binding results in an active repressor complex, which binds to the operator region of the appropriate **operon**. In bacteria, the genes for pathways of amino acid synthesis often follow a single promoter, so that all enzymes of the pathway are synthesized together. This type of gene arrangement is an operon. As long as the co-repressor is bound to the operator, none of the genes of the operon will be expressed, and none of the enzymes will be synthesized. When cytoplasmic levels of the amino acid fall, the repressor complex disassociates, and the promoter can then interact with RNA polymerase to initiate transcription.

3. **Attenuation**. This complex mode of regulation acts during translation. If a particular amino acid is present at high concentrations in the cytoplasm, then transfer RNA (tRNA) charged with that amino acid will also exist at high levels. For some amino acids, this leads to specific disruption of translation of the operon responsible for synthesis of that amino acid.

In *E. coli*, overproduction of an amino acid can occur only if *all of these levels of regulation are overcome*. This is possible and has been successfully accomplished in some cases, but, in most cases, better results have been obtained using other bacteria.

Serratia marcescens is much easier than *E. coli* to convert to an amino acid overproducer. It is a member of the Enterobacteriaceae family, and its major claim to fame is its distinct red pigmentation. In this bacterium, feedback inhibition is the most important mechanism of regulation of amino acid synthesis. Why does *E. coli* have such complex mechanisms of regulation, but for *S. marcescens* simple regulation of enzyme activity is sufficient? The answer lies in the very different environments of

A. Proline is present

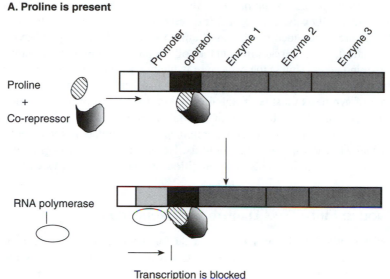

Transcription is blocked

B. Low levels of proline

Transcription is unblocked

FIGURE 7.13 Regulation of amino acid synthesis by co-repression. (A) When the amino acid is present in the cytoplasm, it combines with a co-repressor protein to form a complex that binds to the operator site of the operon. That site contains genes coding for enzymes that are responsible for synthesis of the amino acid. When the complex is bound to the operator, RNA polymerase is unable to transcribe the genes of the operon. (B) When the amino acid is present at low levels, the co-repressor alone is unable to block transcription, which proceeds through the operon.

these bacteria. *E. coli* lives in the mammalian GI tract, an environment of extreme fluctuations in amino acid supply. Bacteria are unable to predict when their hosts will eat, so they must have the ability to rapidly adjust their rates of amino acid synthesis. As an example, consider a human who eats a bowl of of pasta primavera. The food contains an abundance of carbohydrates, but relatively few proteins and amino acids. If *E. coli* is to grow and compete with other bacteria, it must rapidly assimilate carbohydrates and synthesize amino acids from these carbohydrates. Rapid *derepression* of anabolic pathways is therefore essential. However, if the human host then consumes several glasses of milk, a rich source of amino acids, then the bacterium must adapt quickly. Amino acid synthesis is no longer necessary and is energetically wasteful, because it diverts energy from other pathways. Hence, *E. coli* requires fine-tuned regulation of amino acid biosynthetic pathways.

In contrast, *S. marcescens* lives in the soil, an environment that is usually starved of free amino acids. Unless the bacterium is lucky enough to find itself in a protein-rich habitat (e.g., a decomposing racoon), it must synthesize its own amino acids. The rate of synthesis must be metabolically tied to the rate of growth, but this can be done quite simply, by feedback inhibition. If the cell is not growing, it has few needs for amino acids; consequently, amino acids accumulate in the cytoplasm and inhibit pathways leading to their synthesis. However, if the cell is growing, it will use up its store of amino acids, and inhibition of anabolic pathways will no longer occur.

C. Proline: Mutagenesis Leads to Overproduction

Proline is synthesized on an industrial scale for use in intravenous infusions. It is produced through a relatively simple pathway that uses glutamate as the initial substrate (see Figure 7.12). In *S. marcescens*, proline biosynthesis is regulated predominantly by feedback inhibition. In developing a strain of *S. marcescens* that overproduces proline, the first obstacle was overcoming feedback inhibition. This was accomplished using a similar approach to that used to overcome catabolite repression in brewing yeasts (this chapter, Section II.G). Mutants were plated onto media containing proline analogs. Wild-type cells died because they were unable to synthesize proline. This occurred because the analogs shut down proline biosynthesis through feedback inhibition. However, mutants that had impaired feedback inhibition were able to synthesize needed proline and were able to survive and grow.

These mutants overproduced proline, but further increases in the rate of proline synthesis were required. The following steps resulted in strains of *S. marcescens* that dramatically overproduced proline:

1. The gene for an enzyme that degrades proline was deleted from the chromosome of *S. marcescens*. This enzyme allows the bacterium to scavenge proline from its environment and use it as a carbon and nitrogen source, but is unnecessary and undesirable in a proline overproducer.

2. It was observed that growth of *S. marcescens* in a high-salt medium results in further increases in the rate of proline production. *Serratia*, like many other bacteria, uses proline as an osmoprotectant when growing in osmotically stressed environments. Osmoprotectants are compounds that can accumulate within cells to high concentration without adversely affecting cellular metabolism. This is beneficial in high-salt (low-water activity) environments because it decreases cellular water activity enough to allow water to flow into the cell through osmosis. This prevents cell dehydration. When grown in a high-salt medium, the bacteria produced 60 to 75 g/L of proline, and ~20% of the carbon flow was directed to proline production.

3. Finally, through recombinant DNA techniques, the genes coding for enzymes required for proline synthesis were cloned and inserted into plasmids that maintain a high copy number in *serratia*. This resulted in a further 50% increase in the rate of proline production.

Proline overproduction, then, was possible through the use of a random process (mutagenesis) combined with exploitation of an aspect of microbial stress physiology (osmoprotectants) and the use of recombinant DNA techniques. Although overproducing strains were achieved, it is clear that the same strategy would not work with other amino acids. One reason for this is that other amino acids are not used as osmoprotectants by microbes, so growth in high-salt media would not result in increased amino acid production. Also, some amino acids (e.g., leucine) are produced by branched pathways, in which several different end products exert feedback inhibition. This makes it much more difficult to obtain mutants deficient in feedback inhibition. Other strategies must be implemented to obtain overproducers of these amino acids.

D. GLUTAMATE: NATURAL OVERPRODUCTION

Glutamate is an interesting amino acid for several reasons. It was the first amino acid to be produced from microbial culture on an industrial scale, and, along with L-alanine and L-valine, is the only amino acid that is overproduced by wild (unmodified by humans) bacterial strains. Furthermore, glutamate is economically the most important amino acid; it is extensively used within the food industry as a flavor enhancer and is a popular flavor ingredient throughout the world. Glutamate holds a special place in sensory science because it is the fifth basic taste. Human taste buds have specific receptors for glutamate, along with receptors for sour, sweet, bitter, and salty tastes. The taste of glutamate (**umami**) is usually described as "brothy."

The flavor-enhancing capabilities of glutamate were utilized in Japan prior to the availability of purified monosodium glutamate. Dried kelp (konbu) has a long history of use as a food seasoning in Japan, and in the early 20th century K. Ikeda, a Japanese scientist, discovered that glutamate was responsible for the flavoring effects of konbu. Then, in the 1950s, came an unusual finding. Bacteria that naturally overproduce glutamate were isolated from the soil. They belonged to the ***corynebacterium–brevibacterium*** group and are still used today to produce glutamate on an industrial scale. Although the taxonomy of these bacteria is still somewhat unclear, isolates used for overproduction of glutamate are usually called *C. glutamicum*.

This isolation of a natural overproducer was fortuitous. As explained previously, microbes do not normally overproduce amino acids, unless they are used as osmoticants in media with a high solute or low water content. *C. glutamicum* does not use glutamate as an osmoticant, and, in any case, the bacteria secrete large amounts of glutamate, an action that would not be taken with an osmoticant (to protect a cell from the adverse effects of low water potential, osmoticants must be retained within the cell). What other reasons could a microbe have for the overproduction of an amino acid?

After much research, this question has been partially answered. We now know that glutamate overproduction in *C. glutamicum* occurs for two reasons: (1) under certain conditions (e.g., biotin starvation) the membranes become highly permeable to glutamate, and (2) under conditions of low availability of oxygen catabolic pathways shift toward the overproduction of glutamate. The combination of these two phenomena results in the secretion of large amounts (100 g/L) of glutamate.

We do not completely understand why the membranes of these bacteria become selectively permeable to glutamate. It was previously thought that this permeability was a direct result of biotin starvation, because addition of biotin to the growth medium prevents the secretion of glutamate. Biotin is required for the synthesis of fatty acids, so biotin deficiency would be expected to interfere with the synthesis of membranes. However, such effects on membrane structure and integrity should have a wide range of effects on membrane permeability, and *should not* specifically change the permeability to glutamate. The answer to this puzzle appears to be related, at least partly, to the presence of specific **transporters** in the membrane of *C. glutamicum* that pump glutamate out of the cell. Perhaps these pumps are specifically activated under conditions of biotin starvation.

The shift in catabolism under conditions of low-oxygen partial pressure is also poorly understood. The respiratory pathway of *C. glutamicum* is unusual. Under aerobic conditions, carbohydrates are completely respired through a **glyoxylate shunt** (Figure 7.14), which is a modification of the conventional tricarboxylic acid (TCA) cycle. Under anaerobic conditions, the bacteria switch to fermentation, but in conditions of low-oxygen partial pressure, fermentation does not occur. The glyoxylate shunt also operates at low levels, probably because there is insufficient oxygen to oxidize the reduced nucleotides (e.g., NADH) that are normally generated by the glyoxylate shunt. The bulk of carbon flow from acetyl-CoA is toward α-ketoglutarate, and then to glutamate, which is secreted from the cell. In a normal TCA cycle, α-ketoglutarate is converted into succinyl-coA, generating NADH. The enzyme that catalyzes this reaction (**α-ketoglutarate dehydrogenase [α-KDH]**) has very low activity in *C. glutamicum*. The cell rids itself of α-ketoglutarate by converting it to glutamate and actively secreting it from the cell.

The presence of specific pumps for glutamate may be a clue that *C. glutamicum* derives an energetic benefit from the secretion of glutamate. Perhaps under conditions of low oxygen availability, when there is insufficient oxygen to fully oxidize carbohydrates to CO_2, there is still enough oxygen to oxidize a small number of NADH molecules. When glutamate is being secreted, NADH is produced through the formation of acetyl-CoA from pyruvate. Oxidation of this NADH by the electron transport chain would presumably give the cell more adenosine triphosphate (ATP) than it would derive from fermentation. Thus, glutamate secretion may be energetically wise for *C. glutamicum*.

Some gaps still exist in our understanding of glutamate secretion in these bacteria. Fortunately, this does prevent us from growing *C. glutamicum* on an industrial scale for the production of glutamate. We are fortunate that the environmental conditions that lead to glutamate overproduction and secretion (i.e., low levels of biotin and oxygen) are easily achieved. Growth media such as molasses contain biotin, but it tends to be used up fairly quickly. Oxygen levels also decline quickly

A. Catabolism of pyruvate in the presence of oxygen

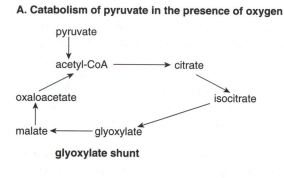

glyoxylate shunt

B. Catabolism of pyruvate under low oxygen availability

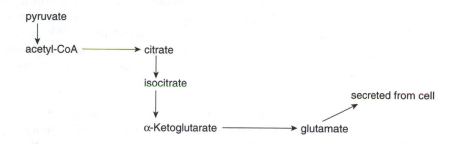

FIGURE 7.14 Carbohydrate catabolism in *Corynebacterium glutamicum*. (A) Under aerobic conditions, *C. glutamicum* operates a glyoxylate shunt, which is a truncated form of the TCA. (B) Under conditions of low oxygen availability, it diverts carbohydrates to glutamate production and secretion.

in dense microbial cultures. Consequently, normal methods of growing microbes lead to the correct environmental conditions for overproduction of glutamate.

E. ASPARTATE: CONVERSION BY IMMOBILIZED ENZYMES

Aspartame, the methyl ester of aspartyl–phenylalanine, is a popular alternative sweetener for soft drinks and a range of products aimed at the calorie-conscious market. Industrially, aspartame is produced through artificial (i.e., without the use of enzymes or cells) synthesis using aspartate and phenylalanine as reactants. Both of these amino acids are industrially produced using bacteria, although in very different ways. Phenylalanine is usually produced by large-scale growth of specific strains of soil bacteria (either *S. marcescens* or *C. glutamicum*). These strains are obtained using a similar strategy as described for obtaining overproducers of proline (i.e., mutagenize and then isolate mutants that have defective feedback inhibition). Phenylalanine can also be produced using immobilized enzymes derived from *E. coli*, but this is not economically feasible because of the high cost of the substrate (*trans*-cinnamic acid).

Although not feasible for phenylalanine, the use of immobilized enzymes is an economic and popular route for the production of aspartate. The enzyme used in this process is **aspartase**, the enzyme that catalyzes the *degradation* of aspartate into ammonia and fumarate. Aspartase is normally used by *E. coli* to scavenge aspartate from the external environment as a source of carbon or energy. However, it will efficiently convert fumarate to aspartate *if the concentration of fumarate and ammonia are high* (2 *M*). Most chemical reactions can proceed in both directions, and the tendency for a reaction to proceed in either direction is affected by the equilibrium constant (K_m) and the concentrations of the reactants and products. In the case of aspartase, if the initial concentration of aspartate is low, and fumarate and ammonia are present at high concentration, the reaction will proceed toward almost total conversion of fumarate to aspartate.

Purified aspartase can be immobilized onto solid supports, but it is equally efficient, and technically easier, to immobilize intact cells of *E. coli*. Carageenan works well as a solid support; cells of *E. coli* are added to molten carageenan, which is then cooled, granulated, and packed into a column. Usually, autolysis of the cells occurs within several days of immobilization. The enzyme is released from cells, but is entrapped within the carageenan, and remains stable for several years. Concentrated fumarate and ammonia are added to the top of the column, and aspartate is collected from the bottom. This process is ideal for the large-scale manufacture of aspartate, for three reasons:

1. The substrates (fumarate and ammonia) are relatively inexpensive.
2. Aspartate is produced in a concentrated form, making purification easy.
3. Complex growth media are not required; therefore, fewer unwanted solutes are present in the effluent from the column, simplifying purification.

Enzyme immobilization is also used industrially to produce L-alanine, L-cysteine, D-*p*-hydroxyphenyl-glycine, L-dihydroxy-phenylalanine, and a number of other amino acids and amino acid derivatives.

F. TRYPTOPHAN AND EOSINOPHILIA–MYALGIA SYNDROME

Because amino acids have specific industrial uses, it is crucial that amino acid production lines end with efficient and effective purification. Contamination of an amino acid by another amino acid or other cellular metabolites would likely interfere with the end use of the amino acid. In the 1980s, an epidemic of **eosinophilia–myalgia syndrome** illustrated tragically that amino acid purification is also crucial to consumer safety. Between 1988 and 1990, 25 people, mostly in Japan, died from eosinophilia–myalgia. Hundreds of other people were injured, and the cause of illness was determined to be the consumption of tryptophan supplements. It was subsequently discovered that the tryptophan consumed by these people came from one company and contained trace levels of several potentially toxic compounds. After a thorough investigation, the presence of these toxins was traced to three factors:

1. A strain of *Bacillus amyloliquefaciens* was used to produce the tryptophan. This strain had recently been modified using recombinant DNA techniques, resulting in overproduction of the product of the first reaction in the synthetic pathway — phosphoribosylpyrophosphate.

2. A step in the purification process was changed. Some of the tryptophan bypassed a reverse osmosis membrane filtration step (see Chapter 8, Section IV.C, for an explanation of reverse osmosis).

3. Lower amounts of activated charcoal were used to remove impurities in the tryptophan.

If only *one* of the preceding factors had been operative, the product may have been safe. However, the combination of increased levels of tryptophan and reduced purification efficiency had lethal consequences. The most important lesson learned from this tragedy is that the overproduction of amino acids can potentially lead to the accumulation of toxins; this is probably related to the reactivity of amino acids, particularly when they are present at high concentrations.

Purification is an expensive component of amino acid production systems. Consequently, there is a powerful temptation for companies to cut corners and allow less-purified products to be sold. The eosinophilia–myalgia scandal amply illustrated that this is an unwise course of action.

Many opponents of recombinant DNA technology use eosinophilia–myalgia as an example of the potential disasters caused by the production of genetically engineered organisms. This is misleading, because this particular disaster could easily have happened if nonrecombinant methods (e.g., mutagenesis) had been used to boost production of tryptophan. Also, it is unclear which of the three causative factors were most important to the appearance and persistence of the toxins. Eosinophilia–myalgia is a potent reminder of the need for regulatory agencies to monitor biotechnology and new methods of producing biotechnological products, regardless of whether recombinant DNA technology is involved.

V. MICROBIAL ENZYMES

A. OVERVIEW

Enzymes have a rich history of use in the food industry. They are used in the production or processing of starch, flour, cheese, fruit juices, artificial sweeteners, and meat, and are also frequently used in brewing and wine making. Enzymes are also used in agriculture; for example, dairy farmers use silage (corn and other crops fermented by LAB) as a nutritive, easily stored feed for cows. Fermentation by LAB lowers the pH of the silage, which inhibits the growth of spoilage organisms, as long as air is eliminated from the silage. Cellulases are often added to silage to increase the amount of fermentable sugars.

Within the food industry, enzymes are particularly useful if a specific chemical transformation is required. The transformation can be a hydrolytic breakdown (e.g., digestion of proteins by proteases), an additive reaction (e.g., the use of aspartase to synthesize aspartate), or a rearrangement (e.g., conversion of glucose to fructose by glucose isomerase). Usually enzymes are used in their "forward" direction (e.g., proteases are used to digest proteins); however, they are sometimes useful in their reverse direction. For example, we saw in the preceeding section of this chapter that aspartame is made from L-phenylalanine and L-aspartate. The industrial process of

TABLE 7.2
Applications of Enzymes to Food Production and Processing

Enzyme	Use	Product	Source of Enzyme
α-Amylase	Starch processing	Dextrins	*Bacillus subtilis*
			Bacillus licheniformis
β-Amylase	Brewing (mashing)	Maltose	*B. subtilis*
Glucoamylase	Starch processing and brewing	Glucose	*Aspergillus niger*
Glucose isomerase	Fructose production	Fructose	*Streptomyces* spp.
Invertase	Candy processing	Glucose + fructose	*Saccharomyces cereviseae*
Pullulanase	Starch processing	Debranched starch	*Klebsiella*
Pectinase	Juice clarification	Galacturonate	*Aspergillus oryzae*
Chymosin	Milk coagulation	Cheese curd	*Kluyveromyces* spp.
Rennin	Milk coagulation	Cheese curd	*Mucor miehei*
β-Glucanase	Brewing (mash)	β-Glucose	*A. niger*
Lipase	Cheese making	Flavor compounds	*Rhizopus oryzae*
Lactase	Dairy processing	Glucose + galactose	*A. niger*

joining these two amino acids to make aspartame involves the use of a protease working "backwards," synthesizing a dipeptide made up of phenylalanine and aspartate. Other methods for this synthesis exist, but enzymes are favored because they preserve the correct optical properties of aspartame. This is important because L-aspartame is sweet, but D-aspartame is bitter!

Theoretically, enzymes could be obtained from animals, plants, or microbes. Indeed, food-grade enzymes are obtained from all of these sources; plant-derived proteases, for example, are preferred over other sources for the purposes of meat tenderizing. However, for most other enzymatic processes, microbial enzymes are preferred, primarily because they are easier and cheaper to obtain. One reason for this is that many bacteria and fungi have the natural ability to secrete enzymes into their environment. This is an important trait for microbes that rely on the digestion of organic matter as a source of energy, carbon, and nutrients. When microbes secrete valuable enzymes, it is relatively easy to separate the enzymes from the cells — simple centrifugation may be sufficient. Another reason for relying on microbes is that several microbes, including the bacteria *Bacillus subtilis* and *Bacillus licheniformis* and the fungus *Aspergillus niger*, have a long history of safe use in the food industry (see Table 7.2 for other important enzyme producers). These microbes are also potent producers of a broad range of useful hydrolytic enzymes. This allows companies to supply enzymes as mixtures; for example, a company might market a product labeled as "proteases" that actually contains a number of different proteases, often produced by the same microbe. Mixtures of related enzymes are often useful, because many applications of enzymes are founded on empirical discoveries. Often, the precise mechanism of enzyme action is unknown, making it difficult to assign specific function to specific enzymes. For this reason, companies prefer to market enzymes as "categories" (e.g., proteases) rather than marketing specific enzymes.

Because the microbes used for enzyme production must be safe for food production, it is often possible to market mixtures of enzymes that have not been highly purified (in other words, purification to remove toxic compounds is unnecessary). This cuts the cost of enzyme production without compromising quality or safety.

The worldwide market for industrial enzymes is around $1 billion; at least 400 companies make enzymes, although 12 companies dominate the industry. Most of the enzymes produced by these companies derive from microbes and are used by brewers, wine makers, starch processors, dairies, and many other food and beverage processors. Many nonfood applications of enzymes also exist, such as the use of proteases and lipases in laundry detergents. We will now examine the major uses of microbial enzymes in the food industry.

B. Amylases

1. Starch Processing

Amylases are arguably the most important enzymes in the food industry. Endogenous amylases in barley are crucial in malting of barley and other cereals (see this chapter, Section II, B and C) and are added to flour to improve baking characteristics. They are the cornerstone of the processing of inexpensive starch to produce a range of value-added products that are used as food ingredients (e.g., glucose).

Starch is a remarkable and useful polymer. It can be transformed into a number of products with many different uses. Some of these transformed products are useful because of their physical properties — for example, maltodextrins are important components of many confections because they give food a chewy texture and are a cheap and bland-tasting filler. Other products of starch processing are valuable sweeteners; the main source of fructose, which is sweeter than sucrose, is from the enzymatic conversion of glucose (derived from starch).

Starch is the most common compound used by plants to store carbohydrates. Plant cells often contain specialized organelles known as **amyloplasts** that are filled with starch grains. Roots and seeds often have a high starch content, as do corms, tubers, bulbs, and fruits. Many of the food crops that humanity is highly dependent on, such as rice, wheat, corn, and potatoes, have high starch content. These are all heavy-yielding plants; consequently, there is an abundant and cheap supply of starch in most regions of the world. Combined with the development of systems for the mass production of inexpensive starch-modifying enzymes, this has led to the establishment of a large and important industry.

Starch is composed of linear chains of α-1,4-linked glucose monomers (Figure 7.15). In **amylose**, the individual chains are unbranched; however, in **amylopectin**, branches occur through the formation of α-1,6 bonds. Amylopectin is more insoluble in water than amylose and is generally more difficult for microbes to fully degrade. The length of each linear chain varies widely but is often in the range of 1000 residues.

A number of enzymes are involved in starch digestion and processing. Some of the following enzymes take part in starch digestion, whereas glucose isomerase and cyclodextrin glycosyltransferases drive the conversion of glucose into other useful compounds:

A. Amylose

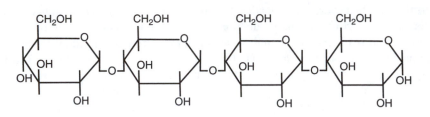

B. Amylopectin

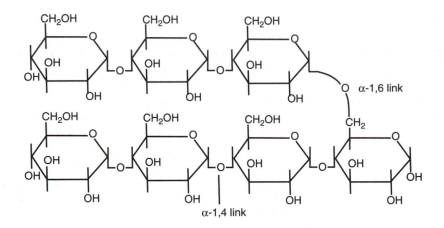

C. Sites of enzyme attack

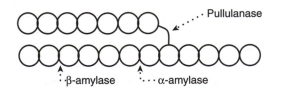

FIGURE 7.15 Structure of amylose (A) and amylopectin (B). Sites of enzymatic digestion of starch are also shown (C), with each circle representing a glucose monomer.

- **α-Amylases** attack α-1,4 links *within* starch. These enzymes increase the solubility of starch because they break the starch into smaller pieces (see Figure 7.15c).
- **β-Amylases** break α-1,4 links at the *nonreducing* ends of starch molecules, releasing maltose (see Figure 7.15c).
- **Pullulanases** are considered to be *debranching* enzymes, because they attack α-1,6 links that join linear starch chains (see Figure 7.15c).

- **Glucoamylases** release glucose monomers from the nonreducing ends of the starch chains.
- **Glucose isomerases** convert glucose to fructose.
- **Cyclodextrin glycosyltransferases** are used to convert dextrins to cyclic compounds. These compounds are not currently widely used in the food industry, but they may be in the future. They form **inclusion complexes** that can enclose "guest" molecules (this is an example of **encapsulation**); this alters properties of the guest molecule and may enhance stability or activity of enzymes or flavor compounds.

These enzymes have distinct uses in starch processing. The nature of the desired end product of the process dictates whether each enzyme is used. Because of the diversity of starch-derived products, it is useful to have a system that allows rapid comparison of compounds. The **dextrose equivalent (DE)** system compares each compound to dextrose (glucose). If a starch molecule is completely hydrolyzed into glucose monomers, this is considered to be **100% dextrose conversion**, or a product with a DE of 100. The range of DE is therefore between 0 and 100; the DE value is a good indicator of the physical nature and potential uses of a modified starch product.

Unmodified starch has a DE of 0, or close to 0. It is very similar to the starch found in the source plant. It has many industrial nonfood applications and is used extensively in the food industry as a thickener (e.g., corn starch). **Maltodextrins** have a DE of less than 20. Like unmodified starch, they are not sweet and have a bland taste. They have a myriad of uses within the food industry. The addition of maltodextrins to a food makes the food less hygroscopic and often gives it a chewy texture. The food is also less susceptible to drying out, and if it is frozen (e.g., ice cream), ice crystal formation is inhibited.

Corn syrup solids are more highly modified than maltodextrins and have a DE of more than 20. They are added to food to protect it from oxidation and to encapsulate flavors. They also affect the physical properties of food and increase its ability to maintain moisture content. Unlike maltodextrins, however, they are moderately sweet. Maltodextrins and corn syrup solids consist of a mixture of dextrins, glucose, and maltose. As the DE increases, the proportion of dextrins decreases and the proportion of maltose and glucose increases.

Glucose (corn) syrup is highly modified starch and has a DE of close to 100. It is extensively used as an inexpensive sweetener in foods. It is also a popular additive to low-calorie foods because it can substitute for fats in certain processed foods. **Maltose-rich syrups** are similar but consist of high levels of maltose and little free glucose. These syrups are moderately sweet and have the typical "malt" flavor that is desirable in breakfast cereals, for example. Finally, **high-fructose syrups** are popular sweeteners in many processed foods and are made by enzymatic conversion of glucose syrups. Fructose is twice as sweet as sucrose.

A range of products can be obtained from starch through the judicious use of microbial enzymes. This requires tight control over the extent and pattern of hydrolysis. In the past, this was done by chemical hydrolysis (e.g., hydrochloric acid). Enzymes are preferred because they give the processor greater control over the process and do not result in large salt residues, as does chemical hydrolysis. A typical enzymatic starch degrading operation consists of the following steps:

1. **Liquefaction.** The starch is heated to 95°C and α-amylase from *B. licheniformis* is added. This heat-stable enzyme solubilizes the starch and releases it from starch grains. After treatment with α-amylase, the starch has a DE of 10 to 12.

2. **Saccharification.** Glucoamylase from *A. niger* generates glucose from nonreducing ends of the starch fragments. Pullulanase from *Bacillus* or *Klebsiella* is also added to break α-1,6 links in the starch. To avoid caramelization, the temperature is reduced to 60°C. The extent of saccharification can be controlled by varying the time that the starch is exposed to glucoamylase. Thus, maltodextrins, corn syrup solids, and glucose syrups are obtained through progressively longer saccharification treatment. If a high-maltose syrup is desired, then plant-derived β-amylase is used instead of glucoamylase.

3. **Purification.** This nonenzymatic step decolorizes the product. Typically, activated charcoal is added, then filtered out, and the filtrate is heated and evaporated to concentrate the solution into syrup. Activated charcoal absorbs many compounds, including those responsible for the brownish color of the glucose syrup.

4. **Isomerization.** This step is taken only if a high-fructose syrup is desired. Immobilized systems using glucose isomerase from *Streptomyces* spp. are popular. Usually, the enzyme is linked to a solid support, which is packed into a column. Using enzymes currently available, only 50% of a batch of glucose can be converted into fructose. We have not yet discovered an enzyme whose natural function is to catalyze the conversion of glucose to fructose *in one step*. Virtually all cells are capable of converting glucose to fructose, but *three enzymes are required*. It is much more desirable to use one enzyme in an industrial operation, so an enzyme is used that is capable of converting glucose to fructose in one step even though this is not the natural function of this enzyme. This enzyme — **xylose isomerase** — has a lower affinity for glucose than its normal substrate (xylose), so it is not possible to convert much more than 50% of the glucose to fructose. Fortunately, it is relatively easy to separate fructose from glucose through differential crystallization. This process is very important to the food industry — greater than 100 billion tons per year of fructose are used as sweeteners.

2. Amylases and Baking

α-Amylases are added to flour to improve baking properties. Flour naturally contains amylases derived from wheat grains, but they are not always present in sufficient amounts. Addition of α-amylases improves the quality of bread and other baked goods through:

- Decreased dough viscosity, leading to easier dough handling, and perhaps contributing to increased bread volume.
- Increased dextrinization during heat treatment, resulting in improved color and flavor.
- Increased supply of fermentable sugars, leading to increased bread volume via increased production of CO_2 by yeasts.
- Improved shelf life, perhaps through the formation of low-molecular-weight branched starch fragments that interfere with starch recrystallization. Recrystallization results in firmer crumb structure and an undesirable texture (hard and excessively chewy) — generally referred to as "bread staling."

The molecular mechanisms behind these improvements in bread quality are unclear but are probably related to the partial digestion of amylose and the generation of dextrins. There is currently great interest in the potential of other enzymes to improve the quality of baked goods. **Hemicellulases**, which release soluble carbohydrates from hemicellulose, one of the insoluble components of the plant cell wall, have received particular attention. Their use could increase the soluble fiber content of dough, resulting in improved functionality (in the health sense) of baked goods. Use of these enzymes may also improve intestinal absorption of nutrients derived from baked goods.

C. LIPASES

Lipases are increasingly popular and important contributors to food processing. Two major applications of lipases are: (1) specific tailoring of lipids to produce desired fatty acid structure, and (2) generation of volatile fatty acids to improve the flavor of butter, cheese, and various other foods. The first application makes use of the synthetic capabilities of lipases, and the second makes use of the hydrolytic properties of these enzymes. Lipid tailoring is achieved by **trans-esterification** (Figure 7.16). Recall that simple lipids are composed of glycerol with ester linkages to three fatty acids. If a simple lipid is exposed to an additional (desired) fatty acid and lipase, the enzyme will catalyze replacement of one or more of the fatty acids by the more desirable fatty acid. Some lipases catalyze this reaction at a specific point in the glycerol molecule (e.g., a 2-specific lipase exchanges fatty acids only at the second carbon of glycerol). Other lipases (e.g., those produced by *Candida rugosa*, a GRAS organism) are nonspecific and can generate several different lipids. Lipases from *C. rugosa* are particularly popular because they can exchange a wide range of fatty acids of different lengths and degree of unsaturation.

Thus, trans-esterification provides precise modification of lipid structure. This is sometimes used to produce specific types of detergents and biosurfactants. It also has great potential to increase the health-promoting aspects of vegetable oils, because it can be used to increase the content of fatty acids such as linolenic acid, which are generally beneficial to human health. This technology can also be used to modify melting points, solubility, and other physical properties of edible oils.

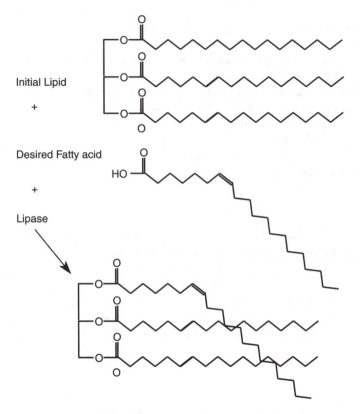

FIGURE 7.16 The use of lipases to drive trans-esterification. One of the fatty acid chains is replaced by a chosen fatty acid.

To "convince" lipases to function as synthetic enzymes, one must decrease the water content to low levels (often <5%). Organic solvents are often used to achieve this, although several immobilization strategies can be used if organic solvent use is undesirable. When lipases are present in the liquid–organic interface, they behave as synthetic enzymes, primarily because of the decreased water concentration (recall that water is a reactant in hydrolytic reactions). These are referred to as **solvent-partitioned** systems.

The hydrolytic function of lipases is also useful because it can be used to generate free fatty acids, which often contribute to food flavor. In cheese making, lipases can be used to shorten the ripening period and improve cheese flavor. They can also be added with LAB during curd formation to generate intensely flavored cheese, which can be converted to a powder and used to add cheddar and other cheese flavors to crackers and other snack foods. These **enzyme-modified cheeses** are an inexpensive way to add flavors characteristic of well-aged cheeses.

D. Polygalacturonase

Pectin is the glue that sticks plant cells together. It is a component of cell walls and is composed of a backbone of α-1,4-linked galacturonic acid, with side chains of arabinanes, rhamnose, and arabinogalactane. Some fruits and vegetables have a high pectin content, which sometimes causes problems. For example, apple flesh has a high content of large pectin polymers, known as protopectin. Normally, as fruit ripens, protopectin is hydrolyzed to shorter polymers. However, the level of protopectin in apples is high enough to cause problems during juice processing. Protopectin is insoluble in water, so it forms large complexes that create haze in apple juice. The easiest way to remove this haze is to treat the juice with **polygalacturonase** (PG) derived from microbial sources. This enzyme converts protopectin to pectin, which is water soluble, thus eliminating the haze problem. A similar process is often used to clarify and increase the yield of grape musts for wine making. PG is usually obtained from *Aspergillus* spp.; it is a relatively common enzyme among both saprophytic and plant pathogenic fungi.

VI. MICROBIAL POLYSACCHARIDES

A. OVERVIEW

As discussed in the previous section, starch and the products of starch processing can be very useful food ingredients. However, starch is not the only polysaccharide used by food processors. Many other polysaccharides derived from plants or algae are popular. For example, carageenan, a complex polymer derived from seaweed, is widely used in ice cream and many other foods. Guar gum and locust gum are examples of plant-derived polysaccharides that are commonly added to food as thickeners or as modifiers of other physical properties of food. These algal and plant products are not usually considered products of biotechnology; however, certain useful polysaccharides *are* considered to be produced through biotechnology, because they are derived from microbial culture. Some of these microbial polysaccharides are used in a relatively uncharacterized and uncontrolled fashion — for example, many of the LAB produce slime layers that are rich in complex polysaccharides and are important contributors to the texture of fermented dairy products.

Other microbial polysaccharides are produced in large-scale culture, are purified, and are sold to food processors as a food additive. The most important of these is **xanthan gum**, produced by *Xanthomonas campestris* (a bacterium). Other microbial polysaccharides used in the food industry include **pullulan gum**, produced by *Aureobasidium pullulans* (a fungus); **gellan**, produced by *Pseudomonas elodea*; **alginate**, produced by brown algae and the bacterium *Azotobacter vinelandii*; and **dextran**, produced by various bacteria (e.g., *L. mesenteroides*).

These polysaccharides are popular food additives because of their tendency to form gels, which help to thicken liquid foods (e.g., sauces, soups, and gravies). Some microbial polysaccharides also exhibit unusual characteristics such as **temperature hysteresis**. Agar is the classic example of this phenomenom — it forms semisolid gels with varying melting points, depending on the *prior* temperature of the solution. When agar is cooled from 90°C, it solidifies between 32 and 39°C; however, if a solution of agar is heated from room temperature, it does not melt until 60 to 90°C.

By far, the most popular microbial polysaccharide in the food industry is xanthan gum. Other microbial polysaccharides have useful characteristics, but their use is restricted by two factors: (1) they are more expensive than xanthan gum or plant- and seaweed-derived gums; and (2) many have not been approved for use in food in most countries. Introduction to the food industry requires expensive toxicity testing and evaluation to satisfy regulatory agencies.

The objectives of this section are to explain the characteristics of microbial polysaccharides that make them useful food additives and to explore the biological roots of these compounds. In other words, why do microbes bother making complex polysaccharides and why are they so useful to the food industry?

B. COMPLEX POLYSACCHARIDES

Before tackling the structure and characteristics of specific microbial polysaccha- rides, let us consider the potential range for variation in polysaccharide structure. Polysaccharide structure can be viewed in a similar way as protein structure. Proteins are polymers of linear chains of amino acids. Similar to proteins, polysaccharides are composed of linear chains of monomers, and both covalent and noncovalent bonds *between chains* affect the overall structure and function of the molecule. Hence, we can ascribe to carbohydrates a **primary, secondary, tertiary, and qua- ternary structure**, conceptually equivalent to the corresponding levels of protein structure. As in proteins, primary structure dictates or heavily influences secondary, tertiary, and quaternary structure of carbohydrates.

Polysaccharides are **homogeneous** (also known as **homopolymers**) if they have only one type of monomer; for example, starch is a homogeneous polysaccharide made up of D-glucose monomers. Yeast β-glucan, pullulan, and dextran are microbial homopolymers used in food. In contrast, **heterogeneous** polysaccharides (also known as **heteropolymers**) may have a backbone that is a repeating sequence of monomers, but they also have covalently bound side chains that contain monomers different than the backbone. Some heteropolymers also have heterogeneous back- bones. Xanthan gum and gellan gum are heterogeneous polymers that are used as food additives.

The primary structure of a heterogeneous polysaccharide is the linear sequence of carbohydrates within the linear chains. A large range of carbohydrate monomers can exist within polysaccharides (Figure 7.17); variation in the carbohydrate content changes the physical and chemical properties of the carbohydrate. A typical monosaccharide (e.g., glucose) can form polymers through bonds of any of the carbon atoms (six, in the case of glucose). Many polysaccharides involve carbons 1 and 4 (see Figure 7.3 for an explanation of carbon numbering), but bonds between carbons 1 and 6 of glucose are common in amylopectin, and carbons 1 and 3 are involved in the backbone of yeast glucan. This variation in bond formation leads to variation in bond angles between monomers, which may profoundly alter secondary structure.

The optical orientation of the polymer (i.e., α vs. β bonds between monomers) also affects the secondary structure. This is best illustrated by a comparison of amylose and cellulose. Both are composed of linear chains of glucose residues; however, they differ in that glucose monomers are connected by α-1,4 bonds in amylose, whereas they are connected by β-1,4 bonds in cellulose (Figure 7.18).

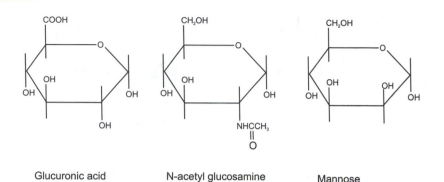

Glucuronic acid   N-acetyl glucosamine   Mannose

FIGURE 7.17 Carbohydrates that are common monomers in microbial polysaccharides.

Cellulose

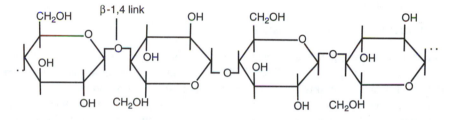

FIGURE 7.18 The structure of cellulose. Glucose residues are connected through β-1,4 links.

Designation of a bond as α or β depends on the rotation of carbon 1 in each monomer. Monosaccharides like glucose, when dissolved in water, can shift irreversibly between a linear chain and a ringed structure. When in the linear form, carbon 1 rotates, resulting in two different orientations for its attached H and OH groups. When the OH is in the equatorial plane (same plane as the ring), it is in an α orientation. In contrast, if the H and OH groups switch places through rotation of carbon 1, then the OH will be above the ring. This is the β rotation. Polymers with α linkages often have very different physical and chemical properties than similar polymers with β bonds. Again, cellulose and amylose illustrate this. Both form coiled structures, but cellulose coils are much more rigid than amylose coils. This rigidity partially explains the ability of cellulose chains to associate through hydrogen bonding to form strong microfibrils. Other possibilities besides coils exist for polysaccharide secondary structure; many polysaccharides form more or less rigid ribbons, for example.

As in proteins, individual polysaccharide molecules can associate with each other, often through noncovalent interactions. These interactions are crucial to the applications of the polysaccharide. For example, in the presence of Ca^{2+}, alginate microfibrils associate and form a gel structure that is essential to the use of alginate as a modifier of food texture.

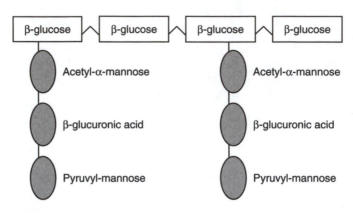

FIGURE 7.19 The primary structure of xanthan gum. The cellulose backbone is covalently linked to trisaccharides. The third monomer of the trisaccharide is sometimes linked to a pyruvyl group.

C. XANTHAN GUM

1. Structure and Characteristics

In economic terms, xanthan gum is the most important microbial polysaccharide. Worldwide production is 10,000 to 20,000 tons/year. It is added to packaged soups and gravies, salad dressings, and prepared sauces, and to nonfood products such as paint. It is used because of its strong effects on the viscosity of foods. Xanthan gum is an excellent thickener and gelling agent that does not change the flavor of food. Because humans are unable to metabolize xanthan gum, it is particularly popular in the preparation of low-calorie foods.

Xanthan gum has a backbone that is composed of a linear chain of β-1,4-linked glucose residues; this is often referred to as the "cellulose backbone" of xanthan gum (Figure 7.19). Every second glucose monomer is covalently linked to a trisaccharide. The trisaccharide is linked to the C3 carbon of the glucose molecule. The trisaccharide is composed of (starting with the residue that is linked to the backbone):

- A mannose that has been modified by addition of an acetyl group
- A glucuronic acid residue
- A mannose residue that has been modified by the addition of a pyruvyl group

When xanthan gum is dissolved in water, pairs of chains associate and form antiparallel double helices, broadly similar to the structure of DNA. Double helices then interact to form microfibrils. These microfibrils have several unique properties. If low concentrations of salts are present, dilute solutions of xanthan gum are very **viscous**; this viscosity is stable between 0 and 100°C and is unaffected by most changes in the chemical environment. For example, the viscosity of a xanthan gum solution is similar between pH 1 and 13.

The shear-thinning properties of xanthan gum solutions are also useful. When solutions are disturbed (e.g., by stirring), the viscosity breaks down instantaneously. However, when the disturbance is removed, viscosity recovers instantaneously. These properties are what make xanthan gum useful as a thickener in foods, because a combination of high viscosity and easy "pourability" is desirable in many processed foods (e.g., salad dressings). Besides being used as a thickener, xanthan gum can also be used as a stabilizer (e.g., foam in beer), adhesive (icings and glazes), flavor encapsulator, emulsifying agent (salad dressings), and ice-crystal inhibitor (ice cream).

2. *Xanthomonas campestris*

Xanthan gum is produced by *Xanthomonas campestris*, a Gram-negative bacterium that is a destructive pathogen of many crops. It causes a vascular wilt that can lead to death of the plant. In such cases, the vascular system of the plant becomes clogged with xanthan gum, and the plant is unable to transport water from the roots to the shoots. Consequently, the above-ground tissues dry out and the plant dies.

X. campestris is relatively easy to grow in large-scale culture, as long as it is well supplied with oxygen. It is usually grown in batch culture (see Chapter 8, Section III.B, for a discussion of the differences between batch and continuous culture). Xanthan gum is produced during all growth phases of the culture, with a burst of production during the exponential growth phase. The major technical problem associated with growing *X. campestris* is the increased viscosity of the medium toward the end of the growth period. Xanthan gum is purified from the batch culture by precipitation with isopropyl alcohol. The precipitated polysaccharide is dried, milled, and packaged.

3. Genetics of Xanthan Gum Biosynthesis

Over the past 20 years, we have learned a great deal about the genes involved in xanthan gum biosynthesis in *X. campestris*, and this has become a model system for heteropolysaccharide synthesis in microbial cells. Much of this research has been driven by the successful application of xanthan gum in the food industry; with an understanding of the mechanism of cellular biosynthesis of xanthan gum, and of the genes involved, perhaps it will be possible to improve the efficiency of xanthan gum synthesis. The pathogenicity of *X. campestris* to vegetable and fruit crops has also driven research into xanthan gum biosynthesis.

The biosynthesis and export of this complex and large molecule seems at first glance to be a formidable problem for a bacterial cell. Remember that prokaryotes lack Golgi-mediated vesicular mechanisms for exporting large molecules. Xanthan gum is synthesized initially as a **pentasaccharide** (chain of five carbohydrate monomers. This pentasaccharide is made up of two glucose residues (that will eventually form part of the "cellulose" backbone), linked to a trisaccharide of acetylated mannose, glucuronic acid, and pyruvyl-mannose. This trisaccharide eventually forms one of the trisaccharide side chains that is linked to the cellulose backbone (see Figure 7.19). Pentasaccharide chains are linked to each other to form xanthan gum.

The three different monosaccharides (glucose, mannose, and glucuronic acid) are synthesized by two enzymes coded for by genes in a 35.3-kb cluster. Another region (*gum*) consists of an operon that contains 12 open reading frames that code for enzymes responsible for the construction of the pentasaccharides, as well as enzymes involved in assembly and secretion of the final polymer. The mechanism of secretion of the polymer remains unclear. The complexity of xanthan gum biosynthesis illustrates the challenges in attempting to modify xanthan gum synthesis through recombinant DNA technology. At least 14 genes are involved in two separate gene clusters. However, the *gum* genes, which take part in construction of the polymer, are all regulated by one promoter, suggesting that modifying their expression would be relatively straightforward. However, activity of the *gum* genes is closely tied to the availability of monosaccharide subunits, whose synthesis is controlled by different genes.

Xanthan gum is the best understood complex microbial polysaccharide. Modification of the genetic aspects of biosynthesis of other heteropolysaccharides, such as alginate, will likely be similarly challenging. Yield increases will probably not occur through genetic modification because production of microbial polysaccharides such as xanthan gum is already highly efficient. However, genetic modification may increase the range of usable substrates for polysaccharide biosynthesis, which may allow the use of cheaper feeds. Currently, a complex mixture of carbon and nitrogen sources is required for efficient xanthan gum production. If this could be replaced by a complex medium such as whey, xanthan gum production would become more cost effective.

VII. CITRIC ACID AND VITAMIN PRODUCTION

Microorganisms produce a diverse range of metabolites that are useful to the food industry, but they are difficult to categorize. They include citric acid (used for flavor and a number of other applications), xylitol (an alternative sweetener), and vitamins (used to enrich and fortify foods). Various means have been used to encourage production of these metabolites, making generalizations difficult. We will examine citric acid production in detail.

A. USES OF CITRIC ACID IN FOOD

Citric acid is widely used (total world production >500,000 tons), particularly in the beverage industry, partly because of its long history of safe use in foods. It is also of considerable historical interest, because commercial citric acid production began in 1923 and was one of the first successful applications of biotechnology. Citric acid is used primarily as an **acidulant** to lower the pH of a food or beverage. Although increased acidity leads to inhibition of most bacteria, citric acid is not normally used as an antimicrobial (acetic acid and lactic acid are more effective). However, citric acid has a number of other properties that make it attractive as an acidulant. It imparts a pleasant sour taste to beverages (e.g., soft drinks) and candies and it enhances other flavors, especially those present in fruits and vegetables.

Citric acid also functions as a metal chelator, and it is used to inhibit off flavor development in fresh and frozen fish. Dimethylamine, produced by endogenous fish enzymes, leads to an unpleasant "fishy" flavor. Dimethylamine production is inhibited by citric acid through chelation of iron and copper ions.

Finally, citric acid is commonly used as an antioxidant, particularly in combination with ascorbic acid. This combination is often used to prevent browning in fruits and vegetables. In this context, chelation of metal ions by citric acid inhibits polyphenol oxidase, a cause of browning.

B. PRODUCTION OF CITRIC ACID BY ASPERGILLUS NIGER

A number of fungi can be used to produce citric acid on an industrial scale, but *A. niger* is currently the most popular. This fungus does not normally produce large amounts of citric acid, but overproduction is possible because of the development of: (1) overproducing strains of *A. niger* (this has mainly been accomplished by extensive screening of isolates), and (2) growth conditions that lead to citric acid overproduction.

To overproduce citric acid, assuming that an efficient strain has been found, three conditions must be met. First, because *A. niger* is a strict aerobe, it must have a sufficient supply of oxygen. Second, the fungus must be **deficient in manganese**. Third, high levels of glucose must be present.

We do not have a complete understanding of the importance of manganese deficiency and high glucose, but details are starting to become clear. Citrate is not the end product of a fermentation or other pathway in *A. niger*. Instead, it is an intermediate in an important cyclic pathway — the TCA cycle. Why would a fungus produce large amounts of such an intermediate compound?

The answer appears to lie in the regulatory function of two key enzymes (Figure 7.20): phosphofructokinase (PFK), a glycolytic enzyme, and α-ketoglutanate dehydrogenase (α-KDH), an enzyme in the TCA cycle. Citric acid accumulation is normally prevented because citric acid allosterically inhibits PFK. Inhibition at this stage results in little flow of glucose through glycolysis and to the TCA cycle. However, manganese deficiency results in the loss of this feedback inhibition. This is probably linked to amino acid metabolism and rates of protein turnover. Manganese-deficient hyphae have increased levels of protein turnover, leading to a requirement for increased intracellular levels of NH_4^+. Increased NH_4^+ appears to be crucial to relief of allosteric inhibition of PFK by manganese deficiency.

α-KDH is the key enzyme regulating catabolism of citrate in the TCA cycle. It is inhibited by oxaloacetate, one of the later intermediates in the TCA cycle. Oxaloacetate production through carbon fixation (pyruvate + CO_2 $\rightarrow$ oxaloacetate) is in turn stimulated by high levels of carbohydrates in the external environment. Thus high levels of glucose or other carbohydrates stimulate oxaloacetate production, which in turn inhibits activity of α-KDH. This model has not been completely verified and may be adjusted in the future. High levels of carbohydrates also act to lower the K_m of PFK, thus increasing its activity and increasing the flux of carbon through glycolysis. This carbon cannot be oxidized by the TCA cycle, so the cell gets rid of it through citric acid secretion.

We are approaching a mechanistic understanding of the empirical observation that high levels of carbohydrates combined with low levels of manganese lead to

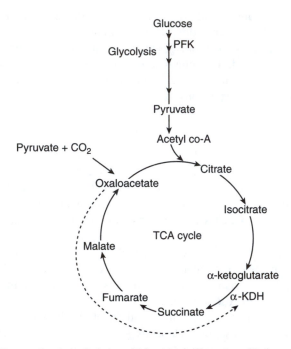

FIGURE 7.20 Overproduction of citric acid in *Aspergillus niger*. Under conditions of high glucose, high oxygen, and low manganese, citric acid is overproduced. High glucose levels result in increased conversion of pyruvate to oxaloacetate. Increased oxaloacetate levels inhibit (dotted line) α-KDH, restricting the flow of carbon through the TCA cycle. High glucose also increases the activity of phosphofructokinase (PFK), which increases the flux of carbon through glycolysis. This extra carbon cannot be oxidized by the TCA and is secreted as citric acid.

citric acid production. But what does the microorganism gain from these biochemical machinations? Is there any cellular benefit to the overproduction of citric acid under these conditions? At present, our limited understanding of the cellular ramifications of manganese deficiency makes it difficult to answer this question.

C. VITAMIN PRODUCTION BY MICROORGANISMS

Vitamin enrichment and fortification is an efficient way of encouraging vitamin sufficiency among consumers. Many people consume vitamin supplements to compensate for poor diets or for other health reasons, or because of conditions such as pregnancy that require increased vitamin supply. Many food processes (e.g., heat treatments) result in losses of vitamins that need to be replaced somehow. In the developing world, vitamin supply is particularly problematic because widespread poverty prevents many people from consuming a diet adequate in all vitamins. Vitamin deficiency leads to a host of health problems (e.g., increased frequency of bacterial infections). Consequently, not only is vitamin production economically important, but it affects global health considerably.

Most vitamins are made through chemical synthesis. However, a number of vitamins, including ergosterol (provitamin D), riboflavin, B_{12}, and B_{13}, are produced through microbial culture. Others, such as vitamin C, are produced through a combination of chemical synthesis and microbial metabolism.

Microbial production of vitamins is mainly achieved through selection of natural (wild) overproducing strains, followed by the design of bioreactor conditions that optimize production. For example, two species of filamentous fungi (*Eremothecium ashbyi* and *Ashbya gossypii*) that are closely related to *S. cereviseae*, naturally overproduce riboflavin. These and other wild overproducers can now be used to produce riboflavin at similar costs as chemical synthesis. This has led several large vitamin companies to phase out chemical synthesis and replace it with microbial synthesis, which also has the advantage of being perceived as a "green" process that uses renewable substrates. In contrast, chemical synthesis relies on nonrenewable fossil fuels as substrates for vitamin production.

One company (Roche) is replacing chemical synthesis of riboflavin with microbial synthesis using mutant strains of *B. subtilis*. This bacterium is not naturally an overproducer of riboflavin, but it is a GRAS organism that is already widely used in the food industry.

We expect that the trend away from chemical synthesis to microbial production will spread to other vitamins, particularly if forecasted shortages of fossil fuels in the mid-20th century drive petroleum prices upward.

VIII. DEVELOPMENT OF NOVEL MICROBIAL PRODUCTS

Although many screening programs have been aimed at finding strains of microbes that "do useful things," undoubtedly there remain a multitude of useful microbes that have not been discovered. This is especially true in reference to microbial polysaccharides. There is enormous range in the structure of these compounds, and future biotechnologists will certainly exploit this variation. However, successful exploitation of a microbe (or cell culture) is not always easy, and is often impossible. To assess the potential success of a microbial venture, it is useful to ask a number of questions about the microbe and its product:

1. *Is there a market for the product?* Obviously, if the answer is no, then there is no reason to continue. If the answer is yes, then several other questions become pertinent. For example, it is useful to estimate the range of prices of the product that would be tolerated by the market.
2. *Can the product be obtained from nonmicrobial sources?* If the product can be obtained from plants or seaweeds, then it is unlikely that microbial production will be able to compete. Plants are fundamentally easier and cheaper to grow than cells or microbes. There are exceptions; for example, vanillin can be produced from large-scale cell culture at comparable costs as from soil-grown plants.
3. *Does the microbe (or cell) produce large amounts of the desired product?* If the answer is no, then a program for strain improvement is required.

Collect and assess a large number of isolates. Expose the best producers to mutagens and screen the resultant mutants. This strategy has been successful in dramatically increasing the production rate of many important microbial compounds. Recombinant procedures can also be applied to most microbes to increase the rate of production, but this requires substantial investment in basic research into the physiology and genetics of biosynthesis of the desired compound. Changes in the bioreactor environment can also strongly affect production and are usually discovered through careful empirical research.

4. *Is the microbe (or cell) easy to grow?* If the microbe requires expensive and well-defined media, then it is unlikely that production on an industrial scale will be feasible. However, if the microorganism can be grown on cheap substrates such as molasses or corn steep liquor, then the chances of success are much greater. If the microbe is difficult to grow, it may be possible to transfer the genetic material responsible for product formation to a microbe that is easier to grow.

5. *Is the microbe safe?* This is a crucial question. If the microbe has been granted **GRAS** status, it can be used in food production. However, if the microbe is not GRAS, you must expend great effort to demonstrate that it is safe. You must clearly show that the microbe is not pathogenic to humans and does not produce any toxins. Furthermore, it must not produce antibiotics, and you must demonstrate that the desired product can be purified without the introduction of harmful contaminants (e.g., organic solvents). Accumulating the evidence required to obtain GRAS status is an expensive process. For this reason, it is vastly preferable to use an organism that already has GRAS status. Consequently, it is not surprising that so many enzymes are obtained from a small number of microbes (e.g., *B. subtilis, A. niger, S. cereviseae*). An alternative to acquiring GRAS status is to transfer the necessary genes to a GRAS organism. However, most countries still require extensive safety testing of the resultant recombinant microbe. This is particularly true if the source of the DNA is a pathogenic, or toxin-producing, microbe. In such cases, it may be impossible to obtain permission to sell the recombinant product.

Once all these questions have been answered satisfactorily, a decision can be made about the potential for the microbial process to be successfully scaled up to an industrial scale. The large number of successful microbial biotechnologies suggests that despite the onerous challenges that are encountered, the large-scale culture of microorganisms is often both desirable and feasible.

Once a suitable microbe has been chosen for a particular process, the biotechnologist is unlikely to be completely satisfied. Microbial biotechnology is always governed by the bottom line; any changes to a microbe or a microbial process that increase efficiency and decrease costs will be desirable. **Strain improvement** is therefore an essential and continuing process. The best example of strain improvement is in the antibiotic industry, where dramatic increases in production efficiency have occurred as a result of intensive improvement programs. The techniques of strain improvement are similar to those used in screening programs. Mutagenesis is often a highly successful strategy, but the current trend is toward exploitation of basic understanding of genetic regulation of product formation, and of interrelationships among metabolic pathways (quantitative physiology). As our understanding of microbial metabolism increases, our power to tailor metabolism to our needs will also grow.

RECOMMENDED READING

1. Glazer, A. N. and Nikaido, H., *Microbial Biotechnology: Fundamentals of Applied Microbiology,* Freeman and Company, New York, 1995.
2. Linko, M., Haikara, A., Ritala, A., and Penttilä, M., Recent advances in the malting and brewing industry, *J. Biotechnol.*, 65, 85, 1998.
3. van Dam, K., Role of glucose signaling in yeast metabolism, *Biotechnol. Bioeng.*, 52, 161, 1996.
4. Randez-Gil, F. and Sanz, P., Construction of industrial baker's yeast strains able to assimilate maltose under catabolite repression conditions, *Appl. Microbiol. Biotechnol.*, 42, 581, 1994.
5. Jensen, L. G., Olsen, O., Kops, O., Wolf, N., Thomsen, K. K., and von Wettstein, D., Transgenic barley expressing a protein-engineered, thermostable (1,3-1,4)-β-glucanase during germination, *Proc. Natl. Acad. Sci. U.S.A.*, 93, 3487, 1996.
6. McCaig, R., McKee, J., Pfisterer, F. A., Hysert, D. W., Munoz, E., and Ingledew, W. M., Very high gravity brewing — laboratory and pilot plant trials, *J. Am. Soc. Brew. Chem.*, 50, 18, 1992.
7. Suihko, M.-L., Vilpola, A., and Linko, M., Pitching rate in high gravity brewing, *J. Inst. Brew.*, 99, 341, 1993.
8. Onnela, M.-L., Suihko, M.-L., Penttilä, M., and Keränen, S., Use of a modified alcohol dehydrogenase, ADH1, promoter in construction of diacetyl non-producing brewer's yeast, *J. Biotechnol.*, 49, 101, 1996.
9. Varnum, A. H., The exploitation of microorganisms in the processing of dairy products, in *Exploitation of Microorganisms*, Jones, D. G., Ed., Chapman & Hall, London, 1993, chap. 11.
10. Chapman, H. R. and Sharpe, M. E., Microbiology of cheese, in *Dairy Microbiology, Vol. 2, The Microbiology of Milk Products*, Elsevier, London, 1990, chap. 5.
11. Martin, M. C., Alonso, J. C., Suárez, J. E., and Alvarez, M. A., Generation of food-grade recombinant lactic acid bacterium strains by site-specific recombination, *Appl. Environ. Microbiol.*, 66, 2599, 2000.
12. Gilbert, C., Robinson, K., Le Page, W. F., and Wells, J. M., Heterologous expression of an immunogenic pneumococcal type 3 capsular polysaccharide in *Lactococcus lactis*, *Infect. Immun.*, 68, 3251, 2000.
13. Hashimoto, S.-I. and Ozaki, A., Whole microbial cell processes for manufacturing amino acids, vitamins or ribonucleotides, *Curr. Opinion Biotechnol.*, 10, 604, 1999.
14. Kumagai, H., Microbial production of amino acids in Japan, *Adv. Biochem. Eng.*, 69, 71, 2000.
15. Whitaker, J. R., Enzymes, in *Food Chemistry*, 3rd ed., Fennema, O. R., Ed., Marcel Dekker, New York, 1996, chap. 7.
16. Poutanen, K., Enzymes: an important tool in the improvement of the quality of cereal foods, *Trends Food Sci. Technol.*, 8, 300, 1997.

17. Benjamin, S. and Pandey, A., *Candida rugosa* lipases: molecular biology and versatility in biotechnology, *Yeast*, 14, 1069, 1998.
18. Lang, C. and Dörnenburg, H., Perspectives in the biological function and the technological application of polygalacturonases, *Appl. Microbiol. Biotechnol.*, 53, 366, 2000.
19. Becker, A., Katzen, F., Pühler, A., and Ielpi, L., Xanthan gum biosynthesis and application: a biochemical/genetic perspective, *Appl. Microbiol. Biotechnol.*, 50, 145, 1998.
20. Roller, S. and Dea, I. C. M., Biotechnology in the production and modification of biopolymers for foods, *Crit. Rev. Biotechnol.*, 12, 261, 1992.
21. Banik, R. M., Kanari, B., and Upadhyay, S. N., Exopolysaccharide of the gellan family: prospects and potential, *World J. Microbiol. Biotechnol.*, 16, 407, 2000.
22. Sutherland, I. W., Microbial polysaccharide products, *Biotechnol. Gen. Eng. Rev.*, 16, 217, 1999.
23. Bigelis, R. and Tsai, S.-P., Microorganisms for organic acid production, in *Food Biotechnology: Microorganisms*, Hui, Y. H. and Khachatourians, G. G., Eds., Wiley, New York, 1994, chap. 6.
24. Vandamme, E. J., Production of vitamins, coenzymes and related biochemicals by biotechnological processes, *J. Chem. Tech. Biotechnol.*, 53, 313, 1992.
25. Stahmann, K.-P., Revuelta, J. L., and Seulberger, H., Three biotechnical processes using *Ashbya gossypii*, *Candida famata*, or *Bacillus subtilis* compete with chemical riboflavin production, *Appl. Microbiol. Biotechnol.*, 53, 509, 2000.
26. Parekh, S., Vinci, V. A., and Strobel, R. J., Improvement of microbial strains and fermentation processes, *Appl. Microbiol. Biotechnol.*, 54, 287, 2000.
27. Steele, D. B. and Stowers, M. D., Techniques for selection of industrially important microorganisms, *Annu. Rev. Microbiol.*, 45, 89, 1991.

8 Industrial Cell Culture

I. SCALE-UP OF CELL CULTURE

Biotechnology often requires the growth of large amounts of microbial cells or of cells derived from plants or animals. In this chapter, the focus is primarily on the methods of and approaches to large-scale growth of bacteria and fungi, because these microbes produce virtually all food-related cell products. When growing bacteria or fungi in an experimental context, we use petri dishes or small volumes of liquid media. In these small-scale systems, nutrients diffuse to cells or, in the case of fungi growing on a petri dish, cells (hyphae) grow toward the nutrients. In a petri dish, all cells are continually exposed to a moist atmosphere made up of 20% oxygen, more than enough to fulfill the needs of a strict aerobe (Figure 8.1). Growing microbes in small volumes of liquid media is also straightforward; mild agitation of the cultures or growth in shallow flasks is sufficient to ensure adequate oxygen supply to the cells.

This type of small-scale culture system is well suited to the experimental manipulation of cells. Much of our understanding of cellular metabolism and behavior has been obtained through small-scale cell culture. However, when the purpose of cell growth is to collect a useful product, and then to sell it, small-scale culture is no longer feasible. The operation must be **scaled-up**, so that the product can be efficiently collected with a minimum of labor materials. A scaled-up culture system is called a **bioreactor**. Bioreactors have traditionally been referred to as **fermenters**. This term is still widely used, but many biotechnologists prefer the term bioreactor, because most applications of large-scale microbial culture rely on aerobic metabolism, not fermentation.

The size of bioreactors for growing plant and animal cells may be relatively modest, particularly in the case of animal cells. It is difficult to grow animal cells in volumes greater than 100 L because of their fragility. Nevertheless, some companies successfully grow animal cells (e.g., hybridomas to produce monoclonal antibodies) in volumes up to 10,000 L. Bacterial and fungal cells, in contrast, are frequently grown in 10,000 to 100,000-L bioreactors. Scale-up of a culture system from a 2-L flask to a 1000-hL bioreactor will invariably result in problems that did not exist in the 2-L flask. Agitation of the bioreactor is not an option. One must find other methods for aerating the culture medium when growing aerobic cells. Diffusion will no longer suffice to supply nutrients throughout the bioreactor. Both of these problems (nutrient and oxygen supply) can be tackled together, through aeration (bubbling oxygen or air into the bioreactor) or agitation by mechanically stirring the liquid medium.

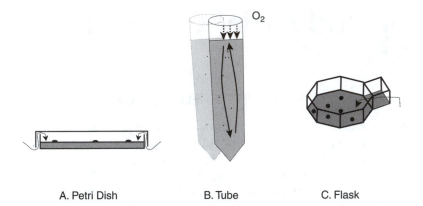

A. Petri Dish   B. Tube   C. Flask

FIGURE 8.1 Oxygen supply to small-scale cell cultures. In petri plates (A), oxygen diffuses through the space between the top and bottom covers of the plate. In small tubes (B), agitation results in mixing of the liquid medium (solid arrows). Oxgen diffuses from the air column above the liquid (dashed arrows). A small flask (C) is often used to grow animal cells. The flask is laid on its side with a thin layer of liquid growth medium. Oxygen diffuses through the loosely attached cap (not shown) and into the flask. Oxygen then diffuses into the liquid.

The temperature within the bioreactor may quickly rise to inhibitory levels, resulting in the need for an efficient cooling system. The biotechnologist may find, to his/her dismay, that after inoculating the bioreactor with the desired organism, appreciable growth in the bioreactor does not occur for several weeks. It may be necessary to prepare large amounts of inoculum in a smaller bioreactor, which can then be used to seed the larger bioreactor, to avoid long delays due to cell division.

As a final insult, the culture may grow well in the bioreactor but fail to make appreciable amounts of the desired product. Microbiologists frequently observe changes in cellular morphology and metabolism upon growth in large bioreactors that may lead to decreased product formation. An even worse scenario is possible — sometimes scale-up leads to frequent contamination by undesirable microbes. If undetected, this could cause major problems for consumers and for the company; contamination can lead to off odors or undesirable flavors, and, if the contaminant is pathogenic or toxigenic, consumers may suffer greatly.

For these reasons, scale-up typically occurs through a gradual increase of bioreactor size. Initially, the cell may be grown in a 100-L container, to find out if this increase in culture volume changes growth rates or rates of product formation. The operation will then be scaled up to the **pilot plant** stage, which may have a volume of 1,000 to 10,000 L. The pilot plant will also incorporate pilot versions of collection and purification systems. When all problems have been resolved in the pilot plant, the company will shift to the final bioreactor volume and perform further tests to ensure that cell growth and product formation occur in the predicted manner.

Scaling up an operation also forces one to carefully consider the growth medium. Is it expensive? Do media exist that offer equivalent growth rates at a fraction of the cost? Are some of the components of the medium unnecessary? Such questions quickly become important as the culture volume increases, and they can make the

TABLE 8.1
Oxygen Requirements of Cells Grown in Large-Scale Systems

Organism	Product	Oxygen Requirements
Aspergillus oryzae	Various enzymes	Aerobic
Bacillus subtilis	Various enzymes	Aerobic
Lactobacillus spp.	Starter cultures	Oxygen not required
Saccharomyces cereviseae	Ethanol	Anaerobic[a]
S. cereviseae	Biomass (baking)	Aerobic
Corynebacterium glutamicum.	Glutamate	Low concentration of oxygen
Zymomonas mobilis	Ethanol	Anaerobic

[a] *S. cereviseae* requires oxygen during the initial phase of ethanol fermentation.

difference between a process that is financially feasible and one that would result in a product that is too expensive for its intended market. It is equally important to accurately assess the marketplace. Will the targeted consumers desire the product, and if so, can it be sold at an affordable price? This is particularly true in the food industry. Most of the microbial products destined for the food industry are food additives. Most have a range of competitors that are produced through alternative means (e.g., chemical synthesis or derived from plants or animals). Consequently, products of cell culture must offer superior performance or must be competitively priced. Economic factors are particularly important for young biotechnological companies. Delays in production caused by scale-up problems can lead to severe cash flow problems, particularly in the current environment where venture capital is difficult to acquire.

We will now discuss aspects of the bioreactor environment that have the most important effects on bioreactor design and function. They are **oxygen supply**, **heat removal**, and **nutrient supply**. Since the original development of industrial-scale microbial culture in the 1920s, scale-up problems associated with these factors have largely been resolved through collaborative efforts of microbiologists, biochemists, and biochemical engineers. In the next section, the major aim is to develop a basic understanding of the approaches taken to ensure that cells in a bioreactor are exposed to an optimal environment for growth and product formation.

II. ENVIRONMENTAL FACTORS

A. Oxygen

One of the basic principles of microbiology is that microbes exhibit great variation in their relationship to oxygen (Table 8.1). **Strict aerobes** require oxygen, whereas others, known as **strict anaerobes**, are killed by oxygen. The industrially important lactic acid bacteria (LAB) are **aerotolerant**, meaning that they do not use oxygen but can tolerate its presence. The amount of oxygen an organism requires affects bioreactor design. Anaerobic organisms are much easier to grow in large volumes than aerobes, because of the low solubility of oxygen in water (8 mg/L at 30°C). Rapidly growing aerobic organisms consume large amounts of oxygen. For example, an actively respiring culture of *Saccharomyces cereviseae* consumes 6 g $O_2/(L \cdot h)$. Oxygen is rapidly depleted, resulting in anaerobic zones around growing cells. Oxygen depletion is relieved by diffusion of oxygen from the surrounding liquid, but this is a slow process. Many factors affect the diffusion rate of a soluble compound; one of the most important is the strength of the concentration gradient. Diffusion is rapid from a region with a high concentration to a region with a low concentration. With oxygen, concentration gradients are relatively small, because of the low solubility of oxygen in water; thus, diffusion of oxygen is relatively slow. Diffusion is also much slower if the *distance of diffusion* is large. Therefore, the biotechnologist who wants to grow large amounts of aerobic cells is faced with two problems: (1) when large amounts of cells are actively metabolizing, the oxygen concentration around the cells quickly falls to low levels; and (2) as the vessel size increases, the distance to the source of oxygen (the surface of the liquid) also increases. The interplay of these factors results in severe stress to aerobic organisms; the lack of oxygen will inhibit or kill them, and the desired product will not be produced.

Why do cells require oxygen? Remember that virtually all eukaryotic organisms produce adenosine triphosphate (ATP) through *respiration* of carbohydrates. Oxygen is the terminal electron acceptor in the electron transport chain that drives ATP formation, and low levels of oxygen severely inhibit respiration in most eukaryotes. Consequently, the supply of oxygen to plant and animal cells is crucial to their growth and metabolism.

Eukaryotic fungi are often more flexible in their oxygen requirements. For example, the yeast *S. cereviseae* is able to survive and grow under conditions of low oxygen concentration because of its ability to ferment carbohydrates such as glucose or sucrose into ethanol. Because fermentation yields less ATP than aerobic respiration, yeasts in anaerobic environments grow more slowly than under aerobic conditions. When ethanol is the desired end product, as is the case in the production of beer or wine, this slow growth is not important, because large amounts of ethanol can be produced by slowly growing cells. However, if we want to produce large amounts of biomass of *S. cereviseae*, we must ensure that the cultures are well supplied with oxygen. Thus, different bioreactor strategies are required by companies that produce and sell baker's yeast (viable cells of *S. cereviseae*) as compared to companies that produce ethanol or ferment beer or wine.

Not all yeasts are able to grow under anaerobic conditions. For example, *Kluyveromyces* is used extensively to produce enzymes such as lactase and recombinant proteins such as chymosin. It is a strict aerobe and thus must be well supplied with oxygen.

Many prokaryotes are capable of anaerobic growth, but many are strict aerobes. *Xanthomonas campestris*, for example, is grown in large quantities to produce xanthan gum and is a strict aerobe. Facultatively anaerobic prokaryotes such as *Escherichia coli* are usually grown aerobically in biotechnological processes, because more rapid growth typically results in better rates of production of the desired compound.

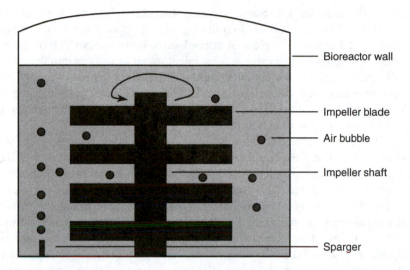

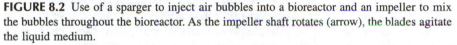

FIGURE 8.2 Use of a sparger to inject air bubbles into a bioreactor and an impeller to mix the bubbles throughout the bioreactor. As the impeller shaft rotates (arrow), the blades agitate the liquid medium.

Some microbes require oxygen, but at low levels. *Corynebacterium glutamicum* produces large amounts of glutamate, an amino acid that is an important flavor enhancer of food. This bacterium is capable of growth under aerobic conditions, but, because of a peculiarity in its carbohydrate catabolism (see Chapter 7, Section IV.D), it overproduces glutamate only when oxygen is present at a low concentration. This bacterium is not a micro-aerophile (organism that grows best at low oxygen levels). It is able to grow well under aerobic conditions, but little glutamate is made when oxygen is abundant.

Prior to the 1940s, most large-scale cell culture involved the growth of facultative or strict anaerobes (citric acid production by the aerobe *Aspergillus niger* is the major exception). However, World War II created an enormous demand for **penicillin**, the newly discovered antibiotic. Efforts to grow large quantities of *Penicillium notatum* were initially unsuccessful because of its oxygen requirements. The development of **surface culture** to grow *P. notatum* solved this problem. The fungus was inoculated onto solid substrates in trays; the fungus grew on the surface of the trays and oxygen was supplied directly from the atmosphere. This method was used until technology was developed that allowed growth of aerobic cells in liquid culture.

The major problem with surface culture is that a large surface area is required. Such extensive surfaces are expensive to construct, and inoculation of each tray is time consuming and expensive. Furthermore, these systems are vulnerable to contamination by airborne propagules of microbes.

Most aerobic systems are based on **liquid culture**, although solid support systems are increasing in popularity. The problem of oxygen supply in a liquid-filled bioreactor is solved by the use of **impellers** (Figure 8.2). Impellers consist of a rotating shaft with blades that extend into the bioreactor; rotation of the impeller mixes liquids within the bioreactor. In most cases, air is also bubbled into the bioreactor. The impeller distributes the air bubbles throughout the bioreactor, and oxygen diffuses from the bubbles into the liquid and into cells. This combination of aeration and impeller is typical of **stirred-tank bioreactors** (STBs), the most common type of bioreactor. In some cases, injecting air or oxygen into the medium provides enough agitation to distribute bubbles throughout the bioreactor; if this is sufficient, then an impeller is unnecessary. Agitation can also mix nutrients through-out the bioreactor, which can help prevent the occurrence of zones that are severely depleted of nutrients due to cell growth.

Impellers aerate large vessels quite well; however, they also generate **shear stress**. The concept of shear stress can be visualized by imagining the flow of liquid within a bioreactor. In the direction perpendicular to the flow, a gradient in velocity exists between moving and stationary liquid. This velocity gradient constitutes shear stress. The flow velocity of the liquid is known as **shear strain** — the moving impeller exerts pressure on the liquid, and this causing strain, or deformation. Liquids cannot be compressed, so deformation leads to flow. As the impeller rotates more quickly, shear strain and shear stress increases. The concept of shear stress is relevant to cell metabolism and viability; some cells (e.g., animal cells) are very sensitive to shear stress, whereas others (e.g., most bacteria) are relatively insensitive to shear stress. Filamentous fungi tolerate shear stress well, but they change in morphology — they typically grow as spherical pellets within STBs. Product for-mation rates are often related to pellet diameter, so it is necessary to maintain a level of shear stress that leads to optimal pellet size.

Because of the undesirable effects of agitation on cells, and because of the expenses involved in agitation and aeration, it is important to be able to quantify the amount of aeration and agitation required to satisfy cellular oxygen requirements. The following formula is often used to model **oxygen supply** (OTR, oxygen transfer rate) to a bioreactor:

$$\text{OTR} = k_L a \, (c^* - c_L) \qquad\qquad (8.1)$$

a is the **specific interfacial area**, the area per unit of culture liquid that participates in oxygen transfer. This can be visualized by imagining a stream of air bubbles injected into a bioreactor. a is correlated with the rate of air flow into the bioreactor, the volume of air injected, and the size of bubbles in the culture liquid. k_L is the mass transfer coefficient. c^* represents the oxygen concentration at the gas–liquid interface (in a bubble of air that is injected into the bioreactor, for example) and c_L is the concentration of oxygen within the culture liquid. Therefore, $(c^* - c_L)$ is a measure of the gradient in oxygen between oxygen-saturated water and the liquid of the bioreactor. It is important to determine the lower (critical) limit of c_L that does not disrupt cell growth or product formation. In most cases cells do not require oxygen-saturated water; the level required will depend on the cells' oxygen con-sumption rate and on other limiting factors within the bioreactor. For example, cellular respiration may be limited by substrate deficiency rather than oxygen supply. In this case, increasing the oxygen transfer rate would not result in increased growth. Some microbes (e.g., *Penicillium* spp.) have relatively high critical oxygen levels; when culturing these organisms, one must pay careful attention to oxygen supply.

$k_L a$, the **volumetric mass transfer coefficient**, is a crucial parameter to Equation 8.1. If $k_L a$ is too small, the oxygen transfer rate will be insufficient to supply oxygen to cells in the bioreactor; if it is too large, energy will be wasted through excessive agitation and aeration. $k_L a$ is affected most strongly by the power applied to rotate the impeller and by the aeration rate. Fortunately, it is possible to calculate the amount of aeration that is required, as long as the following information is available:

- The cell's growth rate (μ_x).
- The size of the initial inoculum (c_x).
- The amount of oxygen needed per kilogram of biomass. This is the **yield coefficient** ($Y_{O/X}$).

These parameters can be used to estimate the **oxygen demand** of a cell culture.

$$\text{Oxygen demand} = \mu_x c_x / Y_{O/X} \tag{8.2}$$

For a steady state to be achieved (i.e., the oxygen level in the bioreactor remains stable and supplies just enough oxygen for the cells), *oxygen supply must equal oxygen demand*. We can state this mathematically by combining Equations 8.1 and 8.2:

$$k_L a \, (c^* - c_L) = \mu_x c_x / Y_{O/X} \tag{8.3}$$

The major task facing the engineer at this point is to determine the type of reactor, operating conditions (e.g., power input to the impeller), and materials that will provide sufficient oxygen supply. Correlations between these bioreactor parameters and $k_L a$ are used to determine the required bioreactor design.

Other aspects of bioreactor function must also be satisfied by the impeller and aeration combination. Mass transfer of nutrients must occur fast enough to feed growing cells (this is usually only a problem with micronutrients), and there must be enough mixing to distribute heat evenly throughout the bioreactor. Cooling systems usually consist of refrigeration or other mechanisms that operate outside of the bioreactor; consequently, heat can be removed only from the outer regions of liquid within a bioreactor.

The mixing efficiency is quantified by the **mixing time**. This is measured by adding a substance to the bioreactor and then recording the time required to achieve uniform distribution of the substance within the bioreactor. Mixing time increases as bioreactor volume increases. For example, 29 s may suffice for an 1,800-L bioreactor, whereas 140 s are required for a 120,000-L fermenter. This requirement may lead to regions within the bioreactor that have low concentration of oxygen or nutrients. Mixing efficiency also decreases as viscosity increases. This is a problem when growing filamentous fungi, because the fungal colonies increase the viscosity of the culture liquid. This problem has an extra element of complexity because filamentous fungi exhibit non-Newtonian fluid properties (non-Newtonian fluids have a complex relationship between viscosity, shear stress, and shear strain). This makes it more difficult to predict optimal levels of gas supply and impeller operation.

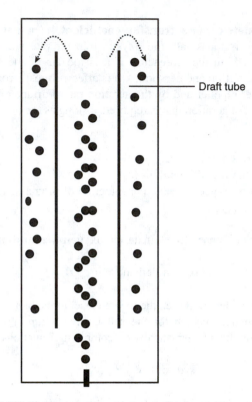

Draft tube

FIGURE 8.3 An air-lift bioreactor with an inner draft tube to encourage mixing throughout the column. Air bubbles up from the bottom of the bioreactor, and, as it rises to the top of the tower, it agitates the culture medium.

The geometry of the bioreactor also affects oxygen supply. It is usually desirable to maintain as much geometric similarity as possible among the different sizes of bioreactor used in the scale-up process. Theoretical and experimental study of bioreactor geometry has led to a number of alternatives. One of these is the **air-lift** bioreactor, which is a cylindrical tower. Oxygen supply and mixing are achieved through bubbling air at the bottom of the tower. As the bubbles rise, they mix and aerate the culture liquid. This type of design is popular in large bioreactors (>100,000 L) where the use of impellers is impractical. Circulation within an air-lift bioreactor can be improved by addition of a **draft tube**, an inner cylinder that directs circulation of the culture liquid. This results in more efficient mixing of large bioreactors (Figure 8.3).

B. pH

Microorganisms are sensitive to changes in pH. Bacteria are particularly sensitive to acidic conditions, and few bacteria (the LAB being notable exceptions) will grow at pH <5.0. Fungi are usually much more tolerant to acidic conditions, but they will not grow well when the pH is lower than 3. Both types of microbes are even more sensitive to alkaline pH, and few can tolerate pH >8.

The strong effects of pH on growth are important to biotechnologists, because cellular metabolism tends to cause changes in the pH of the external environment. The clearest example of this can be found in many fermentative bacteria; LAB such as *Lactococcus lactis* are incapable of respiration and rely on the fermentation of carbohydrates into lactic acid as a source of energy. Thus, large amounts of lactic acid are secreted by LAB, often leading to decreased pH in their external environment.

Other types of cells can decrease the pH of their growth medium. For example, mammalian cells must be grown in well-buffered solutions; growth and metabolism results in the production of acidic end products that rapidly kill mammalian cells if the medium is unbuffered. The extreme sensitivity of these cells to pH changes is not surprising, because of the narrow range of pH that exists in blood (pH 7.36 to 7.42). Mammalian cells in their natural environment are constantly protected from changes in pH by several distinct buffer systems.

Most culture media contain a number of compounds with buffering ability that help to stabilize the pH. However, it is advisable to monitor pH, either by taking samples periodically or by using a continuous pH detector with a probe within the bioreactor. Such systems are usually directly connected to pumps that are activated when fluctuations occur. When the pH decreases below a threshold level, a pump is activated that releases a base into the bioreactor; if the pH increases, another pump is activated that releases an acid. This type of system can effectively regulate pH in large bioreactors.

C. TEMPERATURE

Temperature has strong effects on the growth rate of cells. Because waste heat is released from growing cells, bioreactors usually have cooling systems that allow regulation of temperature, preferably a temperature that allows optimal growth rates. This is an important component of bioreactor control, because most cells have a narrow range of temperature tolerance. Another reason for incorporating efficient cooling into bioreactor design is that steam is frequently used to sterilize bioreactors and culture liquids before inoculation. After such a treatment, rapid cooling to avoid extensive down time is desirable.

Cooling is usually achieved by one of three mechanisms (Figure 8.4): (1) a water-filled jacket that surrounds the bioreactor; (2) coils of water pipes that surround the bioreactor or are enclosed within the bioreactor; or (3) flow of culture fluid through a heat exchanger. Heat exchangers are also required for options (1) and (2), to cool water flowing from the jacket or coils. Cooling jackets are usually restricted to laboratory-scale bioreactors (<100 L) because they are unable to cope with the heat load of larger bioreactors. Internal cooling coils are used for intermediate-size bioreactors, but large bioreactors (e.g., 100,000-L capacity) typically require a system that allows circulation of culture liquid through a heat exchanger (option 3).

Chemical engineers need to be able to predict the cooling requirements for bioreactors. How is this done? The first step is to determine the amount of heat that will be generated within the bioreactor. The **steady-state** formula for heat transfer within a bioreactor is

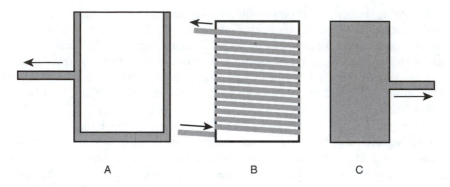

A   B   C

FIGURE 8.4 Systems for cooling bioreactors. (A) A water jacket is used to circulate water or coolant around the bioreactor. (B) A coil of tubes surrounds the inside or outside of the bioreactor. (C) The growth media is circulated through a heat exchanger. In (A) and (B) the coolant is also circulated to a heat exchanger.

$$\text{Heat generation} = hA\ \Delta T \tag{8.5}$$

ΔT refers to the temperature difference between the culture liquid and the liquid used for cooling; h is the **overall heat-transfer coefficient**, which quantifies the ability of the culture liquid and the cooling liquid to conduct heat; and A is the surface area over which heat transfer occurs. "Heat generation" is also a measure of the amount of heat that must be removed from the bioreactor in order to maintain a stable temperature (i.e., a steady state). So, if we examine Equation 8.5, we see that the left side represents the amount of heat that must be removed from the bioreactor, and the right side is a description of how the heat will be removed, in terms of the area of heat transfer and the efficiency of heat conductance to the cooling system.

Estimation of h is challenging and requires careful consideration of heat transfer through the bioreactor wall and tube walls of the cooling system. Estimation of the amount of heat that needs to be removed from the bioreactor is done through an overall energy balance. All mechanisms of heat generation and removal are included in this energy balance, giving an equation such as

$$Q_{met} + Q_{ag} + Q_{gas} = Q_{acc} + Q_{exch} + Q_{evap} + Q_{sen} \tag{8.6}$$

The left side of the equation quantifies the amount of heat generated in the bioreactor, and the right side quantifies where the heat goes. Q_{met} is the amount of heat generated by cell growth and maintenance, the major source of heat in a bioreactor. Q_{ag} and Q_{gas} refer, respectively, to heat gained from the power input to the bioreactor through agitation and gas injection (aeration). Q_{acc} represents the amount of heat that accumulates within the bioreactor. It is usually set at zero, because heat accumulation within the bioreactor results in temperature increase, which usually inhibits microbial growth. Q_{evap} and Q_{sen} are usually quite small in conventional stirred-tank reactors; they represent heat lost through evaporation into the bioreactor headspace (this is the region between the top of the liquid column and the top of the bioreactor chamber and is usually small in volume) and heat lost as materials enter and exit the bioreactor.

We can easily rearrange Equation 8.6 to get an estimate of Q_{exch}, the amount of heat that must be removed by the cooling system in order to maintain a steady state (i.e., constant temperature):

$$Q_{exch} = Q_{met} + Q_{ag} + Q_{gas} - (Q_{evap} + Q_{sen}) \tag{8.7}$$

The amount of heat generated through cellular metabolism can be measured in the laboratory, and the remaining variables can be estimated from known correlations between bioreactor design parameters (e.g., size and rotation speed of impellers) and heat generation. Thus a biochemical engineer can derive an understanding of the magnitude of the heat load that will have to be removed by the cooling system. The remaining step is to implement a cooling design that can remove this heat load. This is where the right side of Equation 8.5 becomes important. Engineers have expended great effort to develop an understanding of correlations between aspects of coolant systems and h, the heat transfer coefficient. These correlations are used to plan appropriate coolant systems. One final point: it is cost effective to use thermophilic bacteria or fungi in bioreactors, if possible. These microbes have high optimal temperatures and thus require less cooling, resulting in considerable cost savings.

D. NUTRIENT SUPPLY

One of the most crucial factors to be considered when designing a biotechnological process that requires the large-scale culture of cells is the type of nutrients that will be incorporated into the growth medium. All cells require nutrients, but there is enormous variation among cells in their nutrient requirements. Animal cells are notoriously demanding in terms of nutrients; RPMI 1640, a medium that is used to culture mammalian lymphocytes, contains 21 amino acids, a range of vitamins, and a number of mineral nutrients (Table 8.2). Furthermore, animal cell culture typically requires **serum**, a mixture of proteins and other compounds obtained from animal blood. These complex nutrient requirements are a major reason for the high costs associated with the large-scale culture of animal cells.

As seen from Table 8.2, plant cell culture also requires an extensive array of nutrients. However, most of the ingredients of plant tissue culture media are relatively inexpensive inorganic compounds. The only organic compounds required are vitamins (e.g., inositol), sucrose, and plant hormones.

In contrast, many microbes have much simpler culture requirements than plant or animal cells. This is not surprising; most bacteria and fungi are adapted to saprotrophic life; to survive, they must be able to efficiently scavenge organic and inorganic nutrients from the environment. Consequently, bacteria such as *E. coli* can be grown on simple media containing a carbon source (e.g., glucose), a nitrogen source (e.g., amino acids), and adequate levels of mineral nutrients (e.g., phosphate). From these building blocks, the whole range of macromolecules, vitamins, and other compounds required for growth are synthesized within the cell. However, biotechnologists

TABLE 8.2
Contents of Media Used in Cell Culture

RPMI 1640 (mammalian cells)		Mirashime-Skoog (plant cells)	LB Broth (E. coli)
Arginine	Biotin	NH_4NO_3	Dextrose
Asparagine	Ca pantothenate	KNO_3	Peptone[b]
Aspartic acid	Choline chloride	$CaCl_2$	NaCl
Cysteine	Folic acid	$MgSO_4$	
Glutamic acid	Inositol	KH_2PO_4	
Glutamine	Nicotinamide	FeNa EDTA[a]	
Glutathione	Aminobenzoic acid	H_3BO_3	
Glycine	Pyridoxine	$MnSO_4$	
Histidine	Riboflavin	$ZnSO_4$	
Hydroxyproline	Thiamine	KI	
Isoleucine	Vitamin B_{12}	Na_2MoO_4	
Leucine	$Ca(NO_3)_2$	$CuSO_4$	
Lysine	KCl	$CoCl_2$	
Methionine	$MgSO_4$	Inositol	
Phenylalanine	NaCl	Thiamine	
Proline	$NaHCO_3$	Sucrose	
Serine	Na_2HPO_4		
Threonine	Dextrose		
Tryptophan	Phenol red		
Tyrosine			
Valine			

[a] EDTA, ethylene diamine tetraacetic acid.
[b] A mixture of amino acids derived by enzymatic digestion of proteins.

have learned that optimal rates of product formation by *E. coli* and similar saprotrophic microbes are seldom possible using defined media (i.e., growth media that contain mixtures of specific compounds). More complex media tend to contain a variety of compounds such as vitamins and nucleotides. Most microbes grow more quickly on complex media than on simple defined media. The reason for this is that cell growth is often limited by the flow of metabolites such as amino acids and glucose toward macromolecule synthesis (e.g., DNA and structural proteins needed for new cells). If metabolites are diverted to pathways involved in vitamin synthesis, then growth slows.

It is sometimes difficult to find a medium that will support the growth of a microbe. They may have unknown nutrient requirements, or they may require the presence of another microorganism, a plant cell, or an animal cell. Such troublesome microorganisms sometimes produce valuable compounds (e.g., novel polysaccharides), but they are unsuitable for biotechnological applications because of the problems associated with their culture. However, if the genes responsible for production of the valuable compound can be cloned and characterized, it may be possible to move the needed genes into a microbe that is better suited for large-scale culture.

After demonstration that a cell can be grown in culture, the next step is to design a growth medium that satisfies nutritional demands and is cost effective. The biotechnologist must pick appropriate **raw materials** that will be used to feed the cells. For each process, a source of **carbon** is required, as well as a source of **nitrogen**.

Some raw materials are relatively well defined; for example, many processes use glucose or sucrose as a raw material. In some applications (e.g., xanthan gum production by *X. campestris*), defined raw materials such as glucose produce more consistent product quality, and so are preferred over less-defined raw materials.

Other raw materials are less well defined and more variable in composition. **Molasses** is an example of a cheap, undefined carbon source that supports the growth of a wide range of fungi and bacteria. It is obtained as a by-product of refining sugar from sugar cane. Sucrose is crystallized from sugar cane juice, and the residue is molasses, which is 30 to 40% sucrose. Because sugar cane grows best in hot tropical climates, molasses is an excellent raw material for biotechnological processes in developing countries. Molasses can also be obtained from sugar beets.

Whey is another example of an undefined carbon source. It is produced in large amounts as a by-product of cheese making. It is rich in lactose, which unfortunately limits its current usefulness. Microbes growing on whey must be able to use lactose as the major source of energy and carbon; many bacteria and fungi are unable to do this.

Nitrogen sources can also be undefined or defined. Often, defined raw materials such as urea or ammonium are used to supply nitrogen. In many processes, though, the cells are less able to use these simple sources of nitrogen and require a more complex form (e.g., amino acids). The solution then is to use undefined nitrogen-rich raw materials such as **corn steep liquor**, a water extract of corn that is 24% protein. Corn steep liquor is a by-product of the starch industry — corn is soaked in water to soften and swell the grains prior to starch extraction.

The choice of carbon and nitrogen sources is based on empirical data (i.e., experimentation with different types of raw materials), known characteristics of the cell, and economic considerations. Some raw materials are commodities that undergo large fluctuations in supply and price. This may force a company to choose an alternative raw material that may be less suitable for the bioprocess itself. It is also necessary to consider **upstream** (before cell inoculation) and **downstream** (after cell growth and product formation has occurred in the bioreactor) processes. The most important upstream process is usually **sterilization** of raw materials, which can be accomplished by steam treatment. The heat applied during steam sterilization may cause undesirable changes or nutrient loss in the raw materials.

For some raw materials (e.g., starch), upstream modification may be necessary — a process similar to mashing (see Chapter 7, Section II.C) may be required to produce usable carbohydrates. Downstream considerations may also be important. For example, some types of raw materials (e.g., whey) may contain compounds that interfere with purification of the product or lead to waste disposal problems.

It is also important to determine the optimal amount of raw materials that need to be added to a bioreactor process. This is usually done empirically, but it can also be achieved through consideration of the nutrient content of substrates and cells.

For example, consider a microbe that is 0.31% nitrogen. If it is desirable to grow this microbe to a final density of 20 g/L, then it will be necessary to have 6.2 mg/L of nitrogen in the bioreactor. If a particular raw material is 15% nitrogen, then 41 g/L of this raw material will be sufficient.

For carbon sources, biotechnologists often use theoretical energy yields as a basis for assessing feed (raw material) requirements. This is usually done by examing the organism's system of ATP generation when supplied with a certain feed. The number of moles of ATP generated from 1 mol of feed is usually directly proportional to the biomass yield. Thus, a fermentative bacterium that obtains 2 mol of ATP from each mole of glucose will have less final biomass than a respiring bacterium that obtains 32 mol of ATP from 1 mol of glucose. To achieve the desired level of biomass, more carbon is required by the fermentative bacterium. Clearly, this type of analysis is easier for simple feeds (e.g., glucose) than for complex feeds (e.g., molasses) that have many fermentable or respirable compounds.

III. TYPES OF BIOREACTORS

A. STIRRED-TANK BIOREACTORS

The STB, which has been the prototype bioreactor for the preceeding discussion, has overwhelmingly dominated industrial microbiology since its introduction in the 1940s. The biggest reason for this dominance is success — STBs can be used to grow a wide variety of cells, and the design of an STB to fit a particular bioprocess has become relatively straightforward.

Typically, STBs are used for cyclical **batch** processes (Figures 8.5 and 8.6): sterile feed is added to the bioreactor, inoculum is added, the cells grow and produce the desired product, product is purified, waste products are disposed of, the bioreactor is cleaned, and the cycle begins anew. A typical STB has a number of pipes that allow import of feeds and inoculants into the bioreactor, as well as pipes (lines) that allow draining (harvesting) of the bioreactor. There should also be a system that allows **cleaning in place**; this typically consists of a series of jets for spraying detergents into the bioreactor between production runs. Steam inlets and outlets can be used to sterilize the bioreactor after cleaning and to sterilize feed.

It is important to be able to monitor the bioprocess. This is achieved through a combination of sensors in the bioreactor (e.g., pH probe and foam detector in the headspace) and periodic sampling through a sampling port. Data from the pH probe often controls the activity of acid and base pumps, so that pH control can be achieved on a continuous basis (i.e., if the pH falls or rises above a specified range, acid or base pumps are activated to return the pH to the desired range). The sampling port may be used to monitor product formation, oxygen concentration, or various other parameters (e.g., nitrogen levels).

If the bioprocess is aerobic, agitation and aeration are required. Depending on the size and geometry of the STB, this may be achieved by air injection through a sparger (outlet into the bioreactor that has numerous small holes for bubble formation) alone or in combination with an impeller. Finally, the STB must have a cooling system of some sort (see previous section).

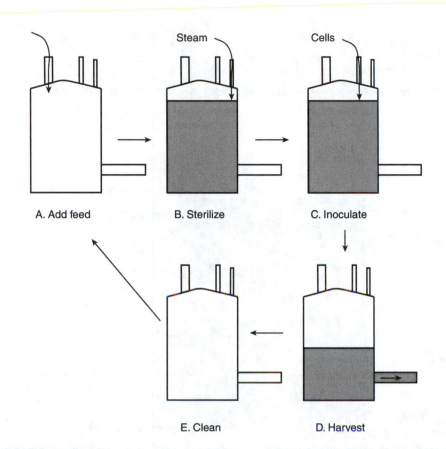

FIGURE 8.5 A batch cycle in a stirred-tank bioreactor. The cycle begins (A) with the addition of the feed (growth medium). The feed is sterilized (B), usually through the injection of steam, and cells are then added to the feed (C). After a suitable growth period, product is harvested from the bioreactor (D). The vessel is then cleaned (E), and the cycle begins again.

This classic type of bioreactor is relatively easy to operate, can be constructed with simple materials, and is protected against microbial and other contamination that may adversely affect the bioprocess. Maintenance is straightforward, and, because of the popularity of this design of bioreactor, replacement parts are easily acquired. However, there are some disadvantages to batch bioreactors. The major problem is related to the typical growth cycle followed by animal, plant, and microbial cells when they are inoculated into a fresh, nutrient-rich growth medium (Figure 8.7). There is an initial lag phase, with little cell division, but much enzyme synthesis as the cells adapt to their new environment. This is followed by a log phase, when the cells begin to divide. Because each cell typically produces two daughter cells with each division, growth is exponential during the log phase. In a bioreactor, the log phase cannot continue indefinitely. Eventually, the cells drive down the nutrient levels and often produce toxic waste products such as ethanol. This leads to a reduction in growth rate, and to the stationary phase, a period of

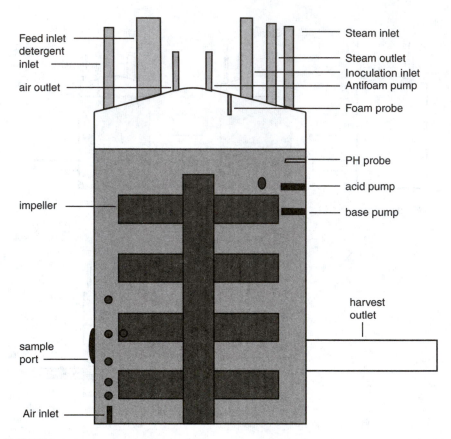

FIGURE 8.6 A stirred-tank bioreactor. Numerous ports (inlets and outlets) are required for monitoring bioreactor performance and for adding and removing various substances. (See text for more explanation.)

zero net increase in cell numbers. The growth cycle ends with the death phase, when most of the cells die and fall to the bottom of the bioreactor.

If the desired product is formed by dividing cells, then there will be little product formation during the lag and stationary phases. On the other hand, if the desired compound is produced by cells only once they have achieved the stationary phase, then the preceding growth phases (lag and log phase) are nonproductive. Thus, regardless of when the product is formed in the cell population, there will be periods of slow product formation in the bioreactor. This can often be alleviated by minimizing the lag phase through the use of large inoculum sizes, and by growing the inoculum in a medium very similar to the bioreactor medium. However, this increases the costs of inoculum production. Indeed, the need to prepare inoculum for each production run is one of the chief disadvantages to STBs.

Finally, the time required for cleaning and sterilization between runs is completely nonproductive. Given all these problems, it may seem surprising that batch-operated STBs have dominated the industry. The main reason for this is that STBs are reliable and relatively trouble free. However, there is great interest in the development of

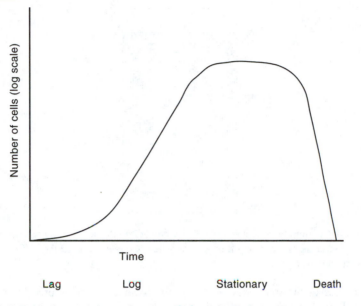

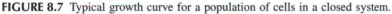

| Lag | Log | Stationary | Death |

FIGURE 8.7 Typical growth curve for a population of cells in a closed system.

alternatives to STBs, particularly in the design of continuous or semicontinuous systems.

B. Continuous Culture

1. Advantages of Continuous Culture

The crucial difference between batch culture and continuous culture is that in continuous culture there is a constant (or periodic) *replenishment of nutrients*. At the same time as new nutrients are added, *spent (used up) media is removed from the bioreactor* (Figure 8.8). Therefore, the volume of media in the bioreactor remains constant throughout the entire culture. Because fresh feed (media) is continuously added to the bioreactor, the cells are able to maintain constant growth rates and do not attain the stationary phase. Production is consequently more efficient than in a batch system. In bioengineeering terms, continuous culture allows maintenance of a **steady-state operation**. This results in increased productivity over the long term, as compared to batch-fed bioreactors. There has been great interest in continuous culture in bioreactor processes that produce relatively inexpensive products (e.g., beer), because corporate profits usually depend on high productivity. In bioprocesses that produce highly valued products, such as pharmaceuticals, there has been less interest in continuous culture.

In continuous systems, cells and product are constantly harvested from the bioreactor. This may or may not be an advantage, depending on the method of extraction and purification of the product. In most cases it is more efficient to have a continuous harvest, because it avoids downtime while the bioreactor is being drained of spent feed and refilled with fresh feed. Equally important, continuous

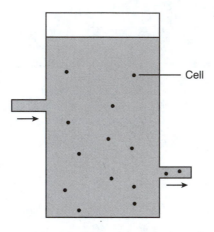

FIGURE 8.8 Continuous operation of a bioreactor. Growth medium is continually fed into the bioreactor, and product is continually harvested. Cells are lost to the harvest stream, so cell growth must balance cell loss.

culture decreases the frequency of inoculum preparation (in engineering terms, **less frequent seed development**). Continuous cultures can be operated for long periods (6 weeks or longer), without the addition of new inoculum. However, periodic shutdown of the bioreactor is necessary, for the following reasons:

- *Mutations.* Mutations occur in all growing cultures of cells. Extended periods of culture in a bioreactor will ultimately result in the occurrence of mutants that have altered, undesirable characteristics. Such mutants may be able to outgrow and eliminate the desired cells. This can happen quickly, because many biotechnological processes require cells to produce compounds in excess of their normal needs. In this sort of environment, mutants that do not produce the desired compound will have a selective advantage.
- *Contamination and biofouling.* Extended periods of continuous culture are unwise, because of the potential for the bioreactor to become contaminated. Periodic cessation of culture and sterilization of the bioreactor helps to avoid this problem. Biofouling is also a problem in continuous culture; for example, biofilms and associated mineral deposits may partially occlude lines.

Not all cells are equally vulnerable to the **genetic instability** that occurs in a bioreactor through the occurrence and growth of mutants. Nevertheless, periodic re-inoculation is a sound practice for both continuous and batch bioreactors. Preserved cell stocks are the usual source of inoculum. In traditional biotechnological systems (e.g., breweries or cheese-making plants), the source of inoculum was the *previous culture*; in other words, a sample of the previous batch of beer (or fermented milk) was added to fresh feed. This is less common now because of genetic instability of cultures and the contamination that is inevitable in long-term cultures. Instead, cultures are preserved. Mammalian cells are susceptible to freezing damage and are best stored in liquid nitrogen in the presence of cryoprotectants such as dimethyl sulfoxide. Bacterial cells, in contrast, are easily preserved at 40 to 80°C, with the addition of glycerol to the culture medium. Fungal cells usually retain their viability when preserved by lyophilization; alternatively, spores can be directly frozen. Biotechnologists prefer these sorts of preservation methods because they do not require periodic passaging (transfer of cells from spent media to fresh media). Each passage carries the risk of contamination or genetic change.

2. Disadvantages of Continuous Culture

Continuous culture is not always advisable or practical. The major disadvantage of this type of bioreactor is the increased chance of contamination. In a batch system, the growing cells are efficiently isolated from contaminating microbes. The only routes of entry for contaminants during the operation of the bioreactor are through the air inlets and the inlets used to chemicals to modify the pH or reduce foam formation. Contamination through these relatively small inlets is not usually a problem, provided that adequate filters are present, ensuring that any air or liquid that enters the bioreactor is sterile. However, continuous systems require the input of large amounts of feed during growth of the culture; the probability of the culture becoming contaminated increases with the addition of multiple feed inputs.

Problems can also occur in downstream processes. For example, one of the reasons for the popularity of batch culture of *X. campestris* (the bacterium that produces xanthan gum) is that few residual carbohydrates are present after bacterial growth. In a batch system, the residual concentration of carbohydrates is very low, because of depletion by the bacteria. However, in a continuous system, the bacteria do not greatly deplete the medium of carbohydrates. The xanthan gum is then more difficult to purify from the culture medium, because of interference from residual carbohydrates.

A more common problem in a continuous system is the potential for **wash-out**. In an ideal continuous culture, the number of cells remains constant; cell division is sufficient to replace cells lost through the harvest port. However, if feed is added (and removed) at an excessive rate, the rate of division may be insufficient to replace cells lost due to harvest. The number of cells in the bioreactor will progressively decline and production will eventually cease.

3. Continuous Culture in Operation

Successful operation of a continuous culture requires careful control of the flow rate through the bioreactor. If the flow rate is too slow, then the number of cells will increase and growth inhibition may occur due to the depletion of nutrients from the feed. As a compromise, many large-scale microbial operations use **semicontinuous** (also known as **semi-batch**) systems. In such a system, feed is added and removed periodically, rather than continuously. These systems can be operated with and without **feedback**. A system without feedback proceeds with a series of programmed feed inputs at specific times during the process run. In a system with feedback, the timing of feed input depends on changes in the environment of the bioreactor. pH can be used in this way, as can turbidity of the culture liquid. Turbidity is related to cell density and is easily monitored. Increased cell density could activate feed input, if a constant level of cell density is desirable (feed input results in cell dilution). This method of monitoring bioreactor function also serves as an alarm for problems such as wash-out. This type of bioreactor is freqently called a **continuous stirred-tank bioreactor** (CSTB).

In continuous and semicontinuous systems, one way to control cell growth is to keep an essential nutrient at a low level. This nutrient will be the **growth-limiting substrate**; cell growth rates can be regulated by adjusting the supply rate of this limiting nutrient. For this to function effectively as a control mechanism, all other nutrients must be present in the bioreactor at greater-than-limiting levels. Although there is a strong theoretical base for continuous culture at the laboratory scale (this is usually referred to as **chemostat** culture), large-scale continuous culture is still undergoing development, and is certainly more complex and difficult to control than fed-batch systems such as the conventional STB.

One common modification of CSTB design is to have a number of CSTBs in **series**. In these systems, culture liquid (containing a mixture of cells and product) flows from one CSTB to the next, and so on. This is a popular option when the prime objective of the bioprocess is to produce biomass. Production of baker's yeast is an example of a biomass process. Up to six CSTBs in series are used to make baker's yeast, with feed added to each bioreactor. Each bioreactor is relatively small (3400 L), and the density of yeast biomass increases in each successive bioreactor in the series. The final bioreactor contains a thick slurry of yeast that can be efficiently dried and packaged. This process is more efficient than having a single batch-fed bioreactor. In a batch system, a large amount of feed would have to be added so that the yeasts could attain a high final cell density. However, this much feed would inhibit growth because of the decreased water potential (activity) of the culture liquid, and it might also induce the Crabtree effect (see Chapter 7, Section II.E), whereby *S. cereviseae* switches to inefficient fermentative metabolism despite the presence of oxygen. In contrast, in a series of CSTBs, small amounts of feed can be added to each bioreactor, avoiding both the Crabtree effect and osmotic inhibition of cell growth.

Continuous processes operating in series are most successful when there is a high degree of conversion of substrate to product. Baker's yeast is typical — most of the substrate (e.g., glucose) added is converted into biomass or CO_2. If there is a low degree of conversion of substrate to product, then it may be more efficient to use a single CSTB, or even a STB.

Sometimes it is practical to use a CSTB followed by a **plug-flow bioreactor**. This is sometimes done, for example, in starch conversion. A plug-flow bioreactor consists of a long coiled tube. Substrates flow through the tube and the bioprocess occurs continually. A true plug has uniform characteristics (temperature, substrate concentration) in the radial direction (i.e., across the diameter of the tube). In the example of starch conversion, each plug (mixture of starch and amylolytic enzymes) would enter the tube mainly as starch and exit the tube in a highly converted state. The combination of CSTB followed by a plug-flow bioreactor is theoretically the

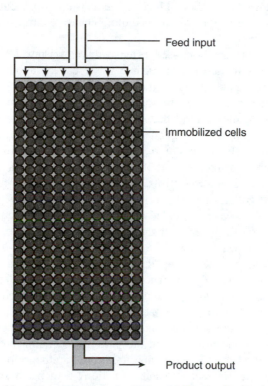

Feed input

Immobilized cells

Product output

FIGURE 8.9 A fixed-bed bioreactor packed with immobilized cells. Feed is input at the top of the column, and as it trickles through the column, it is converted to product.

most efficient bioreactor configuration for many bioprocesses. One of the drawbacks to this design, though, is that it has not proven possible to design an **aerobic** plug-flow bioreactor.

C. Immobilized Cells and Enzymes

1. Immobilized Cells

So far, we have examined only large-scale systems that involve the growth of cells suspended in liquid media. Alternatives do exist. A variety of processes involve immobilization of cells onto supports. This is most often done in a columnar biore-actor (Figure 8.9), which operates as a continuous system — i.e., raw materials are continuously fed to the cells at the top of the column and product is collected at the bottom. This type of bioreactor is only suited to processes where the product is secreted from the cell. There are several advantages to using immobilized cells to produce valuable products. The cells do not need to be harvested in this type of system, and, in fact, it is more efficient to collect secreted product directly, rather than harvest cells plus secreted product. If cells are present, energy must be expended to separate the cells from the liquid containing the secreted product, whereas with immobilized cells, all that is required is to purify the desired product from the culture liquid. Another advantage of immobilized systems is that wash-out is not a problem. This simplifies design of the culture medium, because control over cell dilution is not required.

Immobilized cells are also used extensively in wastewater remediation. For example, water contaminated with organic waste can be applied to the top of a column of immobilized cells, and as the solution trickles through the column, cells catabolize organic compounds in the solution, converting them mainly to CO_2. Thus the solution leaving the column has much less organic contamination. In these systems (unlike most product-oriented systems), the biotechnologist usually lets the microbes immobilize themselves onto a solid substrate. Many microbes can adhere well to solid surfaces, using their carbohydrate-rich glycocalyces. In **trickling-filter** types of wastewater treatment plants, rocks, sand, or gravel form a broad and relatively shallow column that supports a complex community of bacteria, fungi, and protozoans. Wastewater is trickled over the rocks, and the active aerobic microbial communities on the surface of the rocks rapidly and efficiently degrade the organic matter in the wastewater.

The same principle (i.e., using the natural ability of microbes to adhere to surfaces) can be used to produce ethanol by *S. cereviseae*, γ-linolenic acid by *Mucor ambiguous* (zygomycete fungus), and capsaicin and hop flavor compounds by plant cells. Inert particles (**biomass support particles**) are added to a bioreactor. These particles are highly porous and are usually composed of complex fiber meshes. Cells adhere to the fibers and grow within the particles. A number of inorganic (e.g., stainless steel knitted mesh) and organic (e.g., polyurethane foam) particles have been developed. The bioreactor usually functions as a batch system, but higher productivity is obtained than with STBs. For example, *M. ambiguous* grown in polyurethane foam particles resulted in 80% greater production of γ-linolenic acid.

Inert particles (e.g., glass fiber beads) are also used in the culture of animal cells. Again, the ability of the cells to adhere to these beads is essential. These beads have a protective role; animal cells lack cell walls and are highly vulnerable to shear stress.

One of the problems associated with the use of "natural" cell adherence for immobilization is that colonies continually release cells into the culture liquid. Consequently, these cells must be separated from product during downstream processing. This has spurred biotechnologists to develop immobilization systems that restrict cell release but do not interfere with cellular metabolism. One strategy for such immobilization is to **entrap** cells within a gel.

Carageenan is a popular polymer for the entrapment of cells. Carageenan is similar in many ways to agar; it is a complex polysaccharide that is produced by certain algae. It is poorly soluble in water until it is heated to near boiling temperatures. After solubilization, if it is cooled, it forms a highly hydrated gel. If this gel is then reheated to near boiling, the gel melts and is resolubilized. Hot solutions of carageenan are cooled until they are close to the temperature of solidification. The solution is then mixed with cells, and the gel is allowed to continue cooling and solidifying. The entrapped cells, however, cannot be directly used in a bioreactor because of the inability to maintain a flow of feed through the solid carageenan gel. **Granulating** the gel solves this problem. The small granules, each of which contains a number of cells, are then packed into a columnar bioreactor. A liquid medium is fed to the bioreactor and allowed to slowly flow through the packed column. This process works well if the desired product is secreted from cells. After secretion, the product flows through the column and is collected and purified from the effluent.

Besides carageenan, **alginate**, **agar**, and **polyacrylamide** gels have been used successfully to entrap cells. Calcium alginate gels have been particularly popular, because entrapment in alginate is achieved through a relatively mild (in terms of cellular stresses) process. If entrapment is not feasible, it is also possible to immobilize cells to inorganic or organic supports (e.g., carboxymethylcellulose). The challenge is to find chemicals that will stick to cells and surfaces without affecting their viability. This strategy is more successful when applied to enzyme immobilization (see below), where activity, rather than viability, is the vital quality that must be maintained. In some cases (e.g., urocanic acid production by *Micrococcus luteus*), viability is less important, as long as enzyme activity within the cell is present after immobilization. The main advantage of direct immobilization to a solid support is that there are fewer diffusional limitations to product formation, because reactants and products do not have to diffuse through a gel to reach the cells.

Surprisingly, entrapped cells often have higher viability than cells in suspension. In some cases, during *and after* release from entrapment, cells retain higher productivity (e.g., *S. cereviseae* and ethanol production). The main reason for this enhanced productivity appears to be higher activity of glycolytic enzymes (e.g., phosphofructokinase). It is perhaps not surprising that cellular metabolism would change; the micro-environment within a gel is less prone to fluctuation than the micro-environment of a cell in suspension. In some cases, immobilized cells retain their viability and can be used for long periods (even years) without regeneration.

Immobilized cell systems have become quite popular, and a number of biotechnological products are produced using this technology, including malic acid by *Brevibacterium flavium*, aspartic acid by *E. coli*, and monoclonal antibodies by hybridoma cells. Cell immobilization is also used to enhance the stability of certain traditional food biotechnologies. For example, a diverse range of sausages and other cured meats undergo an initial fermentation with LAB. Immobilized cells have successfully been used for these meat fermentations; meat is chopped or comminuted and then incubated with cells of *Pediococcus cereviseae* or other LAB that have been entrapped in alginate gels. Lactic acid production results in acidification of the meat, which inhibits growth of spoilage microbes and pathogens, while allowing the growth of LAB and micrococci (e.g., *Micrococcus varians*) that contribute desirable flavors. The use of immobilized cells leads to greater stability of the fermentative characteristics of the LAB, probably related to increased stability of the micro-environment surrounding the bacteria.

The major application of immobilized cells to food biotechnology is in the production of enzymes that are in turn used for biotechnological processes. The actual bioreactor design used for enzyme production, meat fermentation, and other immobilized cell processes is flexible. Some processes use STBs, but several alternative designs are more popular. STBs are effective when cell viability is required. This refers to both aerobic and anaerobic cells, because anaerobic cells usually produce gaseous end products (e.g., CO_2) that must be removed from the bioreactor. However, if cell viability is irrelevant to the process, then a **fixed-bed** reactor (see

Figure 8.9) can be used. This, the most popular bioreactor for immobilized cells, consists of a bed of cell-bound particles that are packed into the bioreactor. Liquid containing reactants flows through the bed, and product is collected from liquid leaving the bioreactor. **Trickle-bed** bioreactors (described previously in the context of wastewater remediation) have a slow rate of flow; in these systems, aerobic conditions can be maintained because the bed is not water saturated. Thus, trickle-bed bioreactors can be used with viable aerobic cells. This type of bioreactor has a great advantage over a STB in that little energy is required to maintain aerobic conditions. In a STB, the major power input is into the aeration and agitation system. Surprisingly, trickle-bed bioreactors are also amenable to highly productive processes when used with anaerobic cells (e.g., acetone–butanol–ethanol production by *Clostridium acetobutylicium*).

Unfortunately, fixed-bed bioreactors have many disadvantages. Bioprocesses that generate a great deal of heat are difficult to use in these systems, because heat transfer is inefficient. However, an alternative is **fluidized-bed bioreactors**. These are considered to be **three-phase** operations (fixed-bed systems are two-phase) because cells are immobilized onto a **solid** support, **liquid** flows through the support bed, and **gases** are actively injected into the bed. Because liquids (substrate) and gases usually flow into the bottom of the bioreactor, the bed is continually expanding and undergoing some degree of mixing. This allows aeration of the bed, resulting in increased oxygen supply, increased removal rates of waste gases, and improved heat transfer rates. The main problem with fluidized-bed bioreactors is that it is difficult to predict and maintain optimal operating conditions. This has restricted the popularity of these reactors, although they are increasingly used in wastewater treatment.

Finally, the use of membranes to entrap and immobilize cells is currently attention much interest in the research community; membrane bioreactors are also used in certain commercial applications. For example, hybridoma cells (see Chapter 6, Section II.E.2, for an explanation of hybridomas and monoclonal antibodies) can be grown in membrane bioreactors. In these systems, membranes are used to physically separate cells from liquid feed containing substrates and nutrients and to separate the cells from the product stream. The membranes must be porous, so that feed and product can flow through them. Membrane bioreactors alleviate two of the major disadvantages of continuous culture. The cells are not prone to wash-out because the membrane pores are not large enough to allow passage of cells. Second, cells are not mixed with product, facilitating downstream processing.

One final strategy for immobilizing cells deserves mention — **solid-state fermentation** (SSF; note that "fermentation" here refers to processes based on fermentation or respiratory metabolism). This refers to the growth of cells on and within solid supports in the absence of free water. In most cases, the solid support is also the substrate used for cell growth. Many traditional food biotechnologies (e.g., tempeh, soy sauce, and Camembert cheese) are produced through SSF, and it is widely used for the production of industrial enzymes. The main advantages of SSF are related to the low water content of the substrate. Enzymes or other products are in a concentrated state, which makes them relatively easy to purify. SSF is often more productive than comparable submerged systems, when compared on the basis of volume (i.e., volume of substrate required to produce a given amount of product).

Such systems also produce less waste effluent, as compared to submerged systems. Bioreactor design is also fairly simple — a tray, drum, or deep trough is usually sufficient for SSF. Contamination is often unimportant, especially when fungi are used. Filamentous fungi are well adapted to the colonization of solid substrates and are often able to exclude other microbes. In contrast, most filamentous fungi do not naturally colonize submerged systems; hence the array of problems that are encountered when trying to persuade fungi to grow in STBs and other forms of submerged culture (see this chapter, Section II.A, for an explanation of shear stress). Despite these advantages, SSF systems are not widely used outside of Asia. This is partly due to the lack of SSF expertise among engineers in the rest of the world and partly due to inherent disadvantages of SSF. For example, one problem is slow heat dissipation, which limits the upper end of the scale of the bioprocess.

2. Immobilized Enzymes

The past 20 years have seen great expansion of the use of **immobilized enzymes** to produce valuable industrial products. The basic principle is the same for immobilized enzymes as for immobilized cells; in the case of enzymes, a **substrate** is introduced into a column that has been packed with immobilized enzymes, and the **transformed product** is collected from the effluent of the column. Immobilized enzymes are particularly attractive to chemical engineers, because of the well-known advantages associated with the use of solid-phase reactants in nonbiological processes (e.g., less reactants lost to waste). Immobilization is essential when the cost of enzymes is high. In a batch process, the loss of enzyme that typically occurs during recovery and repurification of the enzyme is unacceptable for expensive enzymes.

Enzymes are usually easier to immobilize than cells, because maintenance of cellular viability is not an issue. Consequently, relatively harsh chemicals such as **glutaraldehyde** can be used to fix enzymes to solid supports. Glutaraldehyde cross-links proteins and can be used to covalently bind enzymes to solid beads. Because this usually results in decreased enzyme activity, enzymes are adsorbed first onto a cellophane membrane or other support before application of glutaraldehyde.

Alternatively, enzymes can be **adsorbed** onto solid supports. Ionic attraction, hydrogen bonds, or van der Waals forces may be involved in adsorption. For example, beads of modified cellulose similar to those used in ion-exchange chromatography (e.g., diethylaminoethyl (**DEAE**)–sephadex) can be used to immobilize enzymes. The enzyme is attached through electrostatic attraction to the beads. One important advantage of these systems is that the supporting beads can be recycled. When enzyme activity declines over time, old enzyme can easily be removed from the beads and new enzyme reabsorbed. This is an important component of economic analysis of the feasibility of an immobilized enzyme system. An enzyme that maintains its activity over a long period of use is crucial.

Enzymes can also be immobilized by entrapment within gels; the principles of use are similar to those used with intact cells. The types of bioreactors that are used for immobilized enzymes and cells are also similar. Hence, STBs, packed-bed reactors, and fluidized-bed reactors can all be used for immobilized enzyme processes.

High-fructose syrups are important ingredients in many foods and are typically produced through the use of enzymes attached to modified cellulose beads. Many other food biotechnological processes are produced using a variety of enzyme immobilization techniques, and the application of immobilization technology is bound to increase in the future. The reader may well wonder when immobilized cells should be used and when immobilized enzymes should be used. In many cases, immobilized cells are used for simple conversions accomplished by one or two enzymes, so it is possible to use either cells or enzymes. Cell immobilization is often preferred, primarily because of increased stability and longevity of enzymes when they are part of a living cell, as compared to purified, immobilized enzymes. The main disadvantage of immobilized cells is that they are complex, metabolically active entities — this may lead to unwanted side reactions that deter from product formation. Also, reactants and products must pass through the cell membrane, which may pose a problem for some processes. Thus, before embarking on an immobilization strategy, one must know a great deal about the biochemistry of the process, the stability of the enzymes involved, and the potential for cellular metabolic processes to inhibit process formation.

It is also possible to combine enzyme and cell immobilization in a bioprocess. For example, *Lactobacillus casei* can be used to convert starch to lactic acid. In this system, amylases are immobilized onto a support and *L. casei* cells are entrapped within carageenan.

IV. DOWNSTREAM PROCESSING

A. The Importance of Downstream Processing

It is time to address the problems associated with **collection, processing,** and **purification** of the products of large-scale processes. This aspect of industrial microbiology is crucial; if the product cannot be efficiently purified at a cost appropriate to the market value of the product, then the process is doomed to failure. Downstream processing represents a substantial proportion of the cost of production of most cell products. In large part, the cost of purification is linked to the concentration of the product. For example, therapeutic proteins produced by animal cell culture often need to be purified when present in the range of 10^{-6} g/L. The cost of purifying these dilute proteins is high; commercial production of these proteins is possible only because of their potency in alleviating human disease. In contrast, citric acid is often present at 100 g/L as it leaves the bioreactor. Purification is relatively easy, which allows companies to maintain a low cost for citric acid. This in turn makes it economical to use citric acid in the production of inexpensive products such as soft drinks.

Waste is the other major issue surrounding downstream processes. The brewing and distillation industries are virtually the only microbiological industries that have markets for their wastes. In the brewing industry, waste from mashing is typically sold as a feed supplement, and waste yeast from fermentation is often reused before being incorporated into feed. In other microbial industries, though, product purification leaves a large amount of waste product. Often, the wastes have a large

biological oxygen demand (BOD), similar to human sewage or animal manures. This means that the wastes cannot be directly dumped into a water source without significant negative effects on fish and other aquatic organisms. In most parts of the world, such dumping is illegal, requiring processing to drive down the BOD. In the case of solid wastes, disposal through burial is also increasingly frowned upon and is often illegal. Thus, many companies are developing processes similar to those used to treat organic garbage (composting) and sewage (e.g., trickling-filter bioreactors) to solve their waste problems. Some industries (e.g., the dairy industry) have also actively sought new technologies to transform wastes (e.g., cheese whey) into value-added products (e.g., additives that could be used in fat replacement).

B. CELL LYSIS

If the desired product is not secreted, then cells must be broken in order to release cytoplasmic contents. This is necessary for a number of enzymes; for example, glucose isomerase, used to make high-fructose syrups, is produced intracellularly by various species of *Streptomyces*. Cells can be disrupted by forcing a cell suspension through a small orifice at high pressure, or by agitating a cell suspension in a vessel containing a large density of small glass beads. In the first case, the explosive decompression experienced by the cells during their discharge from the small orifice causes cellular disruption, whereas in the second case, the cells are physically broken up by the glass beads.

Nonmechanical processes can also be used to break cells. These processes rely on enzymes, detergents, or stresses imposed by desiccation to break open cells. Lysozyme is effective in breaking down the cell wall of Gram-positive bacteria and is used industrially to break open cells. Other enzymes, such as zymolase, also have potential for industrial use.

C. SEPARATING SOLIDS FROM LIQUIDS

In STBs and CTSBs, the effluent will contain cells suspended in a liquid. How does one separate cells (a form of suspended solid) from liquids? Many cells (e.g., brewing yeast strains) naturally **flocculate** (i.e., clump together) and sediment to the bottom of the bioreactor. Elimination of the cells is then straightforward; simply remove the liquid over the settled cells. If the cells do not flocculate, then additional steps are required. Large-scale **centrifugation** is commonly used to separate cells and is especially useful if the aim of the process is to harvest biomass (e.g., production of baker's yeast). The main disadvantage of centrifugation is that it produces a thick slurry or a paste; if a dry product is required, further processing is necessary (e.g., filtration).

The principle of centrifugation is the same in industrial-scale centrifuges as in laboratory centrifuges: objects spinning around a central point have a force exerted on them. This force is effective at increasing the sedimentation rate of particles, resulting in clarification of the liquid. A number of large-scale centrifuge designs are available. Most operate as batch systems — slurry from a bioreactor is fed into the centrifuge, the slurry is spun, clarified supernatant is removed, and sedimented particles are removed (often manually) from the centrifuge. Some centrifuges can operate in a continuous or semi-continuous manner. For example, **decanter** centrifuges have an internal conveyor system that slowly removes sedimented solids as the chamber spins. Clarified liquid leaves through the opposite end of the chamber. This type of centrifuge can be used to continuously separate solids from liquids and is effective when either solid or liquid constitute the desired product.

Filtration is another option for separating solids from liquids. This is achieved by passing a slurry against a porous barrier that retains solids. The slurry then forms a dry **cake**, and the filtrate is clear of suspended cells or other solids. Filtration often follows a centrifugation step, and it is an effective method for **dewatering** a slurry. The material used as the solid barrier is crucial to filtration; it must retain solids, but it must also allow flow of liquid. Filters are often made from synthetic compounds such as nylon or polypropylene and are often clamped between metal plates that have ports to allow flow of filtrate from the filter.

When a high degree of purification is desirable (e.g., amino acid production), **membrane filtration** is an excellent strategy. This is different from conventional filtration in that it is applied to clarified liquids and not to slurries. Membrane filters are made of polymers such as cellulose, cellulose acetate, or polyacrylonitrile. A number of membrane filtration technologies are used industrially; we will focus on ultrafiltration and reverse osmosis.

Ultrafiltration can be used to separate macromolecules such as enzymes from liquid. The effluent is forced at high pressure through membranes with extremely small pores (1 to 50 nm). Membranes with a large surface area are usually required for industrial ultrafiltration. Thus, the geometry of the membrane support is crucial to the efficiency of ultrafiltration. This process can be used to remove unwanted proteins from a liquid (e.g., whey). The membrane retains proteins and other macromolecules, and only low-molecular-weight compounds (e.g., glucose, lactose, water) pass through the membrane.

Reverse osmosis also uses membranes, but the pore size is smaller than in membranes used for ultrafiltration. Reverse osmosis is a useful way to concentrate solutions of low-molecular-weight compounds. Consider normal osmosis, a process that drives water flow from solutions of differing solute content that are separated by a semipermeable membrane (Figure 8.10). Normally, the flow of water is toward the solution with higher solute content. However, if a pressure is exerted on one of the solutions, the direction of osmotic flow can be reversed. This reverse osmosis can be used to concentrate small molecules such as amino acids. It can also be used to produce low-alcohol beer — water and ethanol can permeate through reverse osmosis membranes, but other compounds are retained. The resulting concentrate will contain flavor compounds characteristic of beer, with very little ethanol. This demonstrates how reverse osmosis can be used to separate low-molecular-weight compounds in solution.

The importance of reverse osmosis was tragically demonstrated by the eosinophilia–myalgia epidemic attributed to tryptophan supplements (see Chapter 7, Section IV.F). Contamination of tryptophan supplements with toxic metabolites occurred partly because a reverse osmosis purification step was partially bypassed. The reverse osmosis membrane retained molecules with a molecular weight greater

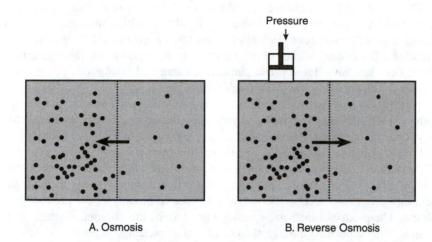

FIGURE 8.10 Reverse osmosis. When two solutions are separated by a semipermeable membrane (A), water normally flows from the solution that has higher water potential. However, if pressure is applied to the solution with lower water potential, water flow is reversed (B).

than 1000 Da. Water and low-molecular-weight compounds, including tryptophan, were able to pass through this membrane. If this process had functioned properly, some of the large molecules that were later implicated as toxins (e.g., 1,1'-ethylidene-*bis*[L-tryptophan]) would have been retained by the membrane. However, some of the toxic compounds are similar in structure to tryptophan and would not have been removed using this reverse osmosis step. The company involved (Showa Denko) also had decreased by 50% the amount of activated charcoal used to remove impurities from the tryptophan. This may have been the crucial mistake that led to accumulation of toxic compounds with the tryptophan. Activated charcoal is able to absorb a large number of low-molecular-weight compounds and is widely used in purification procedures. For example, it is used to produce clear glucose syrup from the brownish syrups produced through starch saccharification.

D. PHASE CHANGES

The final step in downstream processing is a crystallization or precipitation step. Both processes yield dry solids that are convenient for packaging and storing. For example, xanthan gum is purified by the addition of large volumes of isopropyl alcohol. Many proteins are also conveniently purified by precipitation. Often, it is desirable to precipitate proteins without the use of organic solvents. Varying the pH or salt content of aqueous solvent and decreasing solubility through temperature modification are potential strategies. Ammonium sulfate is often used to **salt out** proteins, because of its high solubility in water.

Crystallization is similar to precipitation, but it yields particles with consistent size and shape. Precipitation, in contrast, yields amorphous particles. Crystallization is not possible for all products; however, if possible, it is attractive, because crystallization yields purer product than precipitation.

The liquid–gas phase change is exploited by the distilling industry; ethanol has a lower boiling point than water, so heating a solution with a moderate amount of ethanol (e.g., 12%) will result in loss of ethanol to the gas phase before much water evaporates. The ethanol is then easily returned to the liquid phase through condensation. Distillers have developed a number of batch and continuous processes for efficient distillation of spirits from fermentation media.

Freezing can also be used to purify desired products or to concentrate fermented products such as beer. The rate and extent of ice crystal formation is carefully controlled, and ice crystals are then removed, leaving a concentrated solution.

Finally, valuable compounds can be separated from contaminants by adsorption (i.e., attachment of particles or solutes to a solid phase, thus separating them from a liquid phase). Purification can be achieved by passing effluent through **ion-exchange columns**. Electrostatic interactions lead to binding of the product to the column. The product can then be easily eluted from the column. Glutamate, for example, can be purified through the use of cation-exchange columns.

A related method involves the use of **affinity** columns, wherein the desired product binds *specifically* to the column. As with ion-exchange columns, the product can easily be eluted from the column after the effluent has completely passed through. Affinity systems are highly desirable, especially if they are applied after a preliminary cleaning-up step such as precipitation. **Monoclonal antibodies** are an example of a cell culture product that is easily and efficiently purified using affinity columns. Antibodies against the Fc (constant region) fragment of the monoclonal antibody are covalently attached to cellulose beads. As the effluent from hybridoma cells passes over the cellulose beads, the attached antibodies bind to the Fc fragments of the monoclonal antibodies and retain them. Subsequent addition of an aqueous solution of NaCl will disrupt the link between the two antibodies, and the monoclonal antibody will be released; it can then be collected in a pure, concentrated solution. Similar systems can be used to purify enzymes — the substrate for the enzyme is immobilized on a solid, and enzymes bind to the substrate via their active site as they pass through. One of the problems with these adsorbent-based systems is that they are difficult to operate continuously, due to the need to treat the solid in some way to release the product.

Large-scale culture of bacteria, fungi, and plant and animal cells is not always easy. However, the large number of advances that occurred in the 20th century have led to an impressive expansion of the cell culture industry. Large-scale cell culture provides essential products for the food production and processing industries, which justifies the continued interest in the methods for growing large volumes of cells under controlled conditions.

RECOMMENDED READING

1. Reisman, H. B., Problems in scale-up of biotechnology production processes, *Crit. Rev. Biotechnol.*, 13, 195, 1993.
2. Merchuk, J. C. and Asenjo, J. A., Fundamentals of bioreactor design, in *Bioreactor System Design*, Asenjo, J. A. and Merchuk, J. C., Eds., Marcel Dekker, New York, 1995, chap. 5.

3. Bailey, J. E. and Ollis, D. F., *Biochemical Engineering Fundamentals*, McGraw-Hill, New York, 1986.
4. Ertola, R. J., Giulietti, A. M., and Castillo, F. J., Design, formulation, and optimization of media, in *Bioreactor System Design*, Asenjo, J. A. and Merchuk, J. C., Eds., Marcel Dekker, New York, 1995, chap. 4.
5. Reuss, M., Stirred tank bioreactors, in *Bioreactor System Design*, Asenjo, J. A. and Merchuk, J. C., Eds., Marcel Dekker, New York, 1995, chap. 6.
6. Atkinson, B. and Mavituna, F., *Biochemical Engineering and Biotechnology Handbook*, Stockton Press, New York, 1991.
7. De Gooijer, C. D., Bakker, W. A. M., Beeftink, H. H., and Tramper, J., Bioreactors in series: an overview of design procedures and practical applications, *Enzym. Microb. Technol.*, 18, 202, 1996.
8. Furusaki, S. and Seki, M., Use and engineering aspects of immobilized cells in biotechnology, *Adv. Biochem. Eng.*, 46, 161, 1992.
9. Fukuda, H., Immobilized microorganism bioreactors, in *Bioreactor System Design*, Asenjo, J. A. and Merchuk, J. C., Eds., Marcel Dekker, New York, 1995, chap. 9.
10. Salmon, P. M. and Robertson, C. R., Membrane reactors, in *Bioreactor System Design*, Asenjo, J. A. and Merchuk, J. C., Eds., Marcel Dekker, New York, 1995, chap. 8.
11. McLoughlin, A. J. and Champagne, C. P., Immobilized cells in meat fermentation, *Crit. Rev. Biotechnol.*, 14, 179, 1994.
12. Pandey, A., Selvakumar, P., Soccol, C. R., and Nigam, P., Solid state fermentation for the production of industrial enzymes, *Curr. Sci.*, 77, 149, 1999.
13. Giorno, L. and Drioli, E., Biocatalytic membrane reactors: applications and perspectives, *Trends Biotechnol.*, 18, 339, 2000.
14. Barrios-González, J. and Mejía, A., Production of secondary metabolites by solid-state fermentation, *Biotechnol. Annu. Rev.*, 2, 85, 1996.
15. Atkinson, B. and Mavituna, F., *Biochemical Engineering and Biotechnology Handbook*, Stockton Press, New York, 1991.
16. Bailey, J. E. and Ollis, D. F., *Biochemical Engineering Fundamentals*, McGraw-Hill, New York, 1986.

9 Ethics, Safety, and Regulation

I. OVERALL PERSPECTIVE

When assessing the ethics of a new technology, it is not a simple question of right vs. wrong. Instead, the benefits of the technology must be assessed along with the risks. The nature and severity of risks varies with the type of biotechnology (Table 9.1), but the risks usually deemed most important are those that are directly linked to human health. Biotechnological products that are consumed (e.g., transgenic crops) have a direct risk of toxicity; in contrast, diagnostic systems present indirect risks. However, indirect risks are also important. The use of unvalidated diagnostic tests could lead to incorrect identification of microbes in food, which could have serious consequences.

Many aspects of food biotechnology are also associated with risks or benefits to the environment. For example, widespread plantings of transgenic crops that produce insecticidal proteins have been shown in some cases (e.g., cotton in some regions of the U.S.) to lead to decreased pesticide usage. This should lead to less environmental damage, through decreased accumulation of toxic pesticides in soil and water and less mortality of nontarget insects, birds, and other wildlife. However, these insecticidal crops also have environmental risks. The furor that began in 1999 over toxic effects of pollen from transgenic corn on monarch butterfly larvae is a good example of environmental risk from a transgenic crop (see Chapter 4, Section IV.A).

In general, regulation of biotechnology occurs at the national or international level (e.g., the European Union [EU]). This puts governments in a delicate situation. Environmental and health risks must be balanced against potential health, environmental, and economic benefits, and consumer concerns must also be addressed. Scientific evidence is usually the most important factor in safety assessment, but as safety assessment is translated into policy and legislation, political, trade, economic, and consumer issues all affect decisions about the direction and nature of regulation of food biotechnology. To further complicate matters, governments also face pressure from the business community to maintain minimal levels of regulatory load. As these various inputs vary in strength, governments tend to respond to the strongest forces. Therefore, different governments may develop vastly different regulatory regimes, despite having access to the same scientific evidence regarding safety assessment. Regulations governing the culture and use of transgenic crops in Europe and the U.S. are clear examples of this phenomenon. In Europe, only two transgenic cultivars can be grown, sold, or imported. In most parts of Europe and the U.K., there is little acreage sown to transgenic crops, and food containing transgenic crops is absent

TABLE 9.1
Benefits and Risks of Food Biotechnology and Appropriate Level of Regulation

Type	Potential Benefits	Potential Risks	Regulation
Recombinant proteins	Improved supply (e.g., chymosin); improved enzyme activity (e.g., recombinant amylases)	Toxicity	National
Transgenic crops	Agronomic improvement (e.g., herbicide resistance); reduced pesticide use; improved nutritional quality	Toxicity; environmental damage (e.g., increased weed problems)	National and international
Diagnostics	Faster detection of pathogens; improved specificity; on-line monitoring	Decreased sensitivity; inconsistent performance	Association (e.g., AOAC[a]); national
Transgenic animals	Improved quality (e.g., modified milks); faster growth (e.g., salmon with boosted growth hormone)	Toxicity; decreased animal welfare; environmental damage (e.g., harm to wild fish populations)	National
Microbial biotechnology	Efficient processing (e.g., starch modification); new food ingredients and additives	Toxicity (e.g., eosinophilia–myalgia syndrome)	National, industry
Functional foods	Improved public health	Toxicity (especially in relation to food supplements); misleading labeling	National

[a] Association of Official Analytical Chemists.

from grocery stores. In contrast, in the U.S., Canada, and several other countries, transgenic crops are grown for food on millions of hectares, and grocery stores sell many foods that contain transgenic corn and soybeans.

Another complication is that many governments are vigorous supporters of biotechnological research; this has led many activist groups, and a number of scientific advisory groups, to caution against the danger of conflict of interest. If a government is a supporter of biotechnology, can it honestly and effectively regulate its application? The answer certainly can be yes, provided that there is a clear separation of responsibility for these two aspects of biotechnology. Another common criticism of biotechnology regulators is that in most countries they operate under a cloak of secrecy. Because of corporate desires to restrict dissemination of experimental data derived from company products, regulatory agencies usually do not publish the data used to assess health and environmental risks. This has led to widespread dissatisfaction among scientists and the public over government and corporate accountability, and has likely been a factor in the low public esteem currently experienced by the food biotechnology industry. One final criticism is worth noting at the outset: in most countries, food biotechnology regulation has been piggybacked onto existing legislation covering food safety and, in some cases, pesticide use. In the U.S., legislation governing food biotechnology is administered by three agencies: the U.S. Department of Agriculture (USDA), the Environmental Protection Agency (EPA), and the Food and Drug Administration (FDA). This complex regulatory web is difficult for companies to navigate, and appears from the outside to be difficult to manage. In Europe, the situation is even more complex, because attempts to harmonize legislation at the level of the E.U. have to acknowledge a variety of national laws and policies that are often antagonistic.

Nonetheless, products of biotechnology are regulated, and the food biotechnology industry has a relatively clean health and environmental safety record. Much of the remainder of this chapter is devoted to an examination of the factors that must be considered when regulating products of food biotechnology, with an emphasis on transgenic crops. These are also known as genetically modified organisms (GMOs), a term that also includes transgenic animals and recombinant microbes. But first, we will examine biotechnology from the consumer's perspective and discuss pertinent ethical issues.

II. CONSUMER PERSPECTIVES AND FOOD BIOTECHNOLOGY

As stated above, regulations governing transgenic crops are much more restrictive in Europe than in North America. Much discussion has ensued in the popular media about the reasons for this. It is not a simple issue of consumer concern. Numerous polls in Europe and North America have shown that substantial proportions of the population in both regions are concerned about safety and environmental risks associated with transgenic crops, but that substantial proportions are also willing to consume transgenic crops. However, it is quite possible that in Europe the concerned segment of the population is much more vocal than in North America. This may be due to three recent events in Europe: (1) the bovine spongiform encephalopathy (BSE) epidemic in Britain, (2) the contamination of eggs and other poultry products with dioxins in Belgium in 1999, and (3) the tainted blood scandal in France (at least 3600 people in France were infected by the human immunodeficiency virus via blood transfusions during the mid-1980s). The governments in the respective countries have received extensive criticism over their roles, particularly with regard to communication of risks to the public. Government policy and regulations also failed to protect the public in the BSE epidemic and the French tainted blood scandal. These violations of public trust are probably major forces behind the European people's current lack of confidence in the ability of government to maintain food safety. Although many North American consumers are concerned about transgenic crops, they do not appear to share the European mistrust of government regulators. Hence, there has not been a popular groundswell of protest over the presence of GMOs in food.

The epidemic of BSE in cows and humans illustrates the importance of government regulation of agricultural and food technology and emphasizes the crucial role of government communication to the public about food safety. In 1996–1997, evidence accumulated that at least 10 and probably more than 80 people in the U.K. had a **variant** form of **Creutzfeldt-Jakob disease** (CJD), a degenerative disease that leads to atrophy of brain tissue and death. Pathological symptoms of CJD include the formation of distinctive tangles of fibrils in the brain. CJD has always been a mysterious disease, and even today its cause has not been firmly established. Most evidence, though, points to an abnormal protein that is able to convert its normal counterpart in the brain to the abnormal form. This abnormal protein is a **prion**, a protein capable of transmitting disease. CJD in humans was previously believed to occur sporadicly, mainly afflicting the elderly. However, the ten cases in the U.K. were unusual in that the persons afflicted were all relatively young. Consequently, these cases were considered to be a variant form of CJD (vCJD).

In 1986, BSE was first detected in cows. Unfortunately, until 1996, the British government's policy toward risk assessment of BSE centered on the idea that BSE posed a health risk to cows, but not to humans. In 1996, though, it became clear that the same strain of prion was responsible for BSE and vCJD. At that time, 80 people had died from vCJD. Although it is not clear exactly how the prion was transmitted from cows to humans, it is likely that it was through the ingestion of beef. In 1996, the British government acknowledged that vCJD was linked to BSE. The British public was outraged because of the previous steadfast conviction of the government that British beef was safe. Much of the current antagonism of British consumers toward foods containing transgenic plants has its roots in the breakdown in public trust of government that resulted from the vCJD epidemic. European confidence in food safety also fell considerably in 1999, when it became clear that Belgian authorities had neglected to inform trading partners of dioxin contamination of eggs and poultry. In 2001, Europe was also hit by the news that BSE-infected cows were present in a number of European countries, including Germany. Thus it is not surprising that Europeans are distrustful of new and potentially dangerous food technologies.

Another element that drives consumer concerns is the gut feeling regarding transgenic crops and animals. Many people feel that recombinant DNA technology allows scientists to make unnatural creations that violate fundamental distinctions between organisms. This is difficult to deny — biotechnologists do indeed take genes and transfer them to distantly related organisms, an act that is virtually impossible using natural processes of mating and recombination. The approach taken by biotechnology supporters has mainly been to attempt to educate consumers that transgenic plants and animals are merely extensions of traditional plant and animal breeding practices and that humans have engaged in genetic modification of organisms for a long time.

Finally, many antibiotechnology activist groups have bemoaned the corporate nature of food biotechnology. Although many independent researchers have been involved in the study of transgenic plants, most commercialization of these modified plants has been done by multinational corporations such as Monsanto, Novartis, and Aventis. This has led to the oft-cited fear that commercial release of transgenic seeds will ultimately result in corporate control over the global food supply. This is a valid fear, to some extent. Transgenic seeds, particularly those with herbicide resistance, have been tremendously popular among producers. This has reduced the diversity of seed usage in North America and increased the economic clout of the corporations that supply the seed. Past experience tells us that corporate monopolies can be dangerous. The prime example of this is the culpability of the major vitamin-producing corporations in a conspiracy to fix prices of vitamins. This conspiracy operated from 1990 to 1999, and prosecution by U.S. courts resulted in fines of $500 million to two of the companies involved (Hoffman-La Roche and BASF) and a 4-month jail sentence for one of the executives involved.

In relation to the seed industry, though, fears that transgenic seeds will lead to corporate domination of seed supply must be assessed with the understanding that this domination has already occurred. Five multinational companies share the vast bulk of global trade in corn and wheat seed; in the case of wheat seed, this corporate domination consists entirely of seed bred using conventional techniques. Therefore, corporate domination of global seed supply is a problem that is related to the whole seed industry, not just the segment involving trade of transgenic seed. Also, one of the unfortunate side effects of public hostility to GMOs has been that transgenic seed development is now virtually impossible for small corporations; the risk of public rejection of developed products is too great. Thus, commercialization of transgenic crops is firmly in the hands of large multinational corporations that have the financial resources to accept the risk.

The biotechnology industry blames negative consumer attitudes toward food biotechnology on nongovernmental organizations (NGOs) such as Greenpeace. This and many other activist groups are defiantly hostile to GMOs. For example, when Lord Melchett, the British spokesperson for Greenpeace, was asked by a committee of the House of Lords whether Greenpeace's opposition to GMOs was "absolute," the answer was, "It is a permanent and definite and complete opposition based on a view that there will always be major uncertainties." Thus, no amount of scientific evidence of safety or benefits of transgenic crops or other GMOs (e.g., recombinant bacteria) would decrease Greenpeace's opposition to this technology. Clearly, the biotechnology industry will never convince such groups that GMOs are safe and useful. However, this does not mean that the public cannot be convinced. Certainly, more studies of the environmental and health safety of GMOs are needed. The results of these studies need to be communicated effectively to journalists, so that consumers can make up their own minds, without being unduly influenced by NGOs *or* the biotechnology industry.

III. SAFETY ASSESSMENT AND REGULATION OF TRANSGENIC CROPS

A. ASSESSMENT STRATEGIES

Few issues in biotechnology have roused as much antipathy from consumers and NGOs as the introduction of transgenic plants. Few would have predicted this in the early 1990s, when the first transgenics (e.g., the Flavr Savr™) were introduced without great controversy. However, everything changed in the mid-1990s. Activist groups throughout the world spurred massive protests that led to restrictive legislation in many countries. People, particularly in Britain and France, were galvanized to take direct action, destroying test plots of transgenic plants. Governments have had a difficult time responding to these protests, because of the varied and complex risk factors associated with transgenic plants (see Table 9.1), and because the rights of consumers must be balanced against the rights of farmers, food processors, food retailers, and biotechnology companies. For most consumers, the prime worry with respect to transgenic crops is that they introduce new and unknown risks to human health. Environmental risks are also an essential and powerful part of risk assessment of transgenic crops. These will not be discussed here, but issues related to environmental assessment of transgenic crops are described in Chapter 4, Sections IV.A and C).

Since the mid-1980s, when it became clear that adequate technology existed to develop transgenic crop plants, governments, organizations (e.g., FAO, the Food and Agriculture Organization of the United Nations), as well as biotechnologists became concerned over potential health risks associated with this new technology. Over the next 10 years, it was generally agreed that the best strategy for assessing health risk is through the concept of **substantial equivalence**. This concept holds that if a novel crop has similar chemical and biological characteristics to the crop from which it was derived, then it should have equivalent risks as the derived crop. There are several reasons for this approach, but the most important is that the normal toxicological approaches to assessing safety of new pharmaceuticals, synthetic food additives, or pesticides are not appropriate for the testing of novel foods.

Conventional toxicological analysis of new chemicals requires some understanding of the dose responsible for toxicity. All chemicals have toxic properties; however, we consider chemicals to be safe if they are toxic only in high concentrations and if it is highly unlikely that consumers will be exposed to these high levels. Consequently, the general approach taken is to expose laboratory animals to increasing concentrations to find out the minimal level that causes toxicity and the level when mortality is frequent (the LD_{50}, for example, represents the level that causes mortality of 50% of test animals). Regulators then set at least a 100× safety level, ensuring that the population will not be exposed to the chemical at levels greater than 0.01× the lowest concentration that causes toxicity. Unfortunately, this approach cannot be taken with novel foods, because foods intrinsically have low toxicity. If animals are fed large amounts of a food, they develop symptoms related to nutritional imbalance (e.g., lack of specific vitamins) that are unrelated to toxic characteristics of the novel food. For this reason, regulators currently do not demand toxicity assessment of novel foods. However, toxicity experiments using the protein expressed by the transgene are usually included in safety assessments. The rationale for this is that, theoretically, this should be the only new compound present in the transgenic plant that is not present in the conventional crop from which it was derived. Antibiotechnology forces see this as inadequate and maintain that the lack of requirements for extensive toxicity testing is the result of government collusion with industry to lower development costs for transgenic crops. This conspiracy theory is difficult to accept, given the public nature of the debate over regulation of biotechnology that occurred in the early 1990s.

It is also worth pointing out that regulators and biotechnologists agree that each **novel crop** should be assessed separately, on the basis of its final characteristics, *not on the basis of the method of inducing genetic change*. Thus a herbicide-resistant crop would be considered a novel food that must be assessed for safety, regardless of whether it was produced by recombinant DNA technology or through traditional breeding. Again, antibiotechnology activists disagree; they believe that recombinant

DNA technology itself creates hazards that are common to all crops produced using this technology.

B. The Substantial Equivalence Debate

Substantial equivalence is assessed through the analysis of various aspects of the chemical structure of a transgenic plant, followed by a comparison with the structure of the parent nontransgenic plant. Nutrient levels in the part of the plant that are consumed are particularly important; developers must demonstrate that the new plant has similar levels of proteins, fats, carbohydrates, vitamins, and so forth as the parent. Many crop plants (e.g., potato, beans) also contain toxins. Developers must test the levels of these toxins to demonstrate that they fall in a similar range as the parent plant. This is a controversial area — some scientists maintain that companies should be bound to statistically demonstrate that the new and parent plants are identical. This is technically more difficult than a simple demonstration that nutrient and toxin levels are in the same *range*, but it is certainly possible. The theory of *statistical power* specifically addresses the strategies required to statistically assess whether two populations *do not differ*. However, most regulatory agencies are satisfied with data that show that the new and parent plants have similar ranges of nutrients and toxins.

Substantial equivalence is a controversial concept; some scientists regard it as an unscientific concept, because of its essential vagueness and the imprecise definitions associated with it. One of the chief criticisms is that testing for substantial equivalence will not reveal the presence of *new* toxins. Theoretically, this could happen as a result of the random nature of insertion of transgenes into plants. Biotechnologists virtually never have control over the new location of transferred genes (see Chapter 4, Sections III.E and F). Sometimes the new DNA is inserted in the middle of a crucial gene, making the plant unviable. Viewed in this context, the transferred gene acts as a mutagen, with the potential of creating new genetic sequences that could have unpredictable effects. Most biotechnologists acknowledge that this can and does happen; debate centers on the probability that new toxins could be formed in this way. Proponents of biotechnology assert that this is unlikely. New toxins have been recorded only a few times in crop plants, and they have always resulted from traditional breeding programs. This does not reassure opponents of biotechnology, who demand extensive and expensive rounds of animal testing before transgenic crops are foisted on the public. This would be similar to the type of testing that is required if a new synthetic chemical is intended for use in foods or as a pharmaceutical. Developers must demonstrate that such chemicals have nonexistent or low levels of acute (short-term) or chronic (long-term) toxicity and are not carcinogens or teratogens (substances that produce birth defects). If required to conduct these tests, biotechnology companies would face long delays in development and vastly increased costs. On the other hand, successful tests of this sort would be a powerful weapon in the fight to influence public opinion about food biotechnology.

A recent chronic toxicity experiment involving long-term feeding of rats with transgenic and nontransgenic potatoes found no evidence of toxicity. This experiment by Japanese researchers was probably intended to test the validity of the widely publicized experiments by Arpad Pusztai in Scotland. The latter study observed differences in the intestinal anatomy of rats that were fed transgenic potatoes, as compared to rats fed nontransgenic potatoes. The transgenic potatoes had an insecticidal lectin gene. Unfortunately, that study had serious experimental design problems. For example, because of the large amount of potato in the rats' diet, they may have become deficient in protein, making interpretation of the data difficult. Also, an important control was missing (e.g., rats fed the lectin alone), and the study did not include important supportive data. For example, in any toxicological study comparing varieties of potato, it is essential to compare levels of alkaloid toxins. For these reasons, the main effect observed cannot be assigned conclusively to the mechanism used to introduce the transgene. Finally, the main effect observed in the published study was intestinal crypt length, a variable that is difficult to measure with precision. It is also difficult to assess the health significance of this variable. In statements after publication of his results, Pusztai has emphasized that his study does not show a toxic effect of the transgenic potatoes, but indicates that they are *different* from nontransgenic potatoes. Unfortunately, the message that was transmitted by the mass media was that the transgenic potatoes had significant toxic effects.

Public criticism of substantial equivalence has been particularly strong in Europe. Recently (July 2001), the EU released a proposal for a new regulatory regime for GMOs. This proposal stated that substantial equivalence would still be a tool in risk assessment, but it emphasized that demonstration of substantial equivalence does not constitute a risk assessment in itself. The proposal does not state or suggest methods that could be used with substantial equivalence to help risk assessment, but certainly animal studies of toxicity could be part of the process. In the U.S. and Canada, the main growers of transgenic crops, regulators continue to use substantial equivalence as the backbone of risk assessment of new transgenic cultivars. Several countries (e.g., the Netherlands) plan to expand the range of data that is used in the assessment of substantial equivalence. One popular idea that is currently a hot research topic is the development of **metabolic fingerprints** of trangenic crops. This would involve a broad survey of as many of the compounds within the plant as possible. With this sort of approach, the likelihood of identifying a new and unpredicted toxin should increase. Unfortunately, this approach has many problems. For example, many of the secondary compounds that might be expected to have toxic properties have wide fluctuations in concentration in plant tissues. These fluctuations are probably related to environmental variability. Comparison of metabolic fingerprints is not a straightforward process.

C. RISK ASSESSMENT OF NOVEL GENETIC ELEMENTS

The substantial equivalence concept is built around the idea that a transgenic cultivar and the cultivar used in its development should differ only in relation to the newly introduced genes. Therefore, safety assessment of the introduced genes is a crucial part of the process. As mentioned previously, the protein produced by the transgene is usually subjected to acute toxicity tests. Purified protein is injected into laboratory animals at a range of doses. The animals' health is then monitored, and after a short period of time, the animals are killed and examined for internal lesions. In more thorough studies, histological samples (e.g., slices of liver) are examined to detect any cellular abnormalities. These sorts of studies usually do not uncover evidence of toxicity, primarily because most proteins are nontoxic and are readily broken down in the gastrointestinal (GI) tract. However, this does not mean that toxicity is unheard of. Many microbial proteins are potent toxins (e.g., botulinum toxin), and toxic proteins (e.g., lectins) are also common in plants. For this reason, biotechnology companies also analyze the DNA sequence of the novel gene. It is compared to known protein toxins, so that any homologous (similar order of bases) sequences can be detected. One of the criticisms of this approach (sequence analysis + acute toxicity tests) is that chronic (long-term) toxicity is not assessed. One reason chronic tests are not required is that known toxic proteins exhibit acute toxicity, unlike many synthetic compounds, which exhibit toxic effects after long-term exposure. Biotechnology companies are firmly hostile to chronic toxicity tests, partly because of the financial costs of such tests and partly because chronic tests are more difficult to interpret than acute toxicity tests. The main reason for this is that effects of age-related degeneration must be disentangled from the effects due to the compound being tested.

Sequence analysis is also used to assess allergenicity in the novel protein. Some of the most dangerous allergens are proteins derived from plants (e.g., the peanut allergen), so caution is necessary when transferring genes from other plants into a crop plant. The novel gene sequence is compared to the sequence of known allergens. Companies and government regulators also cast a close eye on the *source* of the novel gene. If the source is a plant or microbe that previously has been implicated in allergy, then both biotechnology companies and regulators will proceed with caution. Unfortunately, it is difficult to test for the presence of new allergens in a food, making it difficult to conclusively show that a novel food is allergen free. This is a common cause of concern for consumer groups and activists. It should be encouraging to all groups that biotechnology companies appear to be concerned about the potential for transfer of allergens. In the early 1990s, development of a soybean with increased levels of methionine (targeted at the animal feed industry) was cancelled when early trials indicated the potential for increased allergenicity of the transgenic soybeans. The source of the gene for the novel methionine-rich protein was Brazil nut, a food that is known to provoke allergic reactions. It turns out that the methionine-rich protein is indeed the allergen. Evidently, extreme caution is necessary when transferring genes from plants that have potent allergens.

Novel genes are never transferred in isolation; they always carry extra baggage. The most controversial of these extra sequences is the gene that is used to select for transformed plants (see Chapter 4, Section III.E and IV.D). The most common gene used is the kanamycin-resistant gene. Application of the antibiotic kanamycin results in death of any untransformed plant cells, whereas any cells that carry the novel gene will also carry the antibiotic-resistant gene and will be unaffected by the antibiotic.

The prevalence of antibiotic-resistant genes in pathogenic bacteria is a serious problem worldwide, and many activist groups have pointed out that the inclusion of antibiotic-resistant genes in novel crops results in a vast increase in the frequency of these genes in the environment. Again, we must look at probabilities to determine if this could lead to an increase in the frequency of antibiotic-resistant pathogens. Several questions must be answered in order to assess the risk. How long will these genes survive in crop residues? Is it possible for soil microbes to acquire these genes from crop residues? Is it possible for these genes to be transferred from soil microbes to pathogenic microbes? Because transgenic crops are destined for consumers' stomachs, additional questions must be asked: Can microbes in the GI tract acquire antibiotic-resistant genes from digested crop residues? Could these genes then give microbes a competitive advantage over other microbes in the GI tract?

We lack the data to conclusively answer most of these questions. It is clear that DNA may survive the digestive process, making it theoretically possible that gut microbes could be exposed to intact antibiotic-resistant genes. We also know that many bacteria are capable of absorbing small pieces of DNA in solution and incorporating them into their own chromosome. However, this process (transformation) does not work with all DNA segments. The process is much more efficient if the DNA is in the form of a plasmid and if it contains certain DNA sequences that allow *competent* bacteria (those capable of transformation) to recognize sequences of DNA from closely related bacteria.

Consequently, government regulators look closely at the DNA that accompanies the novel gene. Would any of the transferred DNA sequences increase the likelihood that the transferred genes could act as a plasmid (especially a *transmissible plasmid*, which can also be taken up by bacteria through *conjugation*)? This is a pertinent question because plasmids are virtually always a key part of the cloning and transformation process. However, transmissible plasmids are *not* used for gene cloning, so to obtain conjugative ability, they would have to acquire the required genes from another plasmid.

Most biotechnologists and government regulators have concluded that there are negligible risks of proliferation of antibiotic-resistant genes in soil or gut bacteria as a result of the widespread culture of transgenic crops. The widespread misuse of antibiotics in animal agriculture and human medicine is regarded to be much more important, because it gives antibiotic-resistant bacteria a competitive advantage over other bacteria. This is crucial because in the absence of the antibiotic, there is little likelihood that the prevalence of antibiotic-resistant bacteria will increase. Nevertheless, most regulators (especially in Europe) strongly advise biotechnology companies to develop transgenic cultivars that do not contain antibiotic-resistant genes.

One final genetic element should be mentioned. Transgenic crops usually contain a *promoter*, which is a sequence of DNA that leads to a high level of transgene expression. One widely used promoter is taken from the cauliflower mosaic virus (CaMV), a DNA virus that commonly infects brassicaceous plants. Activists claim that this is a dangerous sequence of DNA because it is theoretically possible for the promoter to pass intact from the GI tract to the bloodstream and then into individual cells. If incorporated into a human genome, it could potentially cause expression of dangerous proteins, such as those involved in control of cell division. This could have serious consequences, perhaps leading to tumor formation.

Biotechnologists and regulators consider this scenario to be extremely unlikely. The major reasons cited are that the promoter has little effect on expression of animal proteins and that CaMV commonly infects a number of food plants. The human population, then, has been exposed to relatively large amounts of the promoter over a long period, without any detected problems. Indeed, there are no records of any sequence of ingested DNA leading to health problems.

D. Starlink Corn and Implications for Labeling and Trade Issues

In October 2000, food biotechnology became a hot topic for discussion in the popular media in North America, because of the finding that a transgenic cultivar (Starlink) of corn was present in taco shells and several other foods containing corn. Starlink had been approved by the FDA for use in animal feeds, but not in food. The reason for this **split decision** was an increased potential for allergenicity in Starlink corn. This cultivar has a *cry* gene derived from *Bacillus thuringiensis* that gives the plant resistance to the European corn borer. Unlike other *cry*-coded proteins, the Starlink protein is somewhat resistant to digestion by pepsin and other proteases in the human gut. This increases the probability that it will retain its structure longer in the small intestine, which could, in theory, lead to an allergic response in some individuals.

As a result of Starlink contamination of taco shells, the Kraft Corporation underwent an expensive food recall. The FDA also changed its regulatory policies, prohibiting split decisions in future assessments of transgenic crops. The U.S. Centers for Disease Control (CDC) also studied a small number of people that had ingested contaminated taco shells. In spring 2001, they released a report stating that symptoms of allergenicity or other health problems were not observed in people exposed to Starlink corn.

The FDA eventually traced Starlink contamination in the taco shells to a corn miller in Texas. Several loads of Starlink corn were mistakenly included in corn destined for use by food processors. This illustrates the difficulties inherent in the segregation of commodities that are functionally identical but differ principally in the presence of a transgene. This has implications for **traceability** and **verification** protocols, which are required for effective voluntary and mandatory labeling regulations. Currently, the EU, Japan, Australia, and numerous other countries have legislation requiring labeling of foods containing GMOs (this is referred to as **mandatory labeling**). The EU, partly because of the Starlink affair, acknowledges that accidents happen that can lead to the presence of GMOs in GMO-free commodities and foods. Consequently, the EU plans to set a 1% limit on allowable "contamination" of GMO-free food by GMOs. For this system to work, it is important that foods can be traced back to their source; the EU has expended much effort in developing strategies to enhance this traceability of commodities. Another consequence of the Starlink affair has been increased interest in the development of diagnostic tests for the detection of transgenes in commodities and food. The polymerase chain reaction (PCR) has emerged as the method of choice for detecting transgenes, partly because of PCR's success in detecting the original Starlink contamination.

International trade was also disrupted by Starlink contamination. Japanese authorities rejected several shipments of corn because of the presence of the Starlink *cry* gene. Several shipments of canola to Europe have also been rejected because of contamination of GMO-free canola by transgenic canola cultivars. These trade issues are controversial because of the absence of documented health or environmental damage from transgenic crops. GMO exporters, such as the U.S. and Canada, are convinced that the 1% limit for GMO contamination is too low (5% is offered as an alternative) and that the EU system for maintaining <1% contamination will prove to be unworkable. This, and other trade issues surrounding GMOs, will likely continue to be a source of conflict between GMO-exporting countries and food-importing countries in the foreseeable future.

IV. ASSESSMENT AND REGULATION OF DIAGNOSTIC TESTS

Whenever it is possible that technology carries significant risk to the public, it is important that governments regulate application of the technology. Most countries regulate biotechnology at the national level (see Table 9.1); however, some forms of biotechnology are regulated by associations. For example, when a biotechnologist or a company develops a new diagnostic technique and decides to market it, the technique must undergo a validation process, whereby it is compared to established diagnostic methods (**standard methods**) in terms of sensitivity, use in different foods, and consistency (precision). This is usually done as a formal study with a relatively large number of independent collaborating laboratories. If this study is equivalent or superior to existing standard methods, then organizations such as the Association of Official Analytical Chemists (AOAC) or the American Public Health Association (APHA) will review the validation data and may add it to their standard methods.

Governments (e.g., FDA) and NGOs (e.g., International Organization for Standardization [ISO]) often specify association-approved standard methods as part of their guidelines or legal requirements. Thus, standards developed by associations often have quasi-legal status. For example, in most countries legal limits apply to the number of microbes that can be present before and after pasteurization. In setting these limits, regulators specify the standard methods that must be used for milk testing. In the U.S., such standard methods are evaluated by the APHA and are incorporated by the Public Health Service and the FDA as they periodically update the **Pasteurized Milk Ordinance**. This document provides statutes that state and municipal governments can use to regulate milk production in their jurisdiction. In this case, standard methods exist for sampling milk directly from the cow's udder and from other points in the system (e.g., bulk tanks). Standards also exist for preparation of culture media, inoculation of media, and assessment of growth. Associations often give a choice among a number of methods that have similar levels of validity and performance.

V. BIOTECHNOLOGY AND THE DEVELOPING WORLD

In the second half of the 20th century, there was a concerted global effort to increase food security in the developing world. Food security in much of the world is low because large numbers of people have insufficient access to food, either in terms of calories consumed or in terms of food diversity, or both. The prevalence of hunger is a strong social force; hunger can lead to malnutrition, with all of its health consequences (e.g., reduced immune function), and it also makes it difficult for people to perform normal activities such as work, looking for work, and so forth. Therefore, it often accentuates and reinforces poverty, which is usually the root cause of hunger.

After 50 years of development experience, we have had some successes, but in many countries a large proportion of the population experiences hunger on a daily basis. Looking back, we can see that some of the popular strategies were relatively unsuccessful. For example, in the 1950s and early 1960s, the **trickle-down** strategy held that supplying technology to developing countries would have a stimulating effect on those countries' economies that would trickle down to both urban and rural populations. For example, if a country's farmers were given large numbers of tractors, that would increase the productivity and wealth of farmers, who would then spend more money, thus enriching other segments of the population. Unfortunately, the infusion of technology without a support structure (e.g., people skilled in tractor maintenance and repair or industries that can supply spare parts) is ineffective and results in continued dependence on industrialized nations.

One of the later fads surrounding international development was the idea that developing nations should boost their economic productivity mainly through food exports to western countries. This strategy is still somewhat popular today, through a fashionable term (**globalization**) that broadens the trade solution to include imports and exports from developing countries in a free market system. Since the last round of meetings of the World Trade Organization in Seattle in 1999, there has been vigorous public opposition to globalization and trade liberalization. This has mostly been driven by a fear that globalization will reinforce current patterns of exploitation of the developing world by industrialized countries. People who work in and study the development process agree that trade is important but that expanded trade alone does not greatly decrease poverty or the prevalence of hunger and malnutrition.

We also know that simply giving food or money to the developing world is not effective in the long term and should be restricted to emergency relief. In some cases, food donations cause problems in local markets, through decreased purchases of locally produced food. Instead, western money is better spent by attempting to get money to those in need through improved access to low-cost credit and by helping to build the roots of an industrial infrastructure.

Another useful way to spend western money is through the encouragement of agricultural research in the developing world. Many regions of the developing world (e.g., China) have been able to increase agricultural productivity to keep pace with increased population levels. This has mainly been possible because of the **green revolution**. This refers to the use of high-yielding cultivars of rice and other crops that were largely developed in western-funded research centers such as the **International Rice Research Institute** (IRRI) in the Philippines. If these high-yielding seeds were not freely available to peasant farmers, particularly in southeast Asia, hunger would be much more widespread. That said, there have been a number of negative effects of the green revolution. During the past 40 years, the agrochemical industry has deservedly been heavily criticized for overly aggressive marketing tactics, particularly in regard to chemical pesticides. The green revolution has also accentuated the gap between rich and poor in many developing countries.

The main question from our perspective is, Can food biotechnology help alleviate hunger and poverty in the developing world? We have seen previously that biotechnology can be used to increase the vitamin content of food staples such as rice (Chapter 4, Section IV.E). Transgenic cultivars can also be developed to combat many of the diseases and insect pests that plague farmers in the developing world, but it is unlikely that we can rely on the major developers of transgenic seed (i.e., western agrochemical companies) to develop these cultivars for distribution to poor farmers at low cost. It is also doubtful that multinational companies will be willing to give up their intellectual property rights (e.g., patents), which would be necessary for broad distribution of improved seed, similar to what happened during the green revolution.

For these reasons, most development workers view biotechnology as a potentially useful force, particularly if it follows a different path than it has in the developed world. Instead, food biotechnology should be guided by local expertise and needs. For example, peasant farmers of upland rice in southeast Asia are in desperate need of high-yielding seed that can cope with the drought stress that is endemic to their agricultural lands and systems. Such stress-resistant cultivars are being developed in various laboratories.

VI. THE FUTURE OF FOOD BIOTECHNOLOGY

This chapter has mainly concentrated on transgenic crops because they are the most regulated and the most controversial sector of food biotechnology. It is difficult to predict the course of the next 20 years with respect to transgenic crops. Public acceptance may slowly increase, assuming that these plants have a continued record of human and environmental safety. It is also possible that new strategies for safety assessment will arise through the increasing research commitment to this problem. However, consumer hostility to transgenic crops and animals may continue well into this century. Food safety is an increasingly strong source of worry for many people, and antibiotechnology activists have successfully convinced great numbers of people that transgenic technology is dangerous when applied to food.

Other aspects of food biotechnology, such as the functional food revolution, will continue to thrive, as we learn more about the ability of specific food components to fight disease. Microbial biotechnology will continue to be a vital economic force, despite being invisible to most of the public, and diagnostic biotechnology will continue on its course, steadily improving the food industry's ability to ensure that food is safe and pathogen free.

RECOMMENDED READING

1. Matravers, P., Bridgeman, J., and Ferguson-Smith, M., *The BSE Inquiry: Findings and Conclusions*, Vol. 1, Return to an Order of the Honourable the House of Commons, 2000, available at http://www.bse.org.uk/.
2. Proposal for a regulation of the European Parliament and of the council on genetically modified food and feed, Commission of the European Communities, Brussels, 2001.

3. Millstone, E., Brunner, E., and Mayer, S., Beyond "substantial equivalence": showing that a genetically modified food is chemically similar to its natural counterpart is not adequate evidence that it is safe for human consumption, *Nature*, 401, 525, 1999.
4. Love, S. L., When does similar mean the same: a case for relaxing standards of substantial equivalence in genetically modified food crops, *HortScience*, 35, 803, 2000.
5. Goldman, K. A., Bioengineered food safety and labelling, *Science*, 290, 457, 2000.
6. Trewavas, A. J. and Leaver, C. J., Is opposition to GM crops science or politics? An investigation into the arguments that GM crops pose a particular threat to the environment, *EMBO Rep.*, 2, 455, 2001.
7. Marshall, R. T. and Peeler, J. T., Standard methods, in *Standard Methods for the Examination of Dairy Products*, 16th ed., Marshall, R. T., Ed., American Public Health Association, Washington, D.C., 1992, chap. 1.

Index

A

α-Acetolactate decarboxylase (α-ALDC), 194–195
N-acetyl glucosamine, 223
Acidulant, 226
Activated charcoal, 213, 261
Activist groups, 14, 16, 79, 107, 266, 273, 274, 278
Adenosine triphosphate (ATP), 173, 174, 236, 246
Adenylate cyclase, 191
Aeromonas, 173
Aerotolerant anaerobes, 235
Affinity of antibodies, 164
Affinity separation, 162, 262
Aflatoxin, 26, 108, 156–158
    detection, 166, 167
Air-lift bioreactor, 240
Agar, 255
Agaricus bisporus, 26
Agrobacterium tumefaciens, 92, 95, 102, 108, 118, 124, 125, 128
    binary vector system, 100, 101
    life cycle, 98
    plant transformation, 98–99, 101
    tumor formation, 98
    vir, 99, 100, 117
    Ti plasmid, 98, 99, 100
Agrochemicals, 92, 93, 131, 132
Albumin, 121
Ales, 184
Algal toxins, 172
Alginate, 221, 223, 255
Allergen, 15
    assessment in transgenic plants, 273
    detection, 172
    reduction, 129–130
American Public Health Association (APHA), 276
Amino acids, 2, 182
    applications, 203
    economic aspects, 202–203
    energy costs, 205
    overproduction, 208–211
        safety, 213
    purification, 212, 213
    regulation of synthesis, 206–208
    stereochemistry, 204
Ampicillin resistance, 55

Amylase, 12, 26, 126
    in barley, 185, 186, 187
    in flour, 218–219
    types, 216–217
    uses, 214, 215
Amylopectin, 126, 215, 216,
Amyloplast, 215
Amylose, 126, 215, 216
Anamorph, 28
Animal
    breeding, 6, 11
    cell culture, 233, 241, 243, 244
    feeds, 123
    welfare, 10
Annealing, 53
Antibiotic
    production, 230
    residues, 172
    resistance, 25, 274
        removal, 115
Antibody, 69, 164
    class, 167
    production, 168–171
Antibody-based diagnostics, 14, 27, 154, 157
    biosensors, 176–177
    ELISA, 165–168
    food adulteration, 158
    latex agglutination, 165
Antigen, 164
Antinutrients, 121, 122, 130
Antioxidants, 4, 8
Antisense, 92, 113, 114, 126
α_1-Antitrypsin, 11, 143,148, 149
Apical meristems, 83
apicomplexans, 152
Aquaculture, 11, 137
    production efficiency, 140
Arabidopsis thaliana, 127, 128, 129
Artificial chromosomes, 70
Ascomycota, 28
Archaea, 23
Argentina, 15
Artificial pancreas, 178
Artisan cultures, 199
Ashbya gossypii, 229
Aspartame, 203, 211
Aspartase, 211–212

Repressor proteins, 36
Reproductive technologies, 11
Respiration, 236
Restriction enzyme, 48, 49, 50–51, 52, 64, 68
    sticky ends, 50, 53
    blunt ends, 50
Restriction site, 53, 66, 69
    introduction by PCR, 64
Retrogradation, 126
Retroviruses, 58
Reverse transcriptase, 58
Reverse osmosis, 213, 260–261
Riboflavin, 229
Ribonucleic acid (RNA)
    double stranded, 113–114
    polymerase, 36
    purification, 40
Rice, transgenic, 7
Risk assessment, 141, 265, 266, 274,
        see also Environmental risks;
        Transgenic plants
Rhizobium, 4, 131–132
Rhizopus, 28
    oligosporus, 182
Ribosomal RNA (rRNA), 160
Ribosome-binding site, 38
Rice, 116, 117
Risk benefit assessment, 265, 266
RNA, see Ribonucleic acid
Rockefeller Foundation, 117
Rough endoplasmic reticulum (RER), 29, 30
Royal Society of Canada, 16
Ruminants, 137
Rust, 26, 108
Ruthenium chloride, 178

S

Saccharomyces cereviseae, 12, 13, 25, 26, 75, 183,
        229, 230
    artificial chromosomes, 53
    brewing
        catabolite repression, 192
        continuous culture, 195
        crabtree effect, 190, 252
        diacetyl, 191, 195
        ethanol tolerance, 193
        fermentation, 189–191
    episomal plasmid (YEp), 75
    oxygen requirements, 235, 236
    recombinant, 195
    respiration, 189
    strain identification, 158
Safety assessment,
    allergenicity tests, 273

sequence analysis, 273
strategies, 278
Saline soil, 120
Salmon, see Transgenic fish
Salmonella contamination, 153
Salmonella detection
    biosensors, 175, 176
    cultural methods, 153, 154, 155,
    DNA based, 159–161
    ELISA, 167, 168, 172
    latex agglutination, 164–165
Salmonella enteritidis, 13, 175
Sampling port, 246, 248
Sandwich ELISA, 165–166, 167
Sanger dideoxy, 43
Secreted proteins, 30
Secretion vectors, 72, 76
Secretory vesicles, 39
Selective media, 153
Semicontinuous culture, 251
Sauerkraut, 21, 23
Screening
    libraries,
    microbial, 23
Seed modification, 121
Segregation of crops, 9, 97, 158, 275
Sense strand, 36
Sequencing vectors, 72
Serratia marcescens, 205, 206, 208–209
Sewage treatment, 153
Shikimate pathway, 110
Shear stress, 238
Shear-thinning, 225
Sheep, see Transgenic animals
Shine–Dalgarno sequence, 38
Shotgun cloning, 47, 57
Shuttle vector, 72
Sigma factor, 36
Signal recognition particle (SRP), 39
Signal sequence, 28, 38, 39, 76
Silage, 213
Site-directed mutagenesis, 46, 126
Solid phase hybridization, 159
Solid phase reactants, 257
Solid-state fermentation, 256–257
Solvent-partitioned systems, 220
Somaclonal variation, 86, 90–91
Somatic cell cloning, 148–149
Southern blotting, 41, 96
Soybean
    herbicide resistance, 110–111
    methionine content, 273
    sensory properties, 124
    storage proteins, 121
Soy sauce, 25, 256